R. Kellner, F. Lottspeich, H. E. Meyer

Microcharacterization of Proteins

Second Edition

 WILEY-VCH

R. Kellner, F. Lottspeich, H. E. Meyer

Microcharacterization of Proteins

Second Edition

 **WILEY-VCH**

Weinheim · New York · Chichester · Brisbane · Singapore · Toronto

Main authors:
Dr. Roland Kellner
Abt. Biomed Fo/GBT
Merck KGaA
D-64271 Darmstadt
Germany

Priv.-Doz. Dr. Helmut E. Meyer
Ruhr University Bochum
D-44780 Bochum
Germany

Priv.-Doz. Dr. Friedrich Lottspeich
Max Planck Institute for Biochemistry
Am Klopferspitz 18a
D-82152 Martinsried
Germany

Library of Congress Card No.: applied for

British Library Cataloguing-in-Publication Data:
A catalogue record for this book
is available from the British Library

Die Deutsche Bibliothek – CIP-Einheitsaufnahme
Microcharacterization of proteins / R. Kellner ...
2., ed. – Weinheim ; New York ; Chichester ; Brisbane ;
Singapore ; Toronto : Wiley-VCH, 1999
ISBN 3-527-30084-8

© WILEY-VCH Verlag GmbH, D-69469 Weinheim (Federal Republic of Germany), 1999

Printed on acid-free and chlorine-free paper

Printing: strauss offsetdruck, D-69509 Mörlenbach
Bookbinding: Großbuchbinderei J. Schäffer, D-67269 Grünstadt
Printed in the Federal Republic of Germany.

Preface to the Second Edition

Protein chemistry is facing a dramatic change these days. And it is not a step to higher sensitivity in protein analysis which pushes the field of proteins – but what might have been expected to be the bottleneck for further developments: It is a change in our attitude to the proteins. They are not any longer treated like individual components, in most cases with long lasting investigations, time and labor intensive. Instead, they are considered as members of pathways or networks, and protein studies may comprise hundreds or even thousands of components at once. Technology-driven developments permit the handling of large numbers of proteins in parallel or with high throughput. Namely, gel electrophoresis, mass spectrometry, and bioinformatics are the key players which open a new period in protein chemistry: 2D-gel electrophoresis can separate highly complex protein mixtures from cells or tissues; automated sample handling followed by mass analysis quickly determines peptide fragments; and finally bioinformatics enables us to identify a protein in the database within seconds. The improvements of various technologies on the one hand, but also their combined effort on the other hand opened the new research strategy called "proteomics".

The buzz word "proteomics" has gained enormous attention, but nevertheless protein chemistry is more than that. There are still other basic techniques, e.g., chromatography and Edman sequencing, which are indispensable to study proteins, their functions and their modifications. All this is the subject of this book in its second edition: the microcharacterization of proteins. We have rearranged the chapters to render the modern working procedures and we have added relevant information with respect to proteome research.

We are really fascinated by the great diversity of proteins and the challenging attempt to explore the protein world.

Darmstadt   Roland Kellner
Martinsried   Friedrich Lottspeich
Bochum   Helmut E. Meyer

November 1998

Contributors

Appel, Dr. Ron D.; Medical Informatic Division, Geneva University Hospital, CH-1211 Geneva 14

Bahr, Dr. Ute; Institute of Analytical Chemistry, University of Frankfurt, D-60590 Frankfurt

Bairoch, Dr. Amos; Swiss Institute of Bioinformatics, CH-1211 Geneva 4

Binz, Dr. Pierre-Alain; Central Clinical Chemistry Laboratory, Geneva University Hospital, CH-1211 Geneva 14

Behnke, Dr. Beate; Institute of Sanitary Engineering, University of Stuttgart, D-70569 Stuttgart

Blüggel, Dr. Martin; Institute for Physiological Chemistry, University of Bochum, D-44780 Bochum

Bold, Dr. Peter; Institute for Laser Medicine, University of Düsseldorf, D-40225 Düsseldorf

Chaurand, Dr. Pierre; Institute for Laser Medicine, University of Düsseldorf, D-40225 Düsseldorf

Eckerskorn, PD Dr. Christoph; Toplab GmbH, D-82152 Martinsried

Frishman, Dr. Dimitrij; MPI for Biochemistry, D-82152 Martinsried

Gasteiger, Dr. Elisabeth; Medical Informatic Division, Geneva University Hospital, CH-1211 Geneva 14

Görg, Prof. Angelika; Technical University of Munich, D-85350 Weihenstephan

Hillenkamp, Prof. Franz; Institute of Medical Physics and Biophysics, University of Münster, D-48149 Münster

Hochstrasser, Prof. Denis F.; Central Clinical Chemistry Laboratory, Geneva University Hospital, CH-1211 Geneva 14

Houthaeve, Dr. Tony; University of Gent, B-9000 Gent

Immler, Dr. Dorian; Institute for Physiological Chemistry, University of Bochum, D-44780 Bochum

Karas, Prof. Michael; Institute of Analytical Chemistry, University of Frankfurt, D-60590 Frankfurt

Kempter, Dr. Christoph; Institute of Sanitary Engineering, University of Stuttgart, D-70569 Stuttgart

Lohaus, Dr. Christiane; Institute for Physiological Chemistry, University of Bochum, D-44780 Bochum

Maierl, Dr. Andreas; MPI for Biochemistry, D-82152 Martinsried

Mehl, Dr. Ehrenfried; MPI for Biochemistry, D-82152 Martinsried

Metzger, Prof. Jörg W.; Institute of Sanitary Engineering, University of Stuttgart, D-70569 Stuttgart

Mewes, Dr. Hans-Werner; MPI for Biochemistry, D-82152 Martinsried

Schägger, Prof. Hermann; Biochemistry Department, Universitiy of Frankfurt, D-60590 Frankfurt

Schwer, Dr. Christine; Schering AG, D-13342 Berlin

Serwe, Dr. Maria; Institute for Physiological Chemistry, University of Bochum, D-44780 Bochum

Siethoff, Dr. Christoph; Institute for Physiological Chemistry, University of Bochum, D-44780 Bochum

Spengler, Prof. Bernhard; Institute for Physical Chemistry, University of Würzburg, D-97074 Würzburg

Westermeier, Dr. Reiner; Amersham Pharmacia Biotech, D-79111 Freiburg

Wilkins, Dr. Marc R.; Australian Proteome Analysis Facility, Macquarie University, Sydney

Contents

II.2   Microseparation Techniques II:
Gel Electrophoresis for Sample Preparation in Protein Chemistry   25
Hermann Schägger

II.3   Microseparation Techniques III:
Electroblotting   35
Christoph Eckerskorn

Section III: Bioanalytical Characterization

III.3 Analyzing Post-Translational Protein Modifications 159
Helmut E. Meyer

III.4 Analysis of Biopolymers by Matrix-Assisted Laser Desorption/Ionization (MALDI) Mass Spectrometry 177
Ute Bahr, Michael Karas and Franz Hillenkamp

III.5   MALDI Postsource Decay Mass Analysis   197
Bernhard Spengler, Pierre Chaurand and Peter Bold

III.6   Electrospray Mass Spectrometry   213
Jörg W. Metzger, Christoph Eckerskorn, Christoph Kempter and
Beate Behnke

III.7   Fourier-Transform Ion Cyclotron Resonance Mass Spectrometry (FT-ICR-MS)   235
Jörg W. Metzger

III.8   Sequence Analysis of Proteins and Peptides by Mass Spectrometry   245
Christoph Siethoff, Christiane Lohaus and Helmut E. Meyer

Section IV: Computer Sequence Analysis

Abbreviations

API	atmospheric pressure ionization
ATZ	anilinothiazolinone
AUFS	absorption units full scale
BLAST	basic local alignment search tool
BNPS	bromine adduct of 2-(2-nitrophenylsulfonyl)-3-indolenine
CAPS	3-cyclohexylamino-1-propanesulfonic acid
CE	capillary electrophoresis
CEC	capillary electrocromatography
CGE	capillary gel electrophoresis
CHAPS	3-(3-cholamidopropyl)dimethylammonio-1-propane sulfonate
CID	collision-induced dissociation
CIEF	capillary isoelectric focusing
CORBA	common object request broker architecture
CTAB	cetyltrimethylammonium bromide
CZE	capillary zone electrophoresis
Da	Dalton
DABS-Cl	4-dimethylaminoazobenzene sulfonyl chloride
DC	direct current
DHB	dihydroxy benzoic acid
DNA	desoxyribonucleic acid
DNP	2,4-dinitrophenol
DMPTU	dimethylphenylthiourea
DPTU	diphenylthiourea
DPU	diphenylurea
DTPA	diethylene triamine-penta-acetic acid
DTT	dithiothreitol

EDTA	ethylenediamine tetraacetic acid
EOF	electroosmotic flow
ESI	electrospray ionisation
EST	expressed sequence tag
FAB	fast atom bombardment
FID	flame ionization detector
FLASH	fast look-up algorithm for string homology
FLEC	1-(9-fluorenyl)ethyl chloroformate
FMOC	9-fluorenylmethyloxycarbonyl
FT	fourier transform
HFA	hexafluoroacetone
HIC	hydrophobic interaction chromatography
HILIC	hydrophilic interaction chromatography
HPLC	high performance liquid chromatography
ICR	ion cyclotron resonance
IEC	ion exchange chromatography
IEF	isoelectric focusing
IPG	immobilized pH gradient
IR	infrared
ITP	isotachophoresis
LC	liquid chromatography
MALDI	matrix assisted laser desorption ionization
MECC	micellar electrokinetic capillary chromatography
MES	2-(N-morpholino)ethanesulfonic acid
MHC	major histocompatibility complex
MoAb	monoclonal antibody
MS	mass spectrometry
NEPHGE	non equilibrium pH gradient gel electrophoresis
NMR	nuclear magnetic resonance
NTCB	2-nitro-5-thiocyanobenzoic acid
OPA	orthophthaldialdehyde
ORF	open reading frame

PAGE	polyacrylamide gel electrophoresis
PCR	polymeric chain rection
PEDANT	protein extraction, description, and analysis tool
PIR	protein identification resource
PITC	phenylisothiocyanate
PSD	post source decay
PTC	phenylthiocarbamyl
PTH	phenylthiohydantoin
PTM	post-translational modification
PVDF	polyvinylidene difluoride
RF	radio frequency
RNA	ribonucleic acid
RPC	reverse phase chromatography
SDS	sodium dodecyl sulfate
SEC	size exclusion chromatography
SWISS-PROT	protein database from the Swiss Institute of Bioinformatics
TCA	trichloroacetic acid
TEMED	N,N,N',N'-tetramethylethylendiamin
TFA	trifluoroacetic acid
TH	thiohydantoin
TIC	total ion current
TOF	time-of-flight
TrEMBL	protein translations from EMBL database
UV	ultraviolett

Section I: Overview

Microcharacterization of Proteins

Friedrich Lottspeich and Roland Kellner

At the beginning of the 20th century it became apparent to biochemists that proteins are macromolecules which consist of amino acid components. Emil Fischer deduced from hydrolyzation experiments and the very first peptide syntheses [1903] that proteins are built up of numerous linked α-amino acids. The nature of different amino acids and the peptide bond linkage became clear. However, small synthetic peptides did not exhibit the properties of proteins. It was not realized until 1950 that all molecules of a given protein have one unique amino acid sequence. Several key developments helped to reveal the molecular architecture of proteins, namely the development of chromatography by Martin [1941] and amino acid analysis by Stein and Moore [1948]. One milestone was the finding of Frederick Sanger, who identified phenylalanine as an end-group of insulin by its reaction with 2,4-dinitrofluorobenzene. The Sanger method forms yellow 2,4-dinitrophenyl (DNP) derivatives with amino groups. At the same time Pehr Edman established the stepwise degradation of proteins using phenylisothiocyanate as a reagent. Then in 1953 Sanger and Tuppy were able to determine the amino acid sequence of insulin, which was the first protein characterized on the molecular level.

And furthermore human insulin became the first biotechnology product in 1982. A synthetic gene was constructed corresponding to the amino acid sequence of insulin and the foreign DNA was inserted into the bacterium Escherichia coli. The bacteria served as a host and expressed the human gene, synthesizing the recombinant protein insulin. Genetic engineering and recombinant proteins revolutionized the classical field of biochemistry. Very successful techniques were established to synthesize oligonucleotides and to analyze cDNA sequences rapidly and with high throughput. The PCR technology allowed to multiply single DNA chains and therefore to overcome sensitivity problems. The opportunity for automatization of genetic techniques initiated giant research activities like the human genome project. Numerous genome projects for smaller species could already be finished and there are only doubts if the complete human genome will be identified in the year 2001 or in 2005 or even faster. Genomics was the key word for life science activities in the last years. And only recently the twin name came up – proteomics. The technology-driven development of protein chemistry permits to investigate new biologically relevant questions with enormous business opportunities, especially in pharmaceutical areas. The characterization of proteins is a very exciting field these days.

1 General Aspects

In the classical view, the reason for a protein purification is the observation of a bio-logical phenomenon and search for its molecular basis. Thus, a single or a few pro-teins which are correlated with observed activity are isolated and characterized at the molecular level. Knowledge of the primary structure of these proteins as well as their modifications and processing sites is of fundamental importance. To acquire this knowledge, particularly when only a small amount of protein is available, may be difficult and laborious. All techniques available at the time have to work in a con-certed action to elucidate the structure of the protein in a reasonable time. The entire amino acid sequence of a protein is now more easily obtained by deduction from its DNA sequence rather than analyzing the total protein by protein chemical amino acid sequence determination. However, some protein chemical information is always required to isolate the DNA or to check the accuracy of the DNA sequence. Important information, like the actual N-terminus or C-terminus of a protein, can only be deter-mined by protein chemical methods. Probably the most important task of protein chemistry determing the post-translational modifications, about which little can be learned from the analysis of DNA sequences, but which often regulate the activity of the proteins. Supplementation with immunologic, crystallographic or NMR data provides additional information which is not available by analysis of the DNA or pro-tein sequences alone. Modern protein chemistry is one of the important members in the orchestra of all the different disciplines in the biosciences, providing the basis for a detailed understanding of structure-function relationships.

In recent years additional areas of responsibility have been allocated to protein chemistry, mainly due to the developments in recombinant DNA techniques. Projects to analyze the human genome or the genomes of other organisms at the DNA or RNA level are the focus of strong international research activities. However, even today it is apparent that the meaningfulness of the genomic analyses will be limited, unless at the same time more attention is paid to the function of the huge number of sequenced or characterized pieces of DNA. The connective link between the genome and the multi-ple cellular functions of an organism are the proteins. The quantitative appearance and certain posttranslational modifications have major influence on the function and activ-ity of a protein. Thus, the analysis of proteins, which are the real players and tools in the cell, is a supplementary and inevitable approach to understand the biological events in a cell. Figure 1 outlines how protein analysis links genomic and protein information at the molecular level.

2 From a Cell to a Protein Sequence

Nowadays it is possible to separate most of the cellular proteins by two dimensional (2D) gel electrophoresis and to check their amount quantitatively. In this way, a pic-ture of the protein expression for a certain cell stage can be obtained. In a kind of sub-tractive approach, different metabolic stages of cells can be compared with the help of computerized image analysis and data handling. The subtle protein changes that arise

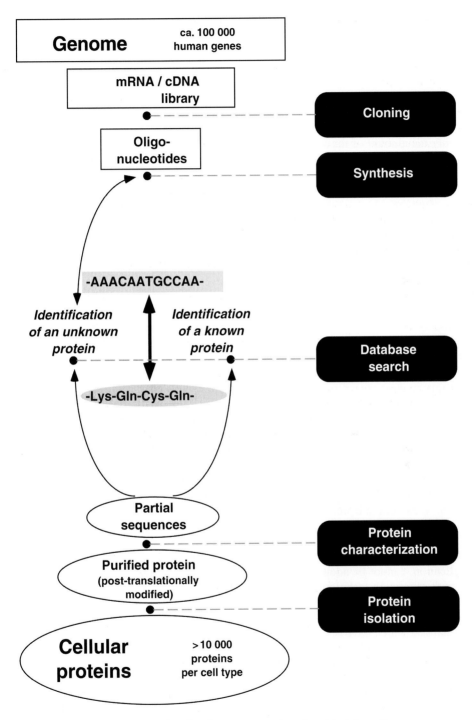

Figure 1. Protein analysis as an interface between genomic and protein information.

in all complex biological systems against a background of constitutive proteins have to be recognized and further analyzed with modern protein chemical micro methods.

The amounts of the most abundant proteins in a 2D gel electrophoretic separation are in the very low picomole or femtomole level, corresponding to few micrograms or even submicrogram amounts of material. It is because the amounts of interesting proteins are so small that it has been so imperative to improve the protein analytical methods towards higher sensitivity and more speed. Sequence analysis, for example, has come to a level where a few picomoles of a protein can be sequenced: that is roughly the amount of the 100–300 most abundant cellular proteins separated in 2D gel electrophoresis. All the other most commonly used methods in protein chemistry, like separation techniques, amino acid analysis or mass spectrometry, today work in a similar sensitivity range. An urgent need also to attack the proteins present in even lower quantities is a strong driving force for further developments in methodology and instrumentation. Improvements in sensitivity to the femtomole level for all the protein chemical methods mentioned is within the reach of modern micro protein chemistry. The consequence will be that thousands of proteins of a cell can be analyzed.

Manipulating these small quantities requires several peculiarities to be taken into consideration. Proteins adsorb strongly to any kind of surface and microgram amounts may be lost in a few minutes to the wall of a vessel. Consequently, several techniques commonly used on the macro scale, like ultrafiltration, dialysis or lyophilization, cannot be recommended when working with micro amounts. Sometimes adsorbed protein can be recovered by treatment with concentrated formic acid or by incubating with detergent solutions. In general a good recovery is achieved for peptide material dissolving it in aqueous trifluoroacetic acid with a few percent of acetonitrile.

At the micro level contamination becomes a major problem. Laboratory dust and impurities in solvents, reagents and equipment are almost inevitable sources of contamination. It is obvious that automated separation and reaction devices which use minimum volumes of solvents and reagents and which keep the sample in a closed environment should be used for preference. However, so far, common laboratory equipment seldom is in accordance with these requirements, and thus instrument development is called for.

As a consequence of the special situation with micro amounts, careful planning of the purification strategy is imperative to minimize all the handling and transfer steps and to keep contamination as low as possible. Often, early fractionation steps by multistep extraction or precipitation techniques, commonly used in large scale protein purification, cannot be adopted at the micro scale. Even at the early stages the most efficient separation techniques available have to be adopted in an optimal sequence. Detection methods like mass spectrometry, or diode array spectroscopy yielding multiple information are strongly recommended. In Figure 2 several strategies are summarized which have been developed and successfully applied in the last years.

Starting materials like a cell or a complex protein mixture already enriched in the protein of interest by conventional purification steps is applied to a high-resolving separation method. In Figure 3 the different fractionation methods are compared according to their capability to separate molecules of different sizes. Immediately it becomes clear why gel electrophoresis is usually the method of choice to separate complex protein mixtures. For small molecules (e.g., amino acids or peptides) chro-

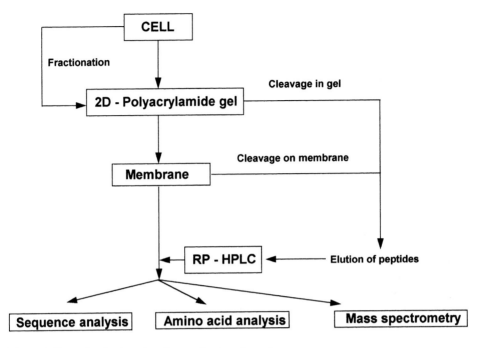

Figure 2. Strategies for the microcharacterization of proteins.

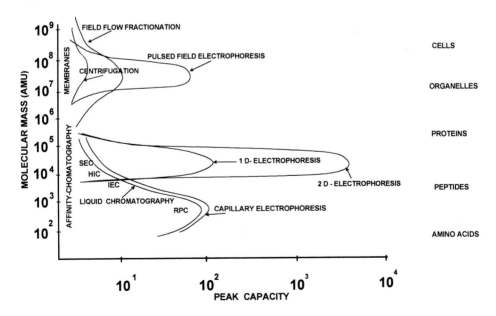

Figure 3. Separation capacity of different methods depending on the molecular mass of the substances.

matography has the best separation power, which is reflected in almost exclusive use of high performance liquid chromatography (HPLC) for amino acid separations or peptide fractionations and peptide maps. For proteins in the molecular weight range of 5000 to 200 000 Dalton gel electrophoresis, particularly in its 2D mode, is capable of separating up to 10 000 proteins in a single analysis.

Capillary electrophoresis is gaining increased attention as a fairly new method in protein chemistry. However, so far it has not been adopted for routine preparative work, due to technical difficulties in applying large quantities of samples and in collecting fractions. Nevertheless, capillary electrophoresis is ideally suited in principle for microscale separations. Samples are eluted at high concentrations and it can be connected on-line to other analysis techniques like mass spectrometry. Furthermore, capillary electrophoresis with its ionic separation mode is ideally complementary to reversed phase micro HPLC, where the separation is caused by hydrophobic interactions of the solutes.

When the individual components of the complex sample are resolved, then they are further characterized by three main methods: amino acid sequence analysis, amino acid composition analysis and mass spectrometry. Here, the immediate goal is to recognize if the protein under investigation has already been sequenced or if it is a new, as yet unknown compound. If no N-terminal sequence can be obtained due to a blocked N-terminus, which happens in almost every second case, a database search with the amino acid composition may give a hint as to the nature of the protein. However, usually with a blocked protein cleavage with enzymes or chemicals has to be performed. The internal fragments produced can be analyzed directly by matrix-assisted laser desorption/ionization (MALDI) mass spectrometry. With the help of computer programs the fragment masses found can be compared with the calculated fragment masses of all the protein sequences stored in protein and DNA databases. Multiple matches will yield significant evidence towards a knowledge of the protein. If the protein is unknown, or if the mass search is not unambiguous, or if modifications are present, the fragments have to be separated by microbore reversed phase HPLC or capillary electrophoresis. Both techniques may be coupled on-line to electrospray mass spectrometric detection, and sequencing will be performed either by Edman degradation or MS/MS. The sequence information obtained is again used for a database search for identities or homologies. Usually one or two short sequences of 6-8 amino acid residues are sufficient to recognize a homology or an identity: in fact, often not more of the fragments than this will be analyzed at the protein level. The identifed peptide fragments usually provide sufficient sequence information for isolating the DNA of the protein of interest or for the production of monospecific antibodies. Once the cDNA is sequenced, the mass of the encoded protein sequence and the actual measured mass of the isolated protein have to be compared. In case of discrepancies the N- and C-termini need to be checked and/or modified amino acid residues must be identified. Determination of the exact position of protein modification is a tremendous challenge when sample amounts are small.

3 Genome and Proteome

Between the stored biological information, the gene, and the functional representative, the protein, several highly complex events take place: transcription of the gene sequence and translation into a protein, transport phenomena, stability criteria, posttranslational modifications. Neither the amount of an active protein in the cell nor the presence or absence of posttranslational modifications can be deduced from the gene level. Both has to be determined by investigating the mature protein. Encouraged by the high-throughput methods of molecular biology and concomitant with the bright success of mass spectrometry for the analysis of peptides the new field of proteome research was initiated. A proteome is best understood as the quantitative protein expression pattern of a cell or an organism under precisely defined conditions. The proteome is a highly dynamic object in contrast to the genome, since any environmental or biological parameter influences protein expression in a complex manner (Fig. 4). These active changes can be used in subtractive approaches, for example, investigating the influence of a pharmacon on a cell or its absence, respectively. After separation of protein mixtures by 2D gel electrophoresis these patterns can be quantitatively compared. The observed differences can identify functional targets for the pharmacon or biochemical pathways which are affected. Most responses are quantitative variations of several proteins and therefore the major issue in proteomics is the maintainance and exact determination of the quantitative ratios of the relevant pro-

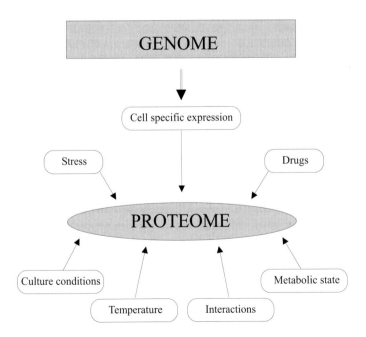

Figure 4. Protein expression is influenced by numerous environmental and biological parameters and a proteome becomes a highly dynamic object.

teins. This requires new techniques in sample preparation, separation, detection and quantification of proteins. Fully automated, high-throughput methods for protein analysis have to be adopted and many of them are described in this book. Bioinformatic methods are inevitable to handle the immense data output and will start via principal component analysis and cluster analysis for functional implementations which are the basis to understand biological phenomenas.

Section II: Microseparation Techniques and Sample Preparation

Microseparation Techniques I:
High Resolution Gel-Electrophoretic Techniques: Qualitative, Quantitative and Micropreparative Applications

Angelika Görg and Reiner Westermeier

1 Introduction

Electrophoretic separation methods in gel matrices offer a number of unique features for protein analysis:

- Gel-electrophoretic techniques are robust concerning contaminants. Crude samples can be applied almost without prepurification steps.
- The resolution is very high for one-dimensional, and extremely high for two-dimensional techniques.
- The separation result is a characterization of the proteins by their physico-chemical parameters: isoelectric points and/or molecular weight.
- With an appropriate staining method the fractions can be quantitated; they can still be further characterized by amino acid analysis, sequencing, or mass spectrometry.

Gel-electrophoretic techniques are applied in different methodological and apparative variants. For further characterization, the proteins are either digested in the gel and then eluted from the matrix, or they are transferred to an immobilizing membrane by blotting procedures.

2 Theory

Electrophoresis is the migration of charged particles in an electric field. Different components migrate with different velocities and form separate zones. The particles, cell organelles, cells, or proteins to be separated must carry charges and be in aqueous solution.

The electrophoretic migration is charactarized as follows:

$$v = \mu \times E$$

- v = migration velocity (cm/s)
- μ = electrophoretic mobility (cm^2/V$\times$s)
- E = electric field strength (V/cm)

The electrophoretic mobility is dependent on charge, size, shape of the molecule as well as viscosity, pore size, buffer pH, ionic strength, and temperature of the medium. The *relative* electrophoretic mobility (m_R) is determined by co-electrophoresis of an ionic dye (e.g. bromophenol blue, xylencyanol, Orange G, pyronine). The values for m_R are very rarely published. *Instead* molecular weights, base pair length, Stokes' radius, and isoelectric points can be found in articles, books, and data bases. The electric field strength is the driving force of electrophoresis.

Electroendosmosis: Wherever solid material is involved, electroendosmotic phenomena may occur. Sulfonic or carboxylic groups in the matrix and silanol groups on the glass surface become negatively charged in a basic buffer: SO_3^-, COO^-, SiO^-. In the electric field they are attracted towards the anode. As they are fixed, they can not migrate, this leads a compensation by the migration of hydrated protons (H_3O^+) towards the cathode. Because electroendosmosis is directed into the opposite direction of the electrophoretic migration, it causes zone widening and blurred bands.

There are three main principles applied in gel-electrophoretic procedures (Fig. 1):

Zone electrophoresis: In a *homogeneous buffer* different sample components migrate with different velocities and form separate zones.

Isotachophoresis (Displacement electrophoresis) in gels is mostly used as the first phase in disc electrophoresis. In a *discontinuous buffer* system the sample is applied between the leading ion with high mobility and a trailing ion with low mobility. All ions – also the sample ions – are forced to migrate with the same velocity (*Greek:* isotachophoresis). This leads to the formation of a stack of the sample components in a field strength gradient. The zones are separated and concentrated, but migrate directly after each other without intermediate space.

Isoelectric Focusing can only be applied for amphoteric samples like proteins or peptides. In a *pH gradient* the sample components migrate towards the anode or the

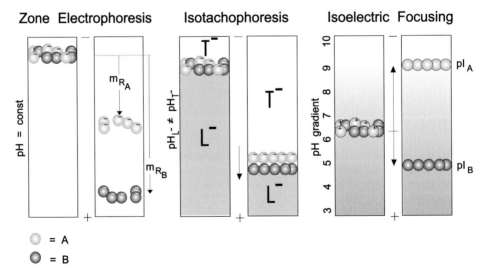

Figure 1. The three electrophoretic methods. See text for explanations [Westermeier 1993].

cathode to the pH values, where their net charges are zero: their *isoelectric points* (pI). Should a protein diffuse away from its pI, it would gain a charge and migrate back: this is the *focusing* effect.

3 Media and Equipment

3.1 Media

Mainly two types of gel matrices are used: agarose and polyacrylamide:

Agarose: Very large pore sizes are obtained with agarose gels, 0.15 μm diameter with a 1 % agarose gel. Negative charges originating from acidic residues (carboxylic and sulfonic groups) in the matrix can cause electroendosmosis: This leads to blurred bands: a lack of resolution and a lack of sensitivity for proteins and peptides. For molecule sizes larger than 1,000 kDa, the submarine technique can be applied: the agarose gel is submerged below a layer of buffer, e.g for the separation of the van Willebrand factor (ca. 1.5 million Da). Isoelectric focusing in agarose is applied more frequently, because the band sharpness is achieved by the focusing effect in the electric field and not by sieving.

Polyacrylamide: These gels have very low electroendosmosis, high mechanical and chemical stability and a clear background. Chemical polymerization with the catalyst system ammonium persulfate and TEMED is preferably used instead of photopolymerization with riboflavin. The pore size is defined by the *T* value: the total monomer concentration (acrylamide + N,N'-methylenbisacrylamide) and the *C* value: the crosslinking factor. For example, a gel with 5 % *T* and 3 % *C* has a pore diameter of 5.3 nm.

More comprehensive information on the modifications of gel matrices are found in Andrews [1986].

3.2 Equipment

Vertical gel rods have been replaced more and more by vertical and horizontal slab gels. In *vertical* systems, the gels are polymerized between glass plates and remain there throughout the entire separation procedure; the samples are loaded at the top into sample wells, which have been formed with a comb. In *horizontal* systems, the gels are run outside of the casting mould with an open surface, the gels are mostly bound to a thin glass plate or a film support; here the samples are loaded into sample wells on the surface. Methodical variants of zone electrophoresis are run on both vertical and horizontal equipment, isoelectric focusing methods are almost exclusively performed in the horizontal plane.

For the dissipation of Joule heat and in order to provide defined temperature conditions, a thermostatic circulator is connected to the apparatus.

4 Gel Electrophoretic Methods

4.1 Disc Electrophoresis

A widely used separation system for proteins is the discontinuous electrophoresis in Tris-chloride/Tris-glycine buffer according to Ornstein [1964] and Davies [1964]. The electrode buffer contains Tris-glycine, a resolving gel is prepared with 10 % *T*, 0.375 mol/L Tris-chloride pH 8.8 and a stacking gel with 4 % *T*, 0.125 mol/L Tris-chloride pH 6.8 (near the pI of glycine). During the *first phase*, the sample components are stacked and concentrated in an isotachophoresis system (because of discontinuity of ions). In the *second phase* the stack arrives at the edge of the resolving gel (small pore size), which is highly restrictive to proteins. A second concentration effect occurs. As the gel is not restrictive to glycine, the glycine ions pass the proteins with constant speed: now proteins are in a homogeneous buffer. In the *third phase* proteins are destacked and separated according to zone electrophoresis conditions. More recipes are found in a book dedicated to disc electrophoresis by Maurer [1968].

Advantages of disc electrophoresis are:

- Slow migration in the beginning prevents protein aggregation while entering the gel matrix.
- Two times concentration of zones lead to sharp bands and high resolution.

4.2 Gradient Gel Electrophoresis

Resolution can be further improved by employing gels with graded porosity [Margolis and Kenrick 1968]. A simple gradient maker containing two chambers and a magnetic stirrer can be used. For linear gradients, the principle of communicating vases is applied: half as much of the dilute solution flows in as of solution flowing out of the mixing chamber. The solution in the mixing chamber contains also glycerol in order to prevent the solutions of mixing in the cassette and to stabilize the gradient.

Advantages of gradient gels are:

- Very sharp bands.
- A wide range of molecular sizes can be included in one gel.
- Molecular sizes of native proteins and glycoproteins can be determined.
- A more even transfer in Western Blotting is obtained.

4.3 Additives in Electrophoresis

When hydrophobic proteins and/or highly complex samples have to be analyzed, additives have to be used in the gel for several reasons:

- Disulfide bonds between polypeptides form aggregates, they can be cleaved by adding *thiol reagents* like dithiothreitol (DTT) or 2-mercaptoethanol. As these reducing agents severely inhibit polymerization, these additives can only be added to prepolymerized gels by rehydrating a dried gel in buffer and additives, or by adding the compounds to the cathodal buffer.

- *Urea* at high concentrations (8 mol/L) highly improves the solubility of samples and prevents aggregation. However, the gels have to be prepared freshly, because urea degrades to isocyanate in solution, which causes carbamylation of proteins. Acid-urea gel systems are employed for the separation of very basic proteins like histones and cereal prolamins.
- *Non-ionic* detergents like Triton X-100 are added for solubilization of hydrophobic proteins. They do not change the net charges on the proteins, and they are often added to urea-gels.
- *Zwitterionic detergents* like CHAPS (3-(3-cholamidopropyl)dimethylammonio-1-propane sulfonate) in many cases further improve the solubility of proteins. They do not change the net charges on the proteins either, and they are also often added to urea gels.
- *Anionic* (SDS: Sodium dodecyl sulphate) and *cationic detergents* (CTAB: Cetyltrimethylammonium bromide) change the charges on the proteins and denature them.
- For membrane proteins, according to Schägger and von Jagow (see chapter Microseparation Techniques II.2 in this book) *Coomassie Brilliant Blue* can be added instead of a detergent.

4.4 SDS Electrophoresis

When the anionic detergent SDS is added to the protein samples and to the buffer, the upper two methods provide a separation according to the molecular weight of the proteins [Shapiro *et al.* 1967, Lämmli 1970]. 1.4 g of SDS per gram of protein are bound quantitatively and form ellipsoid micelles. All the micelles have a negative charge which is proportional to the mass:

- Individual charge differences of the proteins are masked.
- Hydrogen bonds are cleaved, the polypeptides are unfolded.
- Hydrophobic interactions are cancelled.
- Aggregation of the proteins is prevented.

The relationship between the logarithm of the molecular weights and the relative mobilities is in principle sigmoidal. Over a limited range, there is a linear relationship for gels with constant *T* values. For porosity gradient gels, the interval of linearity is wider. The molecular weight of the proteins can be estimated with a calibration curve using marker proteins. For the separation of low molecular weight peptides, the buffer system has to be modified according to Schägger and von Jagow (see chapter Microseparation Techniques II.2 in this book).

Sample preparation. Non-reducing SDS treatment: Sometimes, samples like physiological fluids such as serum or urine are simply incubated with a 1 % SDS buffer without reducing agent, because one does not want to destroy the quaternary structure of the immunoglobulins. The disulfide bridges are not cleaved by this treatment and so the protein is not fully unfolded. However, this is not a proper molecular weight measurement: albumin (68 kDa) shows an apparent molecular weight of 54 kDa when it is run under nonreduced conditions.

Reducing SDS treatment: Proteins are completely unfolded, when DTT or 2-mercaptoethanol is added and the sample is heated at 95 °C for 3 min. Using DTT instead of 2-mercaptoethanol has several advantages:

- DTT is less volatile: it does not smell so strongly.
- Reduced and nonreduced samples can be separated in the same gel, because DTT does not diffuse into the traces of nonreduced proteins.

Reducing SDS treatment with alkylation: The SH groups are better and more durably protected by an alkylation with iodoacetamide. Sharper bands result; in proteins containing many amino acids with sulphur groups a slight increase in molecular weight can be observed. In addition, the appearance of artifact lines during silver staining is prevented because iodoacetamide traps the excess DTT. Alkylation with iodoacetamide works best with a sample buffer pH 8.8.

Advantages of SDS electrophoresis:

- SDS solubilizes almost all proteins, even very hydrophobic and denatured proteins
- all proteins migrate in one direction, towards the anode
- together with reduction and alkylation a complete unfolding of the polypeptide
- measurement of the molecular weight
- quicker than most native separations
- high resolution and sharp zones
- bands are easy to fix, no strong acids are necessary
- increased sensitivity
- after blotting SDS can be removed.

Staining. Traditionally, many labs use TCA for fixation and alcohol-acetic acid for staining and destaining with Coomassie Blue. This is not necessary after SDS and native PAGE.

Fixing and staining of proteins can be performed in one step by placing the gel into a solution of 0.02 % (w/v) Coomassie Brilliant Blue R-350 in 10 % (v/v) acetic acid at 50 °C for 15 min. Destaining is performed in 10 % (v/v) acetic acid at room temperature. Note, that the Coomassie Blue concentration is very low. Fixation is caused by the dye at acidic pH at high temperature. This alcohol-free staining method has several advantages:

- cheap (no methanol or ethanol needed, no problems with disposal)
- can be reused several times, is better for the environment
- destaining solution can be permanently recycled (with activated charcoal)
- it does not destain proteins during background-destaining (e.g. collagen derivatives could never be stained with the traditional staining technique)
- does not elute prolamines from the gel (alcohol soluble fraction of proteins).

With silver staining a 20 to 100 fold higher sensitivity is achieved. As this is a multistep procedure with a number of washing steps, the fixation of the proteins and peptides has to be irreversible. Mostly glutardialdehyde is employed for crosslinking the polypeptides with the gel matrix. After some washing steps, the gel is immersed in silver nitrate solution for a certain time period. Then the gel is placed in a formaldehyde/sodium carbonate solution. At this high pH value the formaldehyde reduces silver nitrate to metallic silver, which has a dark brown to black color. As polyacrylamide shifts the redox potential of this reaction, it does only occur on the surface of the protein or DNA fractions and not in the background. The reaction is stopped by placing the gel into an acidic solution, mostly diluted acetic acid. In Figure 2 a SDS electrophoresis pherogram of plant proteins is shown.

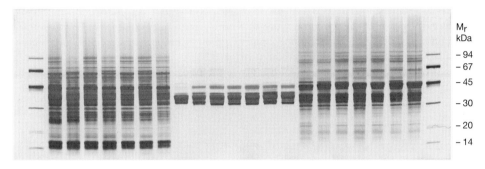

Figure 2. SDS polyacrylamide gel electrophoresis (10–15 % *T*) of albumin/globulin, gliadin and glutenin proteins from seven wheat cultivars. Protein detection with silver staining [Weiss et al. 1993].

4.5 Isoelectric Focusing

In principle isoelectric focusing is an equilibrium method. The proteins are driven to their isoelectric points. The isoelectric points of the proteins can be estimated with a calibration curve using marker proteins. Native or denaturing conditions can be chosen by omitting or adding high amounts of urea. The proteins are highly concentrated at their pI. Thus, there is also a high sensitivity for detection. Small charge differences can be differentiated. For improvement of resolution, narrow gradients can be employed. The method is also suitable for preparative applications, when high amounts of proteins – one hundred and more microgram – have to be purified.

There are different pH gradient concepts:

Free carrier ampholytes: Mixtures of 600 to 700 different homologs of amphoteric compounds with a spectrum of isoelectric points between 3 and 10 form a pH gradient under the influence of the electric field. These substances have high buffering capacities at their isoelectric points.

Immobilized pH gradients: Acrylamido buffers with carboxylic and tertiary amino groups are copolymerized with the polyacrylamide network. A linear gradient of monomer solutions has to be cast analogous to a porosity gradient. With 6 to 10 individuals of Immobilines, all types of gradients can be produced: from wide to very nar-

19

row gradients. Immobilized pH gradient gels are cast on film support, washed with distilled water, dried down, and rehydrated before use with water or urea-detergent solution, or sample solution.

Staining. After isoelectric focusing, the proteins have to be fixed irreversibly in the matrix with trichloroacetic acid (TCA). Most frequently a colloidal staining method according to Diezel and Kopperschläger [1972] is employed: 2 g Coomassie Brilliant Blue G-250 are dissolved in 1 L distilled water and – under permanent stirring – 1 L of 1 mol/L sulphuric acid is added. After 3 h of stirring, the solution is filtered, 220 mL of 10 mol/L sodium hydroxide solution are added to the brown filtrate. Then 310 mL of a 100 % (w/v) TCA is added and mixed thoroughly, the solution becomes green. Fixing and staining is performed for 3 h at 50 °C or over night at room temperature in the colloidal solution. The acid is washed out for 1 to 2 h in tap water, the green bands turn into blue and become more intense.

Figure 3 shows the silver stained gel of carot proteins, separated by isoelectric focusing in an immobilized pH gradient pH 4-9.

The resolving power of the electrophoretic methods described above is very high, when compared to chromatography procedures. However, a sample very often contains more than 50 different proteins, sometimes even several thousands. No one-dimensional method is capable of resolving such mixtures in one go.

Since the scientific community became aware that DNA sequence data alone can not give information about the level of the gene expressions, isoforms of enzymes, posttranslational modification of proteins, and the cellular and subcellular distribution of the proteins, the development of powerful protein analysis techniques reached a high priority. The "Proteome" analysis [Wilkins 1996] depends on the ability to separate complex protein mixtures rapidly and relies on two basic technologies:

1. Two-dimensional electrophoresis of all cell or tissue proteins.
2. Identification or sequencing of individual proteins.

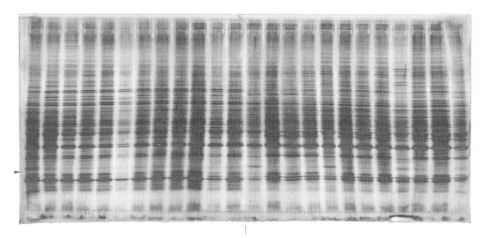

Figure 3. Isoelectric focusing of carot seed proteins in an immobilized pH gradient 4–9. Protein detection with silver staining [Posch et al. 1995].

In the following section we describe the technical aspects of two-dimensional electrophoresis.

4.6 Two-Dimensional Electrophoresis (2-DE)

When isoelectric focusing and SDS electrophoresis are combined to a two-dimensional technique, up to a magnitude of several hundreds to more than thousand proteins can be separated in one gel. The method has an extremely high resolving power, because it separates according to the physico-chemical parameters charge and size, which are completely independent from each other.

Traditional technique. The principle of high resolution 2-DE was introduced by P.H. O'Farrell [1975] and J. Klose [1975]. The main steps were:
- Extraction or dilution of proteins with lysis buffer containing SDS and 2-mercaptoethanol (all proteins should go into solution, prevention of protein aggregates and hydrophobic interactions, prevention of different oxidation steps).
- Isoelectric focusing in 1 to 3 mm thin 4–5 %T polyacrylamide gel rods inside glass tubes under denaturing conditions (9 M urea and 2 % non-ionic detergent). Samples were normally loaded at the cathodal side.
- Equilibration of the gels in 1 % SDS buffer for ca. 10 min.
- SDS PAGE on a 1.5 mm thick vertical slab gel for a separation according to the molecular weights.
- Detection of spots by staining, immunoblotting, autoradiography and fluorography.

However, this procedure has shown a number of limitations:

- Lack of reproducibility, mostly because of problems with the first dimension, the isoelectric focusing step, due to batch to batch reproducibility of the carrier ampholytes and gradient drift.
- Loss of basic and acidic proteins due to gradient drift.
- Sample components influence gradient profile.
- Low loading capacity.
- Personal skill influences results.

NEPHGE (non equilibrium pH gradient gel electrophoresis) with the sample loaded on the anodal side has been introduced as a modification for including the basic proteins [O'Farrell et al. 1977]. However, this method is not very reproducible because it is not a steady-state technique.

For monitoring the complete protein compositions, performing inter-laboratory comparisons and evaluations over long periods, a higher reproducibility of the method and the inclusion of the acid and basic proteins are an urgent demand [Aebersold 1991]. Also the loading capacity had to be increased, in order to use the power of the method for micropreparative separations.

New technique. Therefore, an alternative procedure has been developed by Görg et al. [1988, 1995]: Dried gels containing immobilized pH gradients (IPG) on film supports are cut to 3–5 mm thin strips. Those are rehydrated to the original thickness in a solution of 8 mol/L urea, 1 % (w/v) CHAPS, 0.2 (w/v) DTT, and 0.25 % (w/v) carrier ampholyte mixture overnight before they are used in the first dimension. With this technique even very basic ribosomal proteins and histones with isoelectric points up to 12 can be analyzed [Görg et al. 1997].

The samples are either applied with applicator cups at the anodal or the cathodal end of the rehydrated strips, or the dry strips are rehydrated with the sample solution [Rabilloud et al. 1994]. This in-gel sample application is particularly beneficial when high volumes have to be loaded for micropreparative applications.

These immobilized pH gradients in film-supported gel strips have a number of advantages over the classical method:

- Higher resolution (Δ pI = 0.001) with ultranarrow pH gradients [Görg et al. 1985].
- Improved reproducibility (interlaboratory comparison) [Blomberg et al. 1995].
- Steady-state isoelectric focusing of alkaline proteins up to pH 12 [Görg et al. 1991, 1997].
- Higher loading capacity (up to 10 mg protein/2-D gel) for micropreparative 2-D [Bjellqvist et al. 1993].

First dimension separation. Isoelectric focusing in immobilized pH gradient strips is performed in a horizontal chamber at 20 °C for several hours to overnight.

Subsequently, the strips are rebuffered in SDS equilibration buffer and transferred to the second dimension SDS gel, or they are stored between two plastic sheets at −60 °C to −80 °C. For rebuffering they are equilibrated two times for 15 min in a solution of 50 mmol/L Tris-HCl pH 8.5, 2 % (w/v) SDS, 30 % (v/v) glycerol, 6 mol/L urea, 1 % (w/v) DTT and a trace of bromophenol blue. The second equilibration solution contains 260 mmol/L iodoacetamide instead of DTT.

Second dimension separation. The SDS electrophoresis step is less critical than isoelectric focusing. For multiple runs, mostly *vertical* set-ups are employed. For an optimal contact from the first to the second dimension gel, the IPG strip is sealed on the edge with 0.5 % agarose containing electrode buffer (25 mmol/L Tris, 192 mmol/L glycine and 0.1 % SDS pH 8.3).

Horizontal gel systems are perfectly suited for the use of ready-made SDS gels, cast on plastic backing. The dissipation of Joule heat is very efficient, because the thin gels are run on a cooling plate. Here, after equilibration, the IPG strip is blotted with filter paper, and applied with the gel side facing down onto the stacking area of the SDS gel. A very important feature of these horizontal gels on carrier films is, that they do not break, swell or deform during staining, drying and storage. This makes computer aided evaluation much easier and faster.

Because of the good reproducibility of the IPG 2-D electrophoresis method, many proteins can be identified by comparison of their positions in 2-D gels using new software with database function, and databases on the World Wide Web. Qualitative and quantitative changes of certain protein spots can be determined and studied.

Also, the proteins can be completely transferred to an activated glass fiber membrane or other material for subsequent amino acid analysis, peptide sequencing, or mass spectrometry (see Chapter III.3 in this book by C. Eckerskorn).

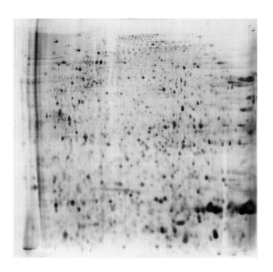

Figure 4. Two-dimensional electrophoresis of mouse liver proteins with isoelectric focusing in IPG 4 to 9 in the first dimension and a vertical SDS gel 13 % *T* in the second dimension. Protein detection with silver staining.

Two-dimensional electrophoresis is the most powerful micropreparative protein purification method [Pennington et al. 1997], because:

1. Proteins can be purified without any prior knowledge of their properties.
2. The crude sample can be loaded: no prepurification (like chromatography) has to be employed.
3. Several hundreds to more than thousand proteins can be purified in one gel.
4. Complex protein mixtures are separated rapidly.
5. Due to the extremely high resolution, in most cases a spot represents a single protein.
6. The proteins are delivered in a completely denatured form, which makes *N*-terminal and internal sequence analysis much easier than after native electrophoresis.
7. The immobilized pH gradient technique allows a much higher sample load than the traditional method; thus also minor components can be identified with the newly developed, very sensitive techniques. Narrow gradients can be used for further increasing of the loading capacity.

Additionally, the post 2-D analysis techniques for protein identification and characterization have been considerably improved:

1. New possibilities in image analysis and databases of 2-D maps with powerful computers and software are available.
2. Microsequencing (Edman degradation) and amino acid composition analysis have become much more sensitive, and many powerful databases with protein and DNA sequence informations are accessible.
3. Mass spectrometry for the determination of the exact protein mass, peptide mass fingerprint, or ladder sequencing provide both, high accuracy and high throughput.

5 References

Aebersold R. in: *Advances in Electrophoresis* (Chrambach A, Dunn MJ, Radola BJ, Eds) Volume 4 (1991) 81–168.

Anderson NL, Anderson NG. *Anal Biochem.* 85 (1978) 341–354.

Anderson NL, Anderson NG. *Proc Nat Acad Sci USA.* 74 (1988) 5421–5425.

Andrews AT. Electrophoresis, *Theory Techniques and Biochemical and Clinical Applications.* Clarendon Press, Oxford (1986).

Bjellqvist B, Sanchez J-C, Pasquali C, Ravier F, Paquet N, Frutiger S, Hughes GJ, Hochstrasser D. *Electrophoresis.* 14 (1993) 1375–1378.

Blomberg A, Blomberg L, Norbeck J, Fey SJ, Larsen PM, Roepstorff P, Degand H, Boutry M, Posch A, Görg A. *Electrophoresis.* 16 (1995) 1935–1945.

Davis BJ. *Ann NY Acad Sci.* 121 (1964) 404–427.

Diezel W, Kopperschläger G, Hofmann E. *Anal Biochem.* 48 (1972) 617–620.

Görg A, Postel W, Weser J, Patutschnick W, Cleve H. *Am J Hum Genet.* 37 (1985) 922–930.

Görg A, Postel W, Günther S. *Electrophoresis.* 9 (1988) 531–546.

Görg A. *Nature.* 349 (1991) 545–546.

Görg A, Boguth G, Obermaier C, Posch A, Weiss W. *Electrophoresis.* 16 (1995) 1079–1086.

Görg A, Obermaier C, Boguth G, Csordas A, Diaz J-J, Madjar J-J. *Electrophoresis.* 18 (1997) 328–337.

Görg, A, Boguth, G, Obermaier, C, Weiss, W. *Electrophoresis* 19 (1998) 1516–1519.

Klose J. *Humangenetik.* 26 (1975) 231–243.

Laemmli UK. *Nature.* 227 (1970) 680–685.

Margolis L, Kenrick KG. *Anal Biochem* 25 (1968) 347–362.

Maurer RH. *Disk-Elektrophorese – Theorie und Praxis der diskontinuierlichen Polyacrylamid-Electrophorese.* W de Gruyter, Berlin (1968).

O'Farrell PH. *J Biol Chem.* 250 (1975) 4007–4021.

O'Farrell PZ, Goodman HM, O'Farrell PH. *Cell* 12 (1977) 1133–1142.

Ornstein L. *Ann NY Acad Sci.* 121 (1964) 321–349.

Pennington SR, Wilkins MR, Hochstrasser DR, Dunn MJ. *Trends Cell Biol.* 7 (1997) 168–173.

Posch A, van den Berg BM, Burg HCJ, Görg A. *Electrophoresis* 16 (1995) 1312–1316.

Rabilloud T, Valette C, Lawrence JJ. *Electrophoresis.* 15 (1994) 1552–1558.

Shapiro AL, Viñuela E, Maizel JV. *Biochem Biophys Res Commun.* 28 (1967) 815–822.

Weiss W, Vogelmeier C, Görg A. *Electrophoresis* 14 (1993) 805–816.

Westermeier R. *Electrophoresis in Practice.* 2nd Ed. VCH Weinheim (1997).

Wilkins MR. *BioTechnology* 14 (1996) 61–65.

Microseparation Techniques II: Gel Electrophoresis for Sample Preparation in Protein Chemistry

Hermann Schägger

1 Introduction

The basic principle of all electrophoretic protein separation techniques is the migration of charged molecules in an electrical field. The many techniques available differ in the ways in which intrinsic or induced charges, charge/mass ratios, or protein sizes are used for separation.

Let us first consider proteins in the native state, and separation techniques making use of intrinsic protein charges.

Proteins with isoelectric points (pI) above and below the buffer pH used will carry negative and positive excess charges, respectively, and only one of the two groups of proteins will enter the gel. The other group of proteins will migrate in opposite direction and will be lost in the electrode buffer. Techniques making use of the protein intrinsic charge are therefore rarely used, except Laemmli-PAGE without SDS, and isoelectric focusing (IEF), where extremes of pH in electrode buffers prevent escape of basic and acidic proteins from the gel.

IEF is used both for separation of native proteins (native IEF) and for the more common separation of proteins denatured by high urea concentrations. The term "IEF" usually means "denaturing IEF". Denaturation by urea, in contrast to denaturation by SDS, does not alter the intrinsic protein charge.

Introduction of charged compounds binding to the proteins leads to "charge shift" protein separation techniques. The added compounds may keep proteins either in the native state (e.g. Coomassie Blue in Blue-Native-PAGE) or denature proteins, as in SDS-PAGE.

Denaturing electrophoretic techniques are much more common for sample preparation in protein chemistry than native techniques. Mostly they are performed on the analytical scale. However, the potency of electrophoretic techniques for isolation of proteins on the preparative scale in SDS-denatured as well as in the native state deserves much more attention than it has received to date. The focus of this chapter therefore is on the preparative use of recently developed techniques that we found to be the most useful and reproducible ones (Table 1).

Table 1. Selection of electrophoretic techniques.

Denaturing techniques	Native techniques
(Blue)-Laemmli-SDS-PAGE	Laemmli-PAGE without SDS
(Blue)-Tricine-SDS-PAGE	Blue-Native-PAGE (BN-PAGE)
electroelution	Native Electroelution
electroblotting	Native Electroblotting
Isoelectric focusing (IEF)	Native IEF (without urea)
2D: IEF/SDS-PAGE	
2D: BN-PAGE/SDS-PAGE	

2 Denaturing Techniques

2.1 Commonly Used SDS-Polyacrylamide Gel Electrophoresis Techniques for Protein Separation

From the many available SDS techniques [Hames 1990], the Laemmli-SDS-PAGE [Laemmli 1970], with advantages for separation of large proteins, and the Tricine-SDS-PAGE [Schägger 1987], with advantages for small proteins and peptides, are selected and are compared in Figure 1 using identical gel types and identical protein samples. These two techniques together cover the whole molecular mass range of proteins. Table 2 provides a guide for selection of the appropriate electrophoresis system and the optimal gel type for particular applications.

Gradient gels are only required if a wide range of molecular masses has to be covered. If there is no protein of interest above 100 kDa or 70 kDa, the Tricine-SDS-PAGE using uniform gels (Figure 2) is recommended, because casting of gels is easier, the low molecular mass range is covered, and there are advantages in electro-

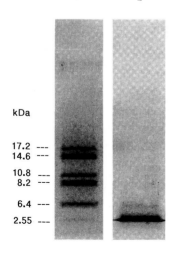

1   2

kDa

17.2 ---
14.6 ---

10.8 ---
8.2 ---

6.4 ---

2.55 ---

Figure 1. Comparison of resolution of Tricine-SDS-PAGE and Laemmli SDS-PAGE. Identical polyacrylamide gel types (10 % T, 3 % C) were used for Tricine-SDS-PAGE (*lane 1*) and Laemmli-SDS-PAGE (*lane 2*), and identical samples (cyanogen bromide fragments of myoglobin) were applied. The advantage of Tricine-SDS-PAGE for resolution of small proteins is obvious. Proteins with molecular masses above 30 kDa would be resolved better by the Laemmli-SDS-PAGE. %T and %C, see Figure 2. Reprinted from [Schägger 1987], with permission from *Analytical Biochemistry*.

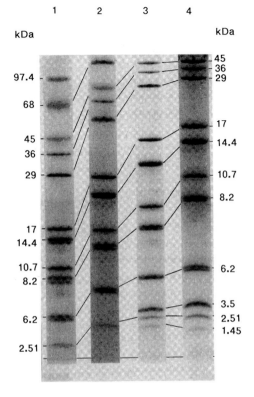

Figure 2. Resolution of Tricine-SDS-PAGE using different gel types.
Lane 1, 10 % T, 3 % C; *lane 2*, 16.5 % T, 3 % C; *lane 3*, 16.5 % T, 6 % C; *lane 4*, 16.5 % T, 6 % C plus 6 M urea. % T, total concentration of both monomers (acrylamide and bisacrylamide), % C, percentage of cross-linker relative to the total concentration. Reprinted from [Schägger 1987], with permission from *Analytical Biochemistry.*

Table 2. Choice of gel type and separation system.

Total range (kDa)	Gel type	System
6 – >250	8–16 % gradient	Laemmli
2–100	10 % uniform	Schägger
1–70	16.5 %	Schägger
Optimal range (kDa)		
50–100	8 %	Laemmli
20– 60	13 %	Laemmli
5– 50	10 %	Schägger
2– 30	16.5 %	Schägger
1– 20	16.5 %[(H)]	Schägger

[(H)]High crosslinker concentration is used.

blotting of large membrane proteins. Uniform acrylamide gels are also preferred for separation of proteins with similar masses. In addition to using the optimal gel types and electrophoresis systems, there are several other possibilities for achieving separation in problematic cases, as described in [Schägger 1994a].

2.2 Blue-SDS-PAGE for Recovery of Membrane Proteins from Gels

Usual protocols to revover proteins from conventional SDS-gels include several sequential steps, e.g.:

- fixation of proteins and staining with Coomassie Blue
- complete removal of acetic acid by several washings with water
- excision of stained bands and crushing of gel pieces
- solubilization by 0.1 % SDS, 50 mM NH_4HCO_3 (pH 9.2; 24 h)
- electroelution according to 2.3.

However, due to the initial fixation step this protocol does not work with membrane proteins. The membrane protein yield after electroelution often is close to zero.

Blue-SDS-PAGE offers an alternative way that works with membrane proteins as well as with water-soluble proteins. Migrating proteins are stained by Coomassie Blue *during* Blue-SDS-PAGE (Fig. 3). Therefore a fixation step is avoided, and the membrane proteins are still in soluble form when the stained bands are excised, and briefly incubated in 0.1 % SDS, 50 mM NH_4HCO_3, pH 9.2. Proteins can immediately be electroeluted.

The basic electrophoretic methods for Blue-SDS-PAGE are either the Laemmli-SDS-PAGE [Laemmli 1970] or the Tricine-SDS-PAGE [Schägger 1987]. In the cathode buffers of both systems the SDS concentration is halved and 25 mg Serva Blue G is added per liter. Under these conditions Coomassie dye can compete with SDS for binding sites on the protein, and proteins migrate as blue bands through the gels.

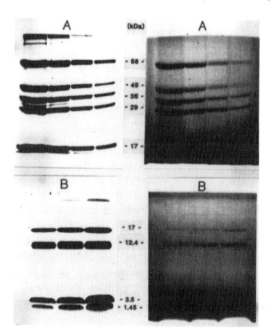

Figure 3. Comparison of protein staining intensity during Blue-SDS-PAGE (*right*) and after conventional restaining of the same gel (*left*). A: Proteins in the molecular mass range from 17-68 kDa were resolved by Blue-Laemmli-SDS-PAGE, using a 12 % T, 3 % C gel. B: Proteins in the molecular mass range from 1.45-17 kDa were resolved by Blue-Tricine-SDS-PAGE, using a 16.5 % T, 6 % C gel. The minimal load was 0.27 µg of each protein per mm^2. The highest load was 2.2 µg/mm^2 in A and 0.82 µg/mm^2 in B. Reprinted from [Schägger 1988], with permission from *Analytical Biochemistry.*

The migration behavior in the Blue Laemmli-SDS-PAGE and in the Blue-Tricine-SDS-PAGE [Schägger 1988] is almost identical to that of the conventional colorless techniques. More than 100 µg per protein band can be resolved on a preparative gel and recovered quantitatively. Use of Blue-SDS-PAGE for separation and recovery of low protein quantities requires some experience (see guidelines), since the staining intensities of protein bands in Blue-SDS-PAGE are considerably lower than after conventional fixation and staining. The minimal protein quantities that can be detected are in the range of 0.2–1 µg per protein band.

Guidelines for Blue-SDS-PAGE:

- Protein bands are detected best after fast runs and short migration distances.
- Staining and detection are better in Blue Laemmli-SDS-gels than in Blue Tricine-SDS-gels.
- Protein spots should be excised directly after the run, because the spots disappear after prolonged standing due to diffusion.

2.3 Electroelution of Proteins after Blue-SDS-PAGE

Electroelution [Schägger 1994a, 1988] was performed with an electroelutor/concentrator made according to Hunkapiller [1983]. A similar apparatus is commercially available from C.B.S. Scientific Company, distributed by ITC Biotechnology, Heidelberg, Germany. The H-shaped electroelutor vessel is composed of two vertical and a connecting horizontal tube. The lower ends of the two vertical arms are sealed by dialysis membranes with a cutoff limit of 2 kDa (Reichelt, Heidelberg). The vessel is placed on the barrier separating the anodic and cathodic compartments which are filled with electrode buffer (0.1 % Na-SDS, 100 mM NH_4HCO_3). The blue protein bands then are squeezed through a syringe directly into the cathodic arm, and incubated for 10 min in electrode buffer. The residual volume of the vessel then is filled with 50 mM NH_4HCO_3 (without SDS) and electroelution is performed at 40 V overnight or at maximally 70 V (considerable heating) for about 5 h. The recovery is nearly quantitative with large membrane proteins as well as with small peptides in the 1–2 kDa range. The proteins collect as a blue solution at the anodic membrane. Since only the SDS present within the elutor vessel (from gel and incubation solution) accumulates at the dialysis membrane, the proteins can be used directly for immunization or protein fragmentation [Schägger 1994a]. Before fragmentation by proteases tolerating low SDS concentrations or by chemical cleavage, e.g. by CNBr, NH_4HCO_3 is removed by lyophilization. Excess NH_4HCO_3 would prevent further resolution of fragments by SDS-PAGE.

2.4 Electroblotting of Blue and Colorless SDS Gels

Electroblotting techniques are discussed in Chapter II.3. One further technique seems to be worth mentioning that helps to retain many peptides in the 1-2 kDa range on PVDF membranes. Essential prerequisites are the use of the Blue-Tricine-SDS-PAGE for separation instead of the colorless SDS-PAGE, and the addition of 20 %

methanol to the anode buffer (Table 3). Except for electroblotting of peptides, the electroblotting of Blue-Tricine-SDS gels offers no advantage. On the contrary, the dye itself occupies binding surface on the PVDF membrane, and at high load causes more protein to pass. Addition of methanol is not recommended for transfer of large proteins.

Table 3. Buffers and electroblotting conditions.

Anode buffer:	300 mM Tris, 100 mM Tricine (+20 % methanol; only for transfer of 1-2 kDa peptides)
Cathode buffer:	300 mM aminocaproic acid, 30 mM Tris

Electrotransfer of proteins from 0.7 mm Tricine-SDS-gels:

10 % acrylamide:	1 mA/cm^2, 3 h
16.5 % acrylamide:	1 mA/cm^2, 5 h
	Voltage: 5–10 V

2.5 Isoelectric Focusing in the Presence of Urea

In contrast to SDS electrophoresis, which separates proteins according to size, iso-electric focusing (IEF) separates according to isoelectric points [Righetti 1990, see also Görg and Westermeier, Microseparation Techniques II.1 in this book]. In IEF using soluble carrier ampholytes a pH gradient develops within the gel during the run. In IEF using ampholytes covalently attached to the gel matrix (IPG-Dalt; Pharmacia) an immobilized pH gradient pre-exists. Besides this technical classification of IEF techniques, IEF can be divided into native (without urea) and denaturing variants. The most commonly applied technique uses urea at a concentration of 8 M or more, and neutral or zwitterionic detergents. The detergents would not necessarily denature proteins; however, 8 M urea does. IEF in the presence of urea is therefore a denaturing technique. It is almost exclusively used as the first step in 2D electro-phoresis [O'Farell 1975, Rickwood 1990, Anderson 1978, Görg 1988].

Advantages of 2D electrophoresis (IEF/SDS-PAGE): Starting from cell homoge-nates thousands of protein spots can be identified, and it is a very valuable technique for comparative analytical studies.
Disadvantages of 2D-electrophoresis:

- Preparative application can be tedious, because collection of spots from many gels might be required to obtain sufficient protein for N-terminal protein sequencing.
- It is not known which fraction of membrane proteins might not enter the IEF gel because of aggregation problems.
- Individual spots in the 2D gel cannot be assigned directly to the physiological protein assemblies. This is quite different to 2D gels using native techniques for the first dimension as described below.

3 Native Techniques

The native electrophoretic techniques are used for purification of proteins instead of or in addition to chromatographic procedures. They are pre-purification steps before denaturing techniques, as described above, are finally applied.

3.1 Colorless-Native-PAGE

Native electrophoresis of water-soluble proteins and protein complexes is preferentially tried with Laemmli-PAGE [Laemmli 1970], but SDS is completely omitted. All proteins with isoelectric points (pI) below 9.5 (running pH in the gel at room temperature) migrate to the anode; however, the applicability of this technique is restricted to rather pH insensitive proteins.

Another colorless native technique (CN-PAGE) [Schägger 1994] separates proteins at pH 7.5 and therefore is restricted to acidic proteins. Steep acrylamide gradient gels, e.g. from 4 to 20 % acrylamide, are used. The migration of proteins stops when appropriate pore sizes of the gel are reached, and migration distances are no longer determined by the intrinsic protein charge, but by the protein size (Figure 4).

3.2 Blue-Native-PAGE

Blue-Native-PAGE (BN-PAGE) is a technique that was initially developed for isolation of enzymatically active membrane proteins at pH 7.5 [Schägger 1991]. One essential component of the electrophoretic system is the dye Coomassie Blue G-250 (Serva Blue G) that is added to the cathode buffer. It competes with the neutral deter-

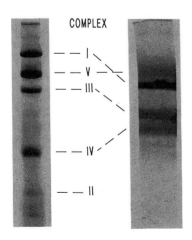

Figure 4. Comparison of resolution of Blue-Native-PAGE and Colorless-Native-PAGE using solubilized bovine heart mitochondria. One hundred micrograms total protein was applied corresponding to about 2-10 µg of the individual membrane protein complexes (complexes I-V) comprising 4 (complex II) to 41 (complex I) protein subunits. The identification of the complexes can be performed by enzymatic assays after native electroelution or by the characteristic polypeptide patterns observed after performing a second-dimension SDS-PAGE from each lane (see Figure 5). Reprinted from [Schägger 1994], with permission from *Analytical Biochemistry*.

gents required for membrane protein solubilization for binding sites on the protein surface. Since Coomassie Blue is negatively charged four main effects are observed:

- All proteins that bind Coomassie Blue (all membrane proteins and many water-soluble proteins) migrate to the anode at the running pH 7.5, even basic proteins. The application range is extended compared to CN-PAGE.
- Negatively charged protein surfaces will repel each other. Therefore the aggregation problem usually observed with membrane proteins is minimized.
- The negatively charged membrane proteins are soluble in detergent-free solution. Therefore any detergent can be omitted from the gel, and the risk of denaturation of detergent-sensitive membrane proteins is reduced.
- Proteins migrate as blue bands through the gel. This facilitates the detection and recovery of native proteins.

The main fields of application of BN-PAGE are:

- The native proteins and protein complexes are separated according to size. BN-PAGE can therefore be used for determination of molecular masses and oligomeric states of microgram amounts of partially purified proteins [Schägger 1994]. The high resolution not only allows discrimination between monomeric and dimeric forms, but also detection of subcomplexes in addition to holocomplexes. In this respect BN-PAGE is superior even to analytical ultracentrifugation.
- Final purification of mg amounts of partially purified membrane proteins from a single preparative gel. The highly pure proteins recovered by native electroelution are suitable, e.g., for immunization, for functional studies, and for protein chemical work [Schägger 1995].
- Purification of membrane proteins directly from biological membranes, e.g. from cell organelles or even from homogenized animal tissue. About 1 μg per protein band is required for detection during BN-PAGE. Proteins are recovered from the gel by native electroelution or native electroblotting [Schägger 1991, 1994b].
- Combined with Tricine-SDS-PAGE, in the second dimension the subunit composition of multiprotein complexes can be studied as shown in Figure 5. The minimal protein load to the first dimension BN-PAGE then only depends on the staining limits of Coomassie or silver staining procedures. This technique is currently used for localization of respiratory chain defects in human diseases, starting from 10 mg skeletal muscle [Schägger 1996]. Since the membrane protein complexes resolved in the native state in the first dimension are denatured only by the second dimension SDS-PAGE, the subunits of the multiprotein complexes can be identified by characteristic polypeptide patterns arranged in a row (Figure 5).

3.3 Native Isoelectric Focusing

The technique of native IEF is similar to the commonly used denaturing IEF, but urea is completely omitted. Due to the omission of urea, proteins are more prone to aggregation, especially when approaching the gel pH corresponding to their isoelectric points. Especially problematic, with respect to the problem of aggregation, is the reso-

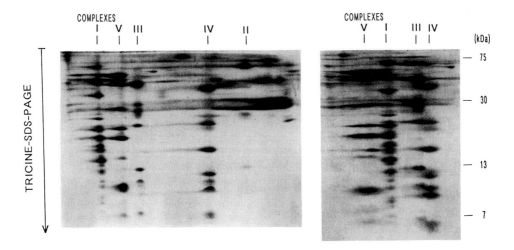

Figure 5. Two-dimensional resolution of the protein subunits of the complexes from bovine heart mitochondria. After first dimension native separation by BN-PAGE (*A*) or CN-PAGE (*B*) as shown in Figure 4, complete lanes were resolved by Tricine-SDS-PAGE in the denaturing second dimension. Reprinted from [Schägger 1994], with permission from *Analytical Biochemistry.*

lution of membrane proteins. The neutral or zwitterionic detergents required for solubilization are usually not very effective in keeping proteins solubilized, and they provoke the risk of dissociation of protein complexes and irreversible denaturation. Nevertheless, native IEF has been used with great success for final purification of special membrane protein complexes on the preparative scale [Tsiotis 1993]. Only after that final purification were the isolated membrane protein complexes suitable for crystallization. The purification was performed by the technique using mobile carrier ampholytes and Ultrodex (Pharmacia) as the gel matrix.

4 References

Anderson, N.G., and Anderson, N.L. (1978) *Anal. Biochem. 85*, 331–354. Two-Dimensional Analysis of Serum and Tissue Proteins: Multiple Isoelectric Focusing.

Görg, A., Postel, W., and Günther, S. (1988) *Electrophoresis 9*, 531–546. The Current State of Two-Dimensional Electrophoresis with Immobilized pH Gradients.

Hames, B.D. (1990). One-dimensional polyacrylamide gel electrophoresis. In: *Gel Electrophoresis of Proteins* (Hames, B.D., Rickwood, D.; eds.), pp. 1–139, IRL Press, Oxford.

Hunkapiller, M.W., Lujan, E., Ostrander, F., and Hood, L.E. (1983). Isolation of Microgram Quantities of Protein from Polyacrylamide Gels for Amino Acid Sequence Analysis. In: *Methods in Enzymology* (Hirs, C.H.W., Timasheff, S.N.; eds.), Vol. 91, pp. 227–236, Academic Press, New York.

Laemmli, U.K. (1970) *Nature 227*, 680–685. Cleavage of Structural Proteins during the Assembly of the Head of Bacteriophage T4.

O'Farell, P.H. (1975) *J. Biol. Chem. 250*, 4007–4021. High Resolution Two-Dimensional Electrophoresis.

Rickwood, D., Chambers, J.A.A., Spragg, S.P. (1990). Two-Dimensional Gel Electrophoresis. In: Gel Electrophoresis of Proteins (Hames, B.D., Rickwood, D.; eds.), pp. 217–272, IRL Press, Oxford.

Righetti, P.G., Gianazza, E., Gelfi, C., and Chiari, M. (1990). Isoelectric Focusing. In: *Gel Electrophoresis of Proteins* (Hames, B.D., Rickwood, D.; eds.), pp. 149–216, IRL Press, Oxford.

Schägger, H., and von Jagow, G. (1987) *Anal. Biochem. 166*, 368–379. Tricine-Sodium Dodecyl Sulfate-Polyacrylamide Gel Electrophoresis for the Separation of Proteins in the Range from 1 to 100 kDa.

Schägger, H., Aquila, H., and von Jagow, G. (1988) *Anal. Biochem. 173*, 201–205. Coomassie Blue-Sodium Dodecyl Sulfate-Polyacrylamide Gel Electrophoresis for Direct Visualization of Polypeptides during Electrophoresis.

Schägger, H., and von Jagow, G. (1991) *Anal. Biochem. 199*, 223–231. Blue Native Electrophoresis for Isolation of Membrane Protein Complexes in Enzymatically Active Form.

Schägger, H., Cramer, W.A., and von Jagow, G. (1994) *Anal. Biochem. 217*, 220–230. Analysis of Molecular Masses and Oligomeric States of Protein Complexes by Blue Native Electrophoresis and Isolation of Membrane Protein Complexes by Two-Dimensional Native Electrophoresis.

Schägger, H. (1994a) Chapter 3: Denaturing Electrophoretic Techniques. In: *A Practical Guide to Membrane Protein Purification* (Von Jagow, G., Schägger, H.; eds.), in press, Academic Press, New York.

Schägger, H. (1994b) Chapter 4: Native Electrophoresis. In: *A Practical Guide to Membrane Protein Purification* (Von Jagow, G., Schägger, H.; eds.), in press, Academic Press, New York.

Schägger, H. (1995) Native Electrophoresis for Isolation of Mitochondrial Oxidative Phosphorylation Protein Complexes. In: *Methods in Enzymology* (Attardi, G.M., Chomyn, A., eds.), Vol. 260, pp. 190–202.

Schägger, H. (1996) Electrophoretic Techniques for Isolation and Quantification of Oxidative Phosphorylation Complexes from Human Tissues. In: *Methods in Enzymology* (Attardi, G.M., Chomyn, A., eds.), Vol. 264, pp. 555–566.

Tsiotis, G., Nitschke, W., Haase, W., and Michel, H. (1993) *Photosynth. Res. 35*, 285–297. Purification and Crystallisation of Photosystem I Complex from a Phycobilisome-less Mutant of the Cyanobacterium Synechococcus PCC 7002.

Microseparation Techniques III: Electroblotting

Christoph Eckerskorn

1 Introduction

The combination of sodium dodecylsulphate polyacrylamide gel electrophoresis (SDS-PAGE) with electroblotting is one of the most versatile methods for the isolation of proteins at the microgram and submicrogram level for further protein chemical analysis (Figure 1). This strategy has several advantages compared to other protein isolation procedures.

- First, the separation method offers the high resolution potential of one and two dimensional (1D and 2D) PAGE. This is advantageous for complex protein mixtures, and especially for the separation of membrane proteins.
- Second, the sample handling steps are minimized due to direct transfer of the separated protein onto a membrane. Then the immobilized protein could be either directly sequenced [Vandekerckhove 1985; Aebersold 1986; Matsudaira 1987; Eckerskorn 1988a] or subjected to amino acid composition analysis [Eckerskorn 1988 a,b; Ploug 1989; Tous 1989; Nakagawa 1989], mass spectrometry [Eckerskorn 1992; Strupat 1994] and chemical or proteolytic cleavage [Aebersold 1987; Scott 1988; Eckerskorn 1989; Jahnen 1990; Patterson 1992].
- Third, proteins purified by this technique can be prepared in a short time relatively free from contamination by other proteins, amino acids and salts.
- Fourth, very small amounts of proteins can be handled with high yields.

The key step in this technique is the quantitative transfer and immobilization of the proteins onto a suitable membrane.

2 Electroblotting

The term "electroblotting" means transfer of electrophoretically separated proteins from the polyacrylamide matrix onto a "protein adsorbing" membrane in an electric field. This technique, described by Renart [1979] and at the same time by Towbin [1979], was first established as "Western blotting" to detect electrophoretically sepa-

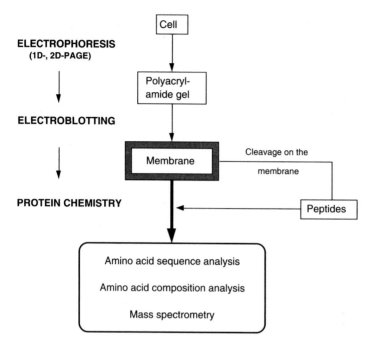

Figure 1. Strategy for obtaining protein chemical information from electrophoretically separated proteins. 1D and 2D gel electrophoresis are the most versatile and highly resolving methods for the separation of proteins. The subsequent isolation of the separated proteins from the gel matrix in a form suitable for protein chemical analysis is a key step in the characterization of these proteins. With the development of efficient blotting techniques and the introduction of chemically inert membranes it is now possible to recover the proteins present in low quantities from the polyacrylamide gel with high yields. The immobilized proteins are suitable for direct sequence analysis, amino acid composition analysis, mass spectrometry and proteolysis. This combination of protein chemical and electrophoretic techniques makes it possible to obtain chemical information from subpicomole quantities of protein, resulting in access to a new set of biologically important proteins.

rated proteins with specific binding properties (antigens, glycoproteins, enzymes, etc.) with antibodies, lectins, substrates, etc. directly on membranes [for review see Beisiegel 1986]. With the development and introduction of sufficient chemically inert membranes, proteins are now directly amenable to protein chemical analysis [for review see Aebersold 1991; Eckerskorn 1990].

2.1 Polyacrylamide Gel Electrophoresis

Proteins which are to be transferred for further analysis onto a membrane can be separated with the most common gel systems according to published protocols [Laemmli 1970; O'Farrell 1975; Görg 1988; Swank 1971; Schägger 1987] with respect to optimal separation of the protein(s) of interest. The concentration of the protein in the gel matrix should be as high as possible (concentrated, salt-reduced sample; small sample slots; thin gels). The concentration of the polyacrylamide and/or the cross-linker should be as low as possible without disturbing the separation of the desired protein. This facilitates the elution of the proteins from the gel matrix during the electrotransfer in the subsequent blotting step.

For any procedure used to enrich or isolate proteins, prevention of chemical modification of the α-amino group (sequence analysis) and the reactive side groups (mass spectrometry, amino acid composition analysis) is a major concern. Sources of N-terminal blocking common to any isolation technique involving polyacrylamide separation include unidentified impurities. These are contained in some batches of buffers or acrylamide and, possibly, amino-reactive moieties generated during polymerization of the acrylamide. For this reason, extensively polymerized gels should be used and each batch of chemicals should be tested with a standardized protein preparation to ensure the absence of blocking impurities. This could be done by comparing sequencing results of aliquots of (radioiodinated) proteins (β-lactoglobulin, trypsin inhibitor) either directly subjected to Edman degradation, or sequenced after gel electrophoresis and electroblotting onto a membrane. We found no statistically significant N-terminal blockage caused by electroblotting when high-purity reagents were used (such as electrophoretic-grade reagents) and allowing the gel to stand several hours at room temperature to complete polymerization. Precautions such as preelectrophoresis prior to sample loading are usually not recommended.

2.2 Blot Systems

2.2.1 Tank Blotting

The standard system for tank blotting is designed according to a construction of Bittner [1980] as vertical buffer tank with platinum wires as electrodes on the side walls. Gel and membrane are assembled between a stack of filter papers and mounted into a grid cassette. Constant slight pressure has to be applied to the blot sandwich to avoid displacement (shifting) of the gel and membrane. This entire "packing procedure" has to be performed in a tube under enough buffer to avoid air bubbles between the sandwich layers. The assembled cassettes are fixed vertically in the buffer tank. Between 2 and 4 l buffer is necessary, depending on the design of the system. Most experiments have been run with a constant voltage of about 50 V to exert uniform electrical power on the charged particles. The initial value of the current is typically 500 mA or higher, depending on the size of the tank and the molarity of the buffer used. During transfer, the value of the current increases due to a continuous increase of the electric resistance of the solution. Under these conditions very efficient cooling is necessary, which is achieved through a vertical cooling insert and sufficient buffer circulation.

2.2.2 Semidry Blotting

The semidry blotting system first described by Kyhse-Andersen [1984] consists of two plates as electrodes in which a blot sandwich made out of filter papers, gel and membrane is mounted horizontally. Compared to tank blotting this assembly is much simpler because no cassettes are required. The filter papers are soaked in buffer, the gel and the blotting membrane are successively arranged in layers on the anode in order (Figure 2). If necessary, air bubbles can easily be removed by rolling a glass rod gently over each layer. The amount of buffer required depends on the size of the blot sandwich and is typically less than 100 ml. Because of the considerably less amount of buffer compared to the tank blotting systems and the shorter transfer times, the proteins are exposed to fewer impurities during the electrotransfer.

Several companies offer different materials for the electrode plates (e.g. graphite, sintered glass carbon, metal covered with platinum, graphite dispersed in a polymeric matrix, conductive polymeric matrices). These differ in their conducting capacity, and their stability against oxidation processes at the anode and at extreme pH values (see below). In most cases, the semidry blotting experiments are performed with constant current (e.g. 1 mA/cm² surface of the blot sandwich), resulting in low voltage values (< 5 V) at the beginning. During transfer, the voltage increases with the electrical resistance depending on the buffer system used, the absolute amount of buffer (= numbers of filter papers and the degree of saturation), the thickness of the gel, and the material of the electrode. After about 3 h, the voltage typically reaches values between 20 V and 50 V. Because of the high electrical conductivities of the electrodes and the relatively low electric power, cooling of the semidry apparatus is not necessary.

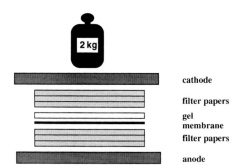

Figure 2. Schematic diagram showing the assembly of the blotting sandwich. Electroblotting can be performed essentially as follows: Filter papers and the blotting membrane should be trimmed exactly to the size of the separation gel. The filter papers should be washed three times for 15 min in transfer buffer (e.g. 50 mM boric acid, adjusted to pH 9.0 with 2 M sodium hydroxide, 20 % methanol for the filter papers on the anodic side and 5 % methanol for the filter papers on the cathodic side). The blotting membrane should be washed three times in 70 % methanol. Immediately after electrophoresis the blot sandwich should be assembled. Electrodes should be rinsed extensively in water before use and the sandwich is built from the lower anode to the upper cathode plate as shown in the schematic diagram above. The thickness of each filter paper stack should be at least 3 mm. Air bubbles between layers can easily be avoided by gently rolling a glass rod over each layer. Typical electrotransfer can be performed with constant current (1 mA/cm²) for 3–5 h.

The electrochemical reaction of water generates a pH gradient (Figure 3) of ca. pH 12 at the cathode (4 H_2O + 4 e^- ——>; $2H_2$ + 4 OH^-) and ca. pH 2 at the anode (6 H_2O ——> O_2 + 4 H_3O^+ + 4 e^-) and a continuous gas evolution. The gas expansion causes the blot sandwich to expand and thereby increases the electrical resistance. This results in a rapid and non reproducible increase of the electrical field (Figure 4). This effect can be reduced by loading a top weight of approximately 2 kg onto the semidry apparatus (Figure 2). The stability of the electrodes varies greatly: all pure graphite electrodes and electrodes with graphite dispersed in a polymeric matrix are oxidized from the oxygen originating at the anode to CO_2. Some of the polymeric matrices are dissolved at the surface due to the high pH at the cathode. Platinum electrodes or electrodes covered with platinum are almost completely stable.

Over the last few years, semidry blotting has become more popular than tank blotting for several reasons: the reduction of buffer consumption and minimization of potential reactive impurities, reduction of air bubbles due to the straightforwardness of assembly, and the significant reduction of transfer time and heat development. In addition to the much greater ease in handling the semidry apparatus, a systematic comparison of the two systems by Tovey and Baldo [1987] showed more efficient protein transfer in a more homogeneous voltage field in semidry blotting. The authors also described slightly more sensitive staining of proteins on the membranes after semidry blotting, because the proteins obviously stack more at the membrane surface under these transfer conditions.

The following experiments all refer to semidry blotting.

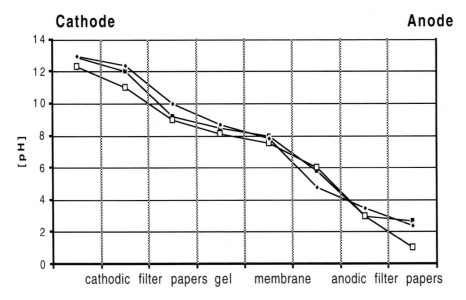

Figure 3. pH values of filter papers in a semidry blotting experiment. The pH values of individual filter papers (thickness 1 mm) were determined after 1 h transfer time. The thickness of the polyacrylamide gel was 1.5 mm. The transfer buffers were (◆): 10 mM 3-cyclohexylamino-1-propanesulfonic acid (CAPS), pH 11, 10 % methanol, (■): 50 mM Na-borate, pH 9, 20 % methanol (□): bidistilled water.

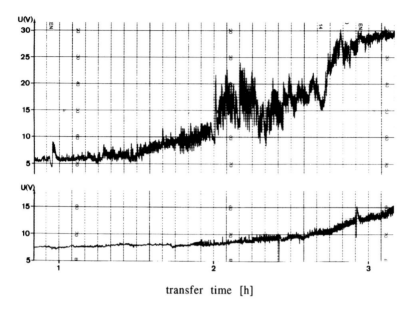

transfer time [h]

Figure 4. Progression of the voltage during a semidry blotting experiment. The progression of the voltage of a standard blotting assembly between pure graphite electrodes (compare Figure 2) was measured without a top weight (*top*) and with a top weight of 2 kg (*bottom*) mounted onto the semidry apparatus. Blotting was performed with constant current ($1\,mA/cm^2$).

2.3 Blotting Parameters

2.3.1 The Blotting Process

During SDS-PAGE proteins migrate in the polyacrylamide matrix as protein-SDS complexes at a velocity of several centimeters per hour. Precipitation is prevented by SDS bound to the proteins. SDS, through its strongly acidic sulphate groups, also provides the main driving force for protein movement through the gel matrix in an electrical field. The required constant reloading of SDS molecules on the proteins is guaranteed by a sufficient concentration of SDS in all electrophoresis buffers.

Unlike the electrophoresis buffers, electroblotting buffers cannot be supplemented with significant amounts of SDS, because even low concentrations of this detergent prevent protein adsorption to the membranes. Therefore, in the blotting process the elution of the protein from the gel matrix towards the membrane surface is driven by the SDS bound during the electrophoresis to the proteins and free SDS molecules from the electrophoresis buffer remaining in the gel volume. During blotting the protein-SDS complexes migrate towards the anode in a medium of continuously decreasing concentration of free SDS. In different protein-SDS complexes, binding of the detergent depends on the SDS-to-protein ratio [Putnam 1945]. Therefore, a decreased SDS concentration will reduce the SDS concentration on the protein surface, disclosing hydrophobic sites which may interact with the hydrophobic membrane. Low molecular weight proteins bind to a lower degree because the protein-SDS complexes migrate faster, so that the amount of SDS in these complexes is not sufficiently decreased. As a result of the dissociation of SDS, migration of the protein-SDS complexes will slow down and some of the proteins, especially high molecular weight proteins, may even become insoluble and precipitate in the gel. The physical properties of the proteins have an additional important influence on the blotting efficiency. The migration velocity is dependent on the size of the proteins. The strength of the interaction between proteins and SDS, polyacrylamide, and membrane surface depends on the amino acid composition and sequence. A detailed analysis of the blotting parameters and a postulated blotting mechanism is described by Jungblut [1990].

2.3.2 Transfer Buffers

A series of blotting buffers is described in the literature:

- 25 mM N-ethylmorpholine, pH 8.3, 0.5 mM dithiothreitol [Aebersold 1988].
- 25 mM Tris/HCl, 10 mM glycine, pH 8.3, 0,5 mM dithiothreitol [Aebersold 1986].
- 50 mM Na-borate, pH 8.0, 0.02 % β-mercaptoethanol, 20 % methanol [Vandekerckhove 1985].
- 10 mM 3-cyclohexylamino-1-propanesulfonic acid (CAPS), pH 11, 10 % methanol [Matsudaira 1987].
- 50 mM Na-borate, pH 9, 20 % methanol [Eckerskorn 1988].

The choice of blotting buffer is not critical if buffers with similar ion strength are used. The influence of the molarity of the transfer buffers was analyzed in detail for Na-borate buffers by Jungblut [1990]. The authors found a significantly decreased transfer yield if the molarity of Na-borate was reduced to less than 10 mM or increased above 100 mM. The pH of the transfer buffer is mainly determined – as described above – with increasing transfer time by the electrochemical reaction of the water at the electrodes. The pH between gel and membrane was maintained around 8.3 due to the (limited) buffer capacity of the running electrophoresis buffer remaining in the gel matrix. The effect of the extreme pH values could be minimized for a gel with a thickness of 1.5 mm if the filter stacks at both electrodes were at least 3 mm and the pH of the Na-borate buffer was $\geq$ 9. Another way to reduce the influence of the electrochemical reactions on the transfer pH is to use discontinuous buffer systems, e.g. 300 mM Tris/ 20 % methanol, pH 10.4 [Kyhse-Andersen 1984] for the filter papers directly contacting the anode to neutralize the protons produced during blotting.

2.3.3 Addition of SDS

In SDS-free transfer buffer hydrophobic proteins, especially membrane proteins with a reduced solubility, and proteins with high molecular weight elute out of the gel matrix in only small amounts. The majority of the proteins remain as precipitates in the gel. The addition of small amounts of SDS (up to 0.01 %) to the transfer buffer for the cathodic site leads to continuous subsequent delivery of SDS. In this way most of these hydrophobic proteins are maintained sufficiently in solution to achieve nearly quantitative elution out of the gel. However, for the reason already discussed above, SDS generally prevents protein adsorption to the membrane, and very low transfer yields were obtained, especially for the hydrophilic and small proteins.

2.3.4 Addition of Methanol

With the addition of methanol, a significant increase of transfer yields to the membranes was obtained. The stability of the SDS-protein complexes is influenced by methanol, as they dissociate more easily at increasing concentrations of methanol. The dissociated SDS molecules are subtracted towards the anode, which leads to an increasing interaction between transferring protein and membrane as well as between protein and polyacrylamide. With the decrease of free SDS molecules during the blotting process, both the solubility and the velocity of protein-SDS migration in the electrical field are influenced. Because of this "retarding" effect of methanol, small proteins and peptides can be more efficiently transferred if methanol is added (up to 40 %) to the cathodic transfer buffer, whereas for hydrophobic or large proteins, methanol on the cathode side should be avoided. The hydrophobic, uncharged blotting membranes require the presence of methanol (10–20 %) to render the interaction between membrane surface and protein possible.

2.3.5 Influence of Protein Concentration

Beside the physico-chemical properties of individual proteins (molecular weight, charge, charge distribution, etc.), the influence of the blotting parameters is also dependent on the concentration of the proteins in a given volume of gel. A low protein concentration in a gel band leads to a high ratio of the constituents of the electrophoretic running buffer (SDS, Tris, glycine) to the protein and vice versa. A (nearly) quantitative transfer can only be expected if the blotting conditions are optimized for the particular protein concentration. In a 2D separation of complex protein mixtures derived from cell lysates, for instance, the ratio of protein to gel matrix is determined by the abundance of the corresponding protein in the cell. To obtain an overview of transfer yields of proteins with different sizes, charges and concentrations, ^{14}C-labelled protein lysates of liver cells were separated in a 2D gel and blotted [Jungblut 1990]. The transfer yields of 50 arbitrarily selected proteins was between 60 % and 100 %, quantified by autoradiography. About 80 % of the evaluated proteins gave transfer yields between 75 % and 85 %.

3 Blotting Membranes

A variety of membranes for blotting have been introduced (Table 1). These supports differ in their composition and texture, being either glass fiber based, modified with positively charged organic groups [Vandekerckhove 1985; Aebersold 1986] or hydrophobic and uncharged [Eckerskorn 1988a], or being pure organic polymers such as polyvinylidene difluoride [Pluskal 1986; Matsudaira 1987] or polypropylene [Lottspeich 1989]. However, with the increasing number of membranes available, determining which membrane is most suitable for further protein chemical analysis becomes more and more important. Many investigations have been performed to elucidate the parameters for high protein transfer onto membranes and to improve the initial and repetitive yields in protein sequence analysis of electroblotted proteins [Xu 1988; Eckerskorn 1988a, 1990, 1991; Moos 1988; Walsh 1988; Lottspeich 1989, 1990; Jungblut 1990; Jacobsen 1990; Lissilour 1990; LeGendre 1990; Baker 1991; Aebersold 1991; Mozdzanowski 1992; Reim 1992]. The key to understanding these divergent results is comprehension of the nature of the membranes and the parameters which influence successful blotting and protein analysis. Structural parameters including specific surface area (Table 2), pore size distribution and pore-volumes (Table 3), and permeabilities of different solvents have been analysed and allow discrimination between membranes relative to their accessible surfaces and membrane densities [Eckerskorn 1993]. Protein binding capacities as well as protein recoveries

Table 1. Hydrophobic blotting membranes suitable for the isolation of proteins from polyacrylamide gels for a subsequent protein chemical analysis.

Name	Description	Manufacturer / Distributor
QA-GF	glass fiber membranes covalently modified with silanes	–
PGCM1	glass fiber membranes coated noncovalently with polybases	Life Science Products, Janssen, Belgium
Glassybond	glass fiber membranes covalently	Biometra, Göttingen, Germany
Immobilon P	Polyvinylidene difluoride	Millipore, Bedford, USA
Immobilon PSQ	Polyvinylidene difluoride	Millipore, Bedford, USA
Fluorotrans	Polyvinylidene difluoride	Pall, Dreieich, Germany
Problott	Polyvinylidene difluoride	PE-ABI, Foster City, USA
Trans Blot	Polyvinylidene difluoride	Bio-Rad, Munich, Germany
Westran	Polyvinylidene difluoride	Schleicher & Scüll, Dassel, Germany
Selex 20	Polypropylene	Schleicher & Schüll, Dassel, Germany
SM 17558	Polypropylene	Satorius, Göttingen, Germany
SM 17507	Polypropylene	Satorius, Göttingen, Germany
SM 17506	Polypropylene	Satorius, Göttingen, Germany

Table 2. Specific surface areas and thickness of blotting membranes. The specific surface area is related to 1 m^2 geometric surface area of the corresponding membrane [Eckerskorn 1993].

Membrane	Specific surface area (m^2)	Thickness (μm)
Trans-Blot	2900	140
Immobilon PSQ	1900	195
Fluorotrans	1600	140
Selex 20	900	145
SM 17507	880	150
SM 17506	810	160
SM 17558	610	100
Westran	570	150
Immobilon P	380	130
Glassybond	130	315

Table 3. Pore size distribution of blotting membranes [Eckerskorn 1993].

Membrane	Pore Size (Distributer) [μm]	Minimum Pore Size [μm]	Maximum Pore Size [μm]	Mean Pore Size [μm]	Pore Volume [%]
SM 17558	0.10	0.107	0.223	0.136	62.1
Immobilon PSQ	0.10	0.169	0.447	0.248	78.1
Selex 20	0.20	0.190	0.533	0.269	72.0
SM 17507	0.20	0.203	0.655	0.278	70.4
Trans-Blot	0.20	0.225	0.532	0.315	75.2
Fluorotrans	0.20	0.232	0.572	0.333	77.1
SM 17506	0.45	0.323	0.880	0.393	80.1
Immobilon P	0.45	0.517	1.136	0.692	68.4
Westran	0.45	0.571	1.361	0.779	62.7
Glassybond	>>	1.921	7.263	2.296	93.3

in electroblotting correlate with these membrane properties. Almost quantitative retention of proteins during electroblotting from gels was obtained for membranes with a high specific surface area and narrow pores (Trans-Blot, Immobilon PSQ, Fluorotrans), whereas membranes with a relatively low specific surface area (Immobilon P, Glassybond) showed reduced recoveries of about 10–20% for the tested proteins [Eckerskorn 1993].

In the same study standardized, radioiodinated protein samples were used to quantitatively assess initial and repetitive sequencing yields of either electroblotted proteins or proteins loaded by direct adsoption. The results showed that for the tested membranes, their different permeabilities for solutions of the Edman chemistry have a major influence on initial yields. The glass fibre based membranes with an extremely low flow restriction produced consistently higher initial yields irrespective of the mode of application of the protein (spotted or electroblotted) or the application of the membranes into the cartridge (discs or small pieces). In contrast, the polymeric membranes showed decreasing initial yields with increasing membrane density for spotted and electroblotted proteins. Yields varied considerably when the membranes were applied as discs into the cartridge. This effect could be minimized if the membranes were cut into pieces as small as possible, as demonstrated for electroblotted proteins.

For amino acid composition analysis no significant influence of the membrane used was observed. All hydrophobic polymer membranes are directly compatible with acid hydrolysis of the immobilized protein and subsequent quantification of the amino acids liberated [Lottspeich 1994]. A special application is shown in Figure 5. Electroblotted proteins were identified by comparison of the experimentally determined amino acid composition with a data set derived from a protein database (Figure 5).

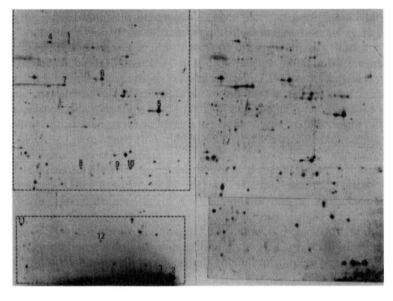

Figure 5. Identification of proteins after 2D-PAGE and electroblotting by amino acid composition analysis [Eckerskorn 1988b]. Patterns of mouse brain proteins obtained by (*A*) 2D-PAGE and (*B*) electroblotting onto siliconized glass fiber membranes. The immobilized proteins were stained with Coomassie Blue, and twelve protein spots were then subjected to both Edman degradation and amino acid analysis. Proteins were identified by comparison of the experimentally determined amino acid composition with a dataset derived from the Protein Identification Resource (PIR) protein database. Eight out of twelve proteins tested were identified by amino acid composition analysis and confirmed by N-terminal sequence analysis.

[*Spot 1*: Serum albumin, *Spot 2*: Hemoglobin, α-chain, *Spot 3*: Hemoglobin β-chain, *Spot 4*: no homology found, *Spot 5*: Glycerinaldehyd-3-phosphat-Dehydrogenase (GAPDH), *Spot 6*: no homology found, *Spot 7*: creatine kinase, *Spot 8*: no homology found, *Spots 9* and *10*: Triosephosphate isomerase, *Spots 11* and *12*: no homologies found.]

In experiments with MALDI-MS (matrix-assisted laser desorption/ionization mass spectrometry) of electroblotted proteins direct from the membrane surface, the choice of the membrane plays an important role [Strupat 1994]. The authors obtained mass spectra of better quality from membranes which exhibit high specific surfaces and low mean pore sizes. The spectra of the tested proteins document that, despite comparable performance with respect to signal intensities, shot-to-shot reproducibility and signal-to-noise ratio, a considerably smaller peak width is observed for a membrane with a high specific surface area.

4 References

Aebersold, R., Teplow, D.B., Hood, L.E., Kent, S.B. (1986) *J.Biol.Chem. 261*, 4229–4239. Electroblotting onto activated glass: High efficiency preparation of proteins from analytical sodium dodecyl sulfate-polyacrylamide gels for direct sequence analysis.

Aebersold, R. H., Leavitt, J., Hood, L. E., Kent, S. H. (1987) *Proc. Natl. Acad. Sci. USA 84*, 6970–6974. Internal amino acid sequence analysis of proteins separated by one- or two-dimensional gel electrophoresis after in situ protease digestion on nitrocellulose.

Aebersold, R. (1991). In: *Advances in Electrophoresis* (Chrambach, A.; Dunn, M. J.; Radola, B.J.; eds.), pp 81–168, VCH Verlag, Weinheim.

Baker, S. C., Dunn, M., Yacoub, M. H. (1991) *Electrophoresis 12*, 342–348. Evaluation of membranes used for electroblotting for direct automated mirosequencing.

Beisiegel, U. (1986) *Electrophoresis 7*, 1–18. Protein blotting.

Bittner, M., Kupferer, P., Morris, C. F. (1980) *Anal.Biochem. 102*, 459–471. Electrophoretic transfer of proteins and nucleic acids from slab gels to diazobenzyloxymethyl cellulose or nitrocellulose sheets.

Eckerskorn, C., Mewes, W., Goretzki, H.W., Lottspeich, F. (1988a) *Eur. J. Biochem. 176*, 509–519. A new siliconized-glass fiber as support for protein chemical analysis of electroblotted proteins.

Eckerskorn, C., Jungblut, P., Mewes, W., Klose, J., Lottspeich, F. (1988b) *Electrophoresis 9*, 830–838. Identification of mouse brain proteins after two-dimensional electrophoresis and electroblotting by microsequence analysis and amino acid composition.

Eckerskorn, C. und Lottspeich, F. (1989) *Chromatographia 28*, 92–94. Internal amino acid sequence analysis of proteins separated by gelelectrophoresis after tryptic digestion in the polyacrylamide matrix.

Eckerskorn, C. und Lottspeich, F. (1990a) *Electrophoresis 11*, 554–561. Combination of two-dimensional gel electrophoresis with microsequencing and amino acid composition analysis: Improvement of speed and sensitivity in protein characterization.

Eckerskorn, C., Lottspeich, F. (1990b) *J. Prot. Chem. 9*, 272–273. The initial yield in automated Edman-degradation depends on the choice of the membrane and the mode of protein application.

Eckerskorn, C., Lottspeich, F. (1991) The initial yield in automated Edman-degradation depends on the choice of the membrane and the mode of protein application. In: *2-D PAGE ^91* (Dunn, M.J., ed.), pp 111–115, Zebra Printing, London.

Eckerskorn, C., Strupat, K., Karas, M., Hillenkamp, F., Lottspeich, F. (1992) *Electrophoresis 13*, 664–665. Matrix-assisted laser desorption/ionisation mass spectrometry of proteins electroblotted after polyacrylamide-gelelectrophoresis.

Görg, A., Postel, W., Günther, S. (1988) E*lectrophoresis 9*, 531–546. The current state of two-dimensional electrophoresis with immobilized pH gradients.

Gültekin, H., Heermann, K. H. (1988) *Anal. Biochem. 172*, 320–329. The use of polyvinylidenefluoride membranes as a general blotting membrane.

Jacobson, G., Karsnäs, P. (1990) *Electrophoresis 11*, 46–52. Important parameters in semi-dry electrophoretic transfer.

Jahnen, W., Ward, L. D., Reid, G. E. Moritz, R. L., Simpson, R. J. (1990) *Biochem. Biophys. Res. Com. 166*, 139–145. Internal amino acid sequencing of proteins by in situ cyanogen bromide cleavage in polyacrylamide gels.

Jungblut., P., Eckerskorn, C., Lottspeich, F., Klose, J. (1990) *Electrophoresis 11*, 581–588. Blotting efficiency investigated by using two-dimensional electrophoresis, hydrophobic membranes and proteins from different sources.

Kyhse-Andersen, J., J. (1984) *Biochem. Biophys. Methods 10*, 203–209. Electroblotting of multiple gels: a simple apparatus without buffer tank for rapid transfer of proteins from polyacrylamide to nitrocellulose.

Laemmli, U. K. (1970) *Nature 227*, 680–685. Cleavage of structural proteins during the assembly of the head of bacteriophage T4.

LeGendre, N. (1990) *Bio Techniques 9*, 788–805. Immobilon-P transfer membrane: Applications and utility in protein biochemical analysis.

Lissilour, S., Godinot, C. (1990) *Bio Techniques 9*, 397–401. Influence of SDS and methanol on protein electrotransfer to Immobilon P membranes in semidry blot systems.

Lottspeich, F., Eckerskorn, C., Grimm, R. (1998) Amino acid analysis on microscale. In: *Cell Biology: A laboratory handbook* (Celis, J.; ed.), pp. 304–309, Academic Press.

Lottspeich, F., Eckerskorn, C. (1989) Two dimensional separated proteins are available for protein chemical analysis on microscale. In: *Electrophoresis forum '89* (Radola, B.; ed.), p. 72–83 Technical University Munich, Munich.

Lottspeich, F. (1990) *J. Prot. Chem. 9*, 268–269. Initial yield in amino acid sequence analysis is a surface dependent phenomen.

Matsudaira, P. (1987) *J. Biol. Chem. 262*, 10035–10038. Sequence from picomole quantities of proteins electroblotted onto polyvinylidene fluoride membranes.

Moos, M., Nguyen, N. Y., Liu, T-Y. (1988) *J. Biol. Chem. 263*, 6005–6008. Reproducible high yield sequencing of proteins electrophoretically separated and transferred to an inert support.

Mozdzanowski, J., Speicher, D. (1992) *Anal. Biochem. 207*, 11–18. Micosequence analysis of electroblotted proteins: Comparison of electroblotting recoveries using different types of PVDF membranes.

Nakagawa, S., Fukuda, T. (1989) *Anal.Biochem. 181*, 75–78. Direct amino acid analysis of proteins electroblotted onto polyvinylidene fluoride membranes from sodium dodecyl sulfate-polyacrylamide gel.

Patterson, S. D., Hess, D., Yungwirth, T., Aebersold, R. (1992) *Anal. Biochem. 202*, 193–203. High-yield recovery of electroblotted proteins and cleavage fragments from a cationic polyvinylidene fluoride-based membrane.

O'Farrell, P. H. (1975) *J. Biol. Chem. 250*, 4007–4021. High resolution two-dimensional electrophoresis of proteins.

Ploug, M., Jensen, A. L., Barkholt, V. (1989) *Anal. Biochem. 181*, 33–39. Determination of amino acid compositions and NH_2-terminal sequences of peptides electroblotted onto PVDF membranes from tricine-sodium dodecyl sulfate-polyacryl-amide gel electrophoresis: Application to peptide mapping of human complement component C3.

Pluskal, M. F., Przekop, M. B., Kavonian, M. R., Vecoli, C., Hicks, D. A. (1986) *BioTechniques 4*, 272–282. A new membrane substrate for western blotting of proteins.

Putnam, F. W., Neurath, H. (1945) *J. Biol. Chem. 159*, 195–209. Interaction between proteins and synthetic detergents II. Electrophoretic analysis of serum albumin-sodium dodecyl sulfate mixtures.

Renart, J., Reiser, J., Stark, G.R. (1979) *Proc. Natl. Acad. Sci. USA 76*, 3116–3120. Transfer of proteins from gels to diazobenzyloxymethyl-paper and detection with antisera: A method for studying antibody specifity and antigen structure.

Scott, M.G., Crimmins, D.L., Mc Court, D.W., Tarrand, J.J., Eyerman, M.C., Nahm, M.H. (1988) *Biochem. Biophys. Res. Com. 155*, 1353–1359. A simple in situ cyanogen bromide cleavage method to obtain internal amino acid sequence of proteins electroblotted to polyvinylidene fluoride membranes.

Schägger, H., von Jagow, G. (1987) *Anal. Biochem. 166*, 368–379. Tricine-sodium dodecyl sulfate-polyacrylamide gel electrophoresis for the separation of proteins in the range from 1 to 100 kDa.

Strupat, K., Karas, M., Hillenkamp, F., Eckerskorn, C., Lottspeich, F. (1994) *Anal.Chem. 66*, 464–470. Matrix-assisted laser desorption/ionisation mass spectrometry of proteins electroblotted after polyacrylamide-gelelectrophoresis.

Swank, R., Munkres, K. (1971) *Anal. Biochem. 39*, 62–477. Molecular weight analysis of oligopeptides by electrophoresis in polyacrylamide gel with sodium dodecyl sulfate.

Tous, G. I., Fausnaugh, J.L., Akinyosoye, O., Lackland, H., Winter-Cash, P., Vitoria, F. J., Stein, S. (1989) *Anal.Biochem. 179*, 50–55. Amino acid analysis on polyvinylidene fluoride membranes.

Tovey, E. R., Baldo, B. A. (1987) *Electrophoresis 8*, 384–387. Comparison of semi-dry and conventional tank-buffer electrophoretic transfer of proteins from polyacrylamide gels to nitocellulose membranes.

Towbin, H., Staehelin, T., Gordon, J. (1979) *Proc. Natl. Acad. Sci. USA 76*, 4350–4356. Electrophoretic transfer of proteins from polyacrylamide gels to nitrocellulose sheets: Procedure and some applications.

Vandekerckhove, J., Bauw, G., Puype, M., Van Damme, J., Van Montagu, M. (1985) *Eur. J. Biochem. 152*, 9–19. Protein-blotting on polybrene-coated glass-fiber sheets: A basis for acid hydrolysis and gas-phase sequencing of picomole quantities of protein previously separated on sodium dodecyl sulfate-polyacrylamide gel.

Walsh, M. J., McDougall, J., Wittmann-Liebold, B. (1988) *Biochemistry 27*, 6867–6876. Extended N-terminal sequencing of proteins of archaebacterial ribosomes blotted from two-dimensional gels onto glass fiber and polyvinylidenefluoride membrane.

Xu, Q.-Y., Shively, E. (1988) *Anal. Biochem. 170*, 19-30. Improved electroblotting of proteins onto membranes and derivatized glass-fiber sheets.

Microseparation Techniques IV:
Analysis of Peptides and Proteins
by Capillary Electrophoresis

Christine Schwer

1 Introduction

The high efficiency, ease of automation and short analysis time (ranging from several ten minutes to less than one minute) has made capillary electrophoresis (CE) a powerful tool in peptide and protein analysis. The use of different techniques applying different separation modes, a wide range of additives and the selection of the buffer pH allows the adjustment of the selectivity for every individual separation problem. Coupling CE to mass spectrometry and micropreparative applications of CE permit a further characterization of the separated compounds.

2 Theory

All electrophoretic techniques have in common that they apply high electrical fields to achieve separation of charged species. To obtain maximum efficiency peak dispersion due to convection caused by temperature gradients (due to Joule heating) in the system has to be kept a minimum. In classical electrophoresis stabilizing media – e.g. paper, cellulose acetate, agarose or polyacrylamide gels – are used to reduce convection. The other approach used in capillary electrophoresis is to use tubings of very small internal diameter (less than 100 μm), as the temperature difference between the center of the capillary and the wall is proportional to the square of the diameter of the capillary [Knox and Grant 1987]. In this way electrical field strengths in the order of several hundred V/cm can be applied, resulting in very fast separations. The basic instrumental setup for CE, as shown in Figure 1, is relatively simple, consisting mainly of a high voltage power supply ($\pm$ 30 kV), a fused-silica tubing as separation capillary and an on-column UV-detector.

Depending on the type and arrangement of the buffer solutions in the capillary different techniques can be distinguished.

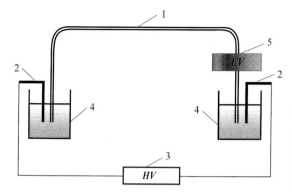

Figure 1. Schematic representation of a capillary electrophoresis instrument. (1) Fused-silica separation capillary, (2) platin electrodes, (3) high-voltage power supply, (4) electrolyte vessels, (5) on-column UV-detector.

2.1 Capillary Isotachophoresis

In isotachophoresis (ITP) two electrolyte solutions are used in the separation system, the leading electrolyte and the terminating electrolyte, which are chosen such that the leading ions have the highest mobility of all ions of interest and the terminating ions the lowest. The sample is injected between these two electrolytes. During separation the electric current is kept constant and the electrical field strength increases according to Ohm's law with decreasing mobility (conductivity) of the consecutive zones. Separation is achieved due to the different velocities v_i of the analyte ions i in the mixed zone formed by the sample:

$$v_i = m_i \cdot E^{(mix)} \tag{1}$$

where $E^{(mix)}$ is the electrical field strength in the mixed zone and m_i is the mobility of the species i in this zone. After the separation of the sample ions into individual zones all zones migrate with the same velocity, each zone having its own electrical field strength (depending on its mobility). The distribution of all ions in the separation system is described for isotachophoresis and zone electrophoresis by the Kohlrausch regulating function [Kohlrausch 1897]:

$$\sum_i \frac{c_i}{m_i} = const \tag{2}$$

where c_i is the concentration of species i.

The concentration profile of the analyte zones in ITP is rectangular, their concentration being adapted to that of the leading zone according to the Kohlrausch regulating function. For the concentration c_A of analyte A it follows:

$$c_A = c_L \cdot \frac{m_A(m_L + m_Q)}{m_L(m_A + m_Q)} \tag{3}$$

where c_L is the concentration of the leading ion L and m_A, m_L and m_Q are the mobilities of analyte A, the leading ion and the counter ion Q in the steady state.

Sharp zone boundaries are maintained because of the "self-sharpening" effect: ions diffusing in a zone behind or in front of them, will be accelerated or slowed down due to the step gradients in the electrical field strength and will therefore return to their own zone.

For dilute samples ITP leads to a concentration of the analytes and can be used therefore as a preconcentration technique for capillary zone electrophoresis.

2.2 Capillary Zone Electrophoresis

In capillary zone electrophoresis (CZE) the separation system is filled with a single electrolyte of relatively high concentration to provide a uniform field strength when an electrical potential is applied to the capillary. Ions i are separated according to their different migration velocities v_i under uniform electrical field strength E:

$$v_i = m_i^{eff} \cdot E \qquad (4)$$

where m_i^{eff} is the effective mobility of an ion of a weak electrolyte, which is related to the mobility m_i of the fully dissociated species i by the degree of dissociation α:

$$m_i^{eff} = m_i \cdot \alpha \qquad (5)$$

In contrast to ITP band broadening caused by diffusion leads to a dilution of the injected sample zone. The concentration profile is approximated by a Gaussian distribution for those cases, where the electrical field is only slightly disturbed by the analyte. High concentrations of analytes with mobilities highly differing from the mobility of the background electrolyte result in a distortion of the electrical field and thus in non-symmetrical peaks [Mikkers et al. 1979]. Under ideal conditions (no contributions to peak dispersion caused by injection, detection, convection due to temperature gradients, distortion of the electrical field, etc.) only longitudinal diffusion leads to band broadening and the width of the concentration distribution, expressed by the spatial variance σ_z^2 is given by:

$$\sigma_z^2 = 2 \cdot D \cdot t \qquad (6)$$

where D is the diffusion coefficient of the sample component, σ_z is the standard deviation of the Gaussian peak and t is the migration time. Using the concept of theoretical plates to describe efficiency, an expression for the number of theoretical plates N can be obtained [Giddings 1969; Jorgenson and Lukacs 1981]:

$$N = \frac{mU}{2\,D} \qquad (7)$$

where U is the applied voltage.

As mobility and diffusion coefficient are related according to the Einstein-Nernst equation, $D/m = kT/ze_0$, equation 7 can be rewritten as [Giddings 1969; Kenndler and Schwer 1991]:

$$N = \frac{ze_0 U}{2\,kT} \tag{8}$$

where k is the Boltzmann constant, T the absolute temperature, e_0 the electronic charge and z the charge number of the analyte. Thus the plate number is only dependent on the charge, but not on the size of the sample ions. This explains the extremely high plate numbers that can be obtained for highly charged solutes as proteins or oligonucleotides.

2.3 Electroosmotic Flow

The most frequently used separation capillaries are made of fused silica. The silanol groups of the inner surface dissociate when in contact with electrolyte solution, leading to the formation of an electrical double layer, which is characterized by the zeta potential ζ. When an electrical potential is applied to the capillary, a bulk flow of the liquid in the direction to the cathode – the electroosmotic flow (EOF) – results, which is related to the zeta potential as follows:

$$v_{EO} = \varepsilon z E / 4\,\pi\eta \tag{9}$$

where v_{EO} is the electroosmotic velocity, ε is the dielectric constant and η the viscosity of the liquid in the electrical double layer. This electroosmotic velocity of the bulk liquid is superimposed on the electrophoretic velocity of the analytes, as shown in Figure 2. The net velocity of the analytes is the vectorial sum of electroosmotic and electrophoretic velocity.

The zeta potential and thus the electroosmotic velocity depends on the ionic strength of the electrolyte solution – it increases with decreasing ionic strength – and on the pH. The pH dependence, as shown in Figure 3, resembles a titration curve with an inflection point at about pH 5.5 [Schwer and Kenndler 1991]. At pH values above 7 the EOF is usually higher than the migration velocity of the anions, so that cations and anions can be detected at the side of the cathode. The EOF shows a plug-like profile in contrast to the parabolic profile of hydrodynamic flow. Its contribution to peak dispersion can therefore be neglected, leading to an expression for the number of theoretical plates under conditions of electroosmotic flow as follows [Jorgenson and Lukacs 1981]:

$$N = \frac{(m + m_{EO})U}{2\,D} = \left[\frac{m_{EO}}{m} + 1\right] \cdot \frac{ze_0 U}{2\,kT} \tag{10}$$

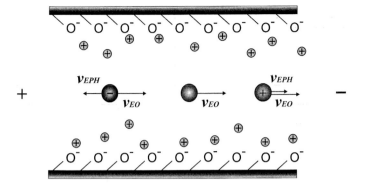

Figure 2. Electroosmotic and electrophoretic velocity, v_{EO} and v_{EPH} : The surface silanol groups are dissociated when in contact with buffer electrolyte, forming, together with the counter ions, the electrical double layer. When an electrical field is applied to the capillary, a flow of the bulk liquid in the direction of the cathode results, which contributes to the overall velocities of all species: as the electroosmotic velocity at high pH is usually higher than the electrophoretic velocity, anions also move in the direction of the cathode; neutral molecules are transported by the electroosmotic flow and can be detected at the side of the cathode.

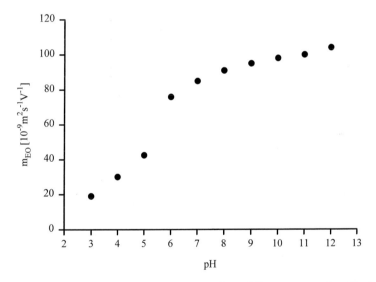

Figure 3. Dependence of the electroosmotic mobility, m_{EO}, on the pH of the buffer electrolyte. The electroosmotic velocities were determined using neutral marker substances in a 10^{-2} mol/L KCl / 10^{-3} mol/L K-phosphate buffer.

By substituting the Einstein-Nernst equation it can be seen that the plate number depends on the charge number of the analyte and the ratio of electrophoretic and electroosmotic mobility [Schwer and Kenndler 1992].

Although the EOF is of advantage in a large number of applications, because anions and cations can be determined in a single run, adsorption of solutes onto the capillary wall can make a modification of the capillary wall necessary. Strong electrostatic interactions between the charged capillary surface and analytes with high charge density, especially proteins, lead to a significant loss in efficiency and unreproducible migration times.

To decrease solute-wall interactions several different approaches have been described. Using buffer electrolytes of extreme pH values results either in a repulsion of anions from the highly negatively charged wall at high pH or no electrostatic interactions because of a non-charged surface at low pH [McCormick 1988; Lauer and McManigill 1986; Waldbroehl and Jorgensen 1989; Lindner et al. 1992]. However, to work at moderate pH other approaches have to be used: permanent coatings of the inner wall by chemically modifying the silanol groups or dynamic modifications by using buffer additives.

The modification of the capillary surface with polyacrylamide via different types of bonding is a widely used permanent coating with relatively good stability at higher pH [Hjertén and Kiessling-Johansson 1991; Cobb et al. 1990; Schmalzing et al. 1993; Hjertén 1985]. Other permanent coatings include methylcellulose, polyethylene glycol, different silanes, ion exchangers, polyvinylalcohol, etc., some of them with limited pH stability and/or effectiveness in suppressing solute-wall interactions [Hjertén 1985; Bruin et al. 1984; Swedberg 1990; Jorgenson and Lukacs 1983; Hjertén and Kubo 1993; Gilges et al. 1992].

Dynamic modifications include the use of high ionic strength buffers, zwitterionic salts, divalent amines, non-ionic surfactants, polymers or charge-reversal reagents [Chen et al. 1992; Bushey and Jorgenson 1989; Bullock and Yuan 1991; Towns and Regmer 1991; Mazzeo and Krull 1991; Hjertén et al. 1989; Emmer et al. 1991; Wiktorowicz and Colburn 1990].

3 Instrumentation

3.1 Injection

Injection in CE is usually performed either hydrodynamically or electrokinetically. In order to obtain maximum efficiency the variance of the injection profile must not contribute significantly to peak broadening. For pressure injection it follows that the injection volume must not exceed a few nanoliter for capillaries of 75 to 100 µm i.d.. As the injected volume with hydrodynamic injection techniques is dependent on viscosity, temperature control is decisive for reproducible injection volumes.

If injection is carried out electrokinetically (the only injection method for gel-filled capillaries) discrimination depending on the mobility of the analytes has to be considered. Further, the actually injected amount is very much dependent on the concentration of matrix components. These effects are the more pronounced, the lower the EOF is.

To increase the injection volume and thus the detection sensitivity sample stacking can be performed. The easiest way of sample preconcentration is achieved, if the sample is dissolved in water or very dilute buffer [Haglund and Tiselius 1950; Mikkers et al. 1979; Burgi and Chien 1991]. The low conductivity in the sample plug results in a high electrical field strength in this zone and consequently to a concentration of the analytes at the front of the zone. This type of sample stacking and modifications of it have been extensively treated by Burgi et al. [Burgi and Chien 1991, 1993; Chien and Burgi 1991; Burgi 1993].

Another approach uses discontinuous buffer systems to achieve an isotachophoretic sample concentration. This can be performed in a dual-column mode [Kaniansky and Marak 1990; Foret et al. 1990; Stegehuis et al. 1991], where an enrichment factor of about 1000 can be reached or in a single-column mode in commercial instruments, where the injection volume can be increased by a factor of 100 to about 1 µL [Schwer and Lottspeich 1992; Schwer et al. 1993; Hjertén et al. 1987; Hjertén 1990; Foret et al. 1992]. Various types of discontinuous buffer systems have been developed for pre-concentration, e.g. the one buffer system, applicable for the concentration of peptides and proteins, where the sample plug is sandwiched between zones of extreme pH, i.e. a zone of OH^- and H^+, as shown in Figure 4. As OH^- and H^+ are migrating towards each other, the analytes are concentrated in between. In Figure 5 this system was applied to the separation of tryptic peptides of β-lactoglobulin. The injection volume was increased to 750 nL, while still maintaining the high separation efficiency.

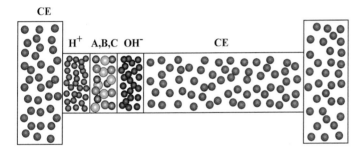

Figure 4. Arrangement of the different solutions in the one-buffer stacking system. CE: carrier electrolyte; A,B,C: sample; H^+, OH^-: solution of an acid or base, respectively. For detailed composition of the electrolytes see Table 1, system I.

A further discontinuos buffer system is shown in Figure 6 and consists of two electrolytes, a leading and a terminating electrolyte like in ITP, but the capillary is filled with terminating electrolyte and only in front of the sample a zone of leading electrolyte is applied. The ends of the capillary are placed in terminating electrolyte. The presence of the leading electrolyte leads first to isotachophoretic conditions for the sample ions and therefore to a concentration adaption to the concentration of the leading electrolyte. However, as the leading electrolyte is preceded by terminating electrolyte, the leading ions migrate zone electrophoretically into the terminating electrolyte, thus migrating away from the sample ions. As the isotachophoretic conditions no longer hold for the sample ions, they are themselves separated by zone electrophoresis in the terminating electrolyte. As an example the separation of 5 basic proteins is shown in Figure 7. The injection volume was increased to 500 nL. The adsorption of the proteins was suppressed by dynamic coating with polyethylene glycol as a buffer additive. The composition of these and other stacking systems is given in Table 1.

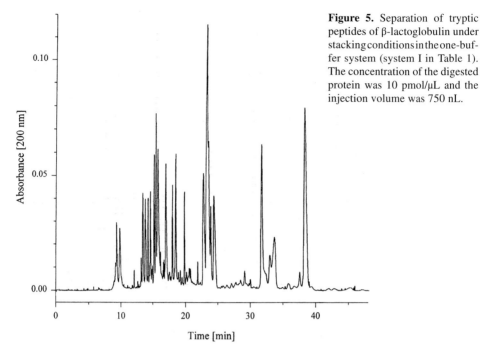

Figure 5. Separation of tryptic peptides of β-lactoglobulin under stacking conditions in the one-buffer system (system I in Table 1). The concentration of the digested protein was 10 pmol/μL and the injection volume was 750 nL.

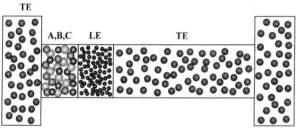

Figure 6. Arrangement of the buffer solutions in the two-buffer stacking system.
LE: leading electrolyte; TE: terminating electrolyte; A,B,C: sample. For detailed composition of the buffer solutions see Table 1, systems IIa-c.

Table 1. Discontinuous buffer systems.

One-Buffer System I	CE:	20 mmol/L Na-phosphate, pH 2.9
	OH⁻:	100 mmol/L NaOH
	H⁺:	100 mmol/L H₃PO₄
Two-Buffer System IIa	LE:	50 mmol/L ammonium acetate
	TE:	50 mmol/L β-alanine/50 mmol/L acetic acid, pH 4.0
Two-Buffer System IIb	LE:	50 mmol/L ammonium acetate
	TE:	50 mmol/L betaine/50 mmol/L acetic acid, pH 3.3
Two-Buffer System IIc	LE:	50 mmol/L HCl/50 mmol/L β-alanine
	TE:	50 mmol/L MES/50 mmol/L β-alanine, pH 4.8

CE: carrier electrolyte; LE: leading electrolyte; TE: terminating electrolyte; MES: 2-(N-morpholino)ethanesulfonic acid. System IIc is used for the concentration of anions, while the other systems are cationic systems.

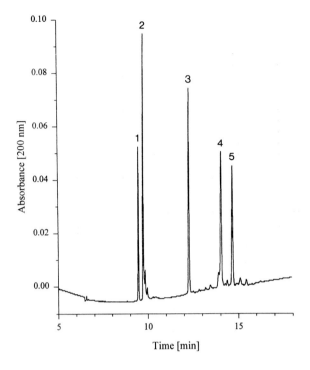

Figure 7. Separation of 5 basic proteins under stacking conditions in a two-buffer system at pH 4 (system IIa in Table 1). The capillary was dynamically coated with polyethylene glycol (0.02 % PEG 5 000 000) as buffer additive. Injection volume: 500 nL. Protein concentration: 10 ng/μL. Peaks: (1) cytochrome c, (2) lysozyme, (3) ribonuclease A, (4) trypsinogen, (5) chymotrypsinogen A.

3.2 Detection

Due to the small optical pathlength in on-column detection relatively high sample concentrations are needed for UV-detection. To increase sensitivity either preconcentration techniques, as described above, can be applied or more sensitive detection methods used, e.g. laser-induced fluorescence [Cheng and Dovichi 1988; Swaile and Sepaniak 1991]. For non or low UV-absorbing analytes indirect UV or indirect fluorescence detection have proven to be a good alternative [Foret et al. 1989; Kuhr and Young 1988; Jandik and Jones 1991]. Other specific detection methods, as e.g. electrochemical, radiometric or conductivity detection have been described in the literature, but are not commercially available yet [Kaniansky et al. 1983; Pentoney et al. 1989; Wallingford and Ewing 1987, 1988].

The coupling of CE to mass spectrometry has already shown to be a very powerful method, especially in peptide and protein chemistry, as it allows a further characterization of the analytes [Olivares et al. 1987; Smith et al. 1988; Moseley et al. 1989]. An example is shown in Figure 8, where CE was coupled to electrospray ionization mass spectrometry to identify tryptic peptides of β-casein. Good agreement between the determined molecular masses and the known fragments of β-casein was found. As an illustration the mass spectra of two selected peaks are depicted in Figure 9.

Further characterization of the separated compounds can be achieved off-line after collection of fractions. Fraction collection has been performed using different techniques. By calculating the time window, when a compound has migrated to the end

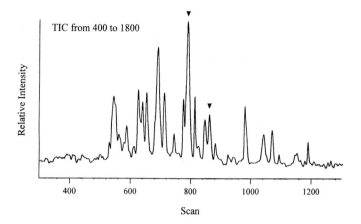

Figure 8. CE-MS analysis of tryptic peptides of β-casein in an ammonium acetate buffer at pH 3.5. A P/ACE 2100 CE-instrument was coupled to a SCIEX API III with a coaxial interface arrangement. As sheath liquid 0.1 % acetic acid in 50 % methanol at a flow rate of 5 µL/min was used. A potential of 25 kV was applied at the inlet side of the CE-instrument, while the electrospray needle was set at a potential of 5 kV. A fused-silica capillary with an internal diameter of 50 µm and a total length of 95 cm was used for separation. The orifice voltage was 75 V.

of the capillary, fractions can be collected by simply changing the vial at the outlet [Guttman et al. 1990; Banke et al. 1991; Bergman et al. 1991; Camilleri et al. 1991; Altria and Dave 1993; Chen et al. 1992]. This method requires a prerun to determine the migration velocities and highly reproducible and constant migration times. Fractions are diluted with ca. 10 µL buffer electrolyte. A non-uniform electrical field strength along the capillary (e.g. under stacking conditions) makes the prediction of the time window for fraction collection difficult.

A continuous approach makes use of a frit structure or a similar construction to obtain electrical contact to the counter electrode [Huang and Zare 1990; Guzman et al. 1991; Fujimoto et al. 1991, 1992]. No contact between electrode and analytes occurs preventing possible decomposition of the analytes by electrode reactions. The solutes are transported by electroosmotic flow to the end of the capillary. Highly reproducible electroosmotic flow rates are required for fractions of high purity.

Using a moving membrane the analytes can be continuously collected without losing resolution. Detection can be performed e.g. by staining, immunological methods or radioactivity measurements [Eriksson et al. 1992; Cheng et al. 1992; Konse et al. 1993].

Making use of a make-up flow a constant flow rate allows fraction collection as in HPLC with the possibility of automation. A schematic representation of the instrumental setup is shown in Figure 10. This method of fraction collection was applied to tryptic peptides of β-casein. The sample solution was applied to a micropreparative system under stacking conditions to increase the injection volume to about 300 nL. Ten fractions were collected and the purity was controlled by reinjection under stack-

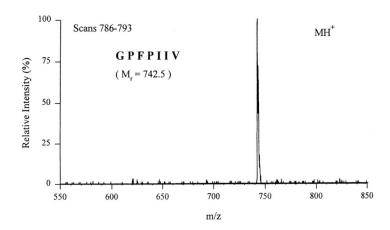

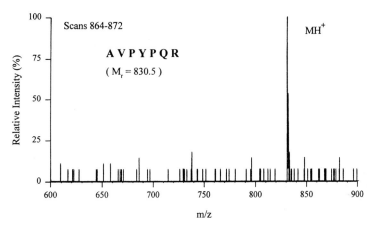

Figure 9. Mass spectra of 2 selected peaks from the CE-MS separation in Figure 8. Known sequences of tryptic fragments could be assigned to the corresponding m/z values of the mass spectra of the individual peaks.

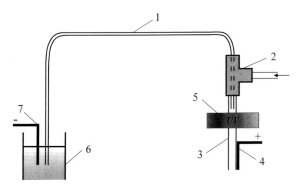

Figure 10. Schematic representation of the arrangement for fraction collection using a make-up flow. (1) separation capillary with an i.d. of 100 μm and an o.d. of 190 μm; (2) make-up flow, delivered by a syringe pump at a flow rate of 5 μL/min; (3) outer fused-silica capillary with an i.d. of 250 μm; (4) counter electrode connected to ground; (5) UV-detector for on-column detection in outer capillary; (6) buffer vial; (7) electrode at inlet.

ing conditions. Even closely migrating peptides were collected in high purity. Amino acid sequence analysis and MALDI-MS were performed from several fractions, proving that fractions from a single run were sufficient for the determination of the molecular mass and sequence.

4 Applications

4.1 Peptide Separations

Peptides are usually separated at low pH of about 2-3, where they all migrate as cations or at high pH under conditions of EOF. The mobility of a peptide can be predicted by the following equation:

$$m = KzM^{-\frac{2}{3}} \tag{11}$$

where K is a constant, z is the valence and M the molecular mass [Offord 1966; Rickard et al. 1991]. Good correlation between migration times and molecular masses were found, if ionization constants, which are typical for peptides, are used to calculate the overall charge of the peptide instead of the pK_a of the free amino acids. A semiempirical model based on charge, size and hydrophobicity was also described to predict the mobilities of peptides [Grossman et al. 1989].

The selectivity of the separation of peptides can be influenced by:

- pH of the buffer electrolyte,
- micelle forming reagents,
- ion-pairing reagents,
- complexing reagents,
- cyclodextrins,
- organic solvents.

CE is routinely applied to the purity control of synthetic peptides or HPLC fractions, as shown in Figure 11. Due to the orthogonal separation principles CE is able to resolve one RP-HPLC fraction into several peaks.

In combination with mass spectrometry CE can be used to gain further structural information of tryptic peptides. A micropreparative application of CE allows the further characterization of peptides by the determination of its amino acid composition or sequence or other off-line methods.

4.2 Protein Separations

Proteins can be separated applying different modes of CE, using different sample properties to achieve separation:

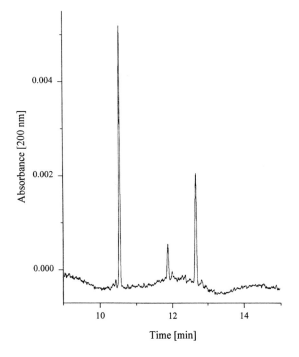

Figure 11. Purity control of an RP-HPLC fraction of tryptic peptides of fetuin. To increase the injection volume and the sensitivity the two-buffer stacking system IIb (see Table 1) was used for separation. Due to the orthogonal separation mechanisms a single RP-HPLC peak can be separated into several components by CZE.

- capillary zone electrophoresis (CZE): mobility,
- capillary isoelectric focusing (CIEF): isoelectric point (pI),
- capillary gel electrophoresis (CGE): molecular mass.

CZE: Separation is obtained due to differences in mobility (depending on size and charge). Loss of efficiency may occur due to interactions between the solutes and the charged wall of the capillary, as discussed previously. In Figure 12 the separation of isoforms of a monoclonal antibody is shown. The inner surface of the capillary was dynamically coated with polyethylene glycol to reduce protein-wall interactions. As the buffer pH is lower than the pIs of the antibody isoforms, all isoforms are positively charged. Thus, isoforms with longer migration times have slightly more acidic pI values.

CIEF: Separation occurs according to the pI of the proteins in a pH-gradient, formed by an ampholyte solution. Usually coated capillaries are used to suppress EOF in order to obtain high resolution. For the detection of analytes under conditions of suppressed EOF, mobilization of the focused proteins must be carried out. This can be achieved either electrophoretically by changing the composition of anolyte or catholyte or hydrodynamically by pressure mobilization, where pressure is applied together with high voltage in order to maintain the high resolution [Hjertén and Zhu 1985; Hjertén et al. 1987; Zhu et al. 1991; Schwer 1995]. Under conditions of EOF no mobilization is necessary, as the focused proteins are transported by the EOF past the detector [Mazzeo and Krull 1991; Mazzeo et al. 1993]. In Figure 13 the separation

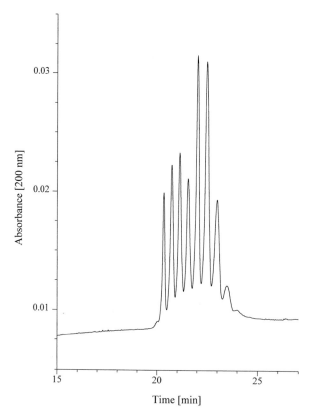

Figure 12. CZE separation of the different isoforms of a monoclonal antibody. Capillary: fused-silica, 75 cm i.d., 57 cm length. Buffer: 0.1 mol/L ε-aminocaproic acid/ acetic acid, pH 4.7. Polyethylene glycol (0.02 % PEG 5 000 000) was added to the buffer electrolyte for dynamic coating of the capillary wall.

of isoforms of a monoclonal antibody by CIEF with cathodic mobilization is shown. The resolution obtained was in the order of 0.02 pI units.

CGE: Proteins are separated as their SDS-complexes according to their molecular mass. As sieving medium either cross-linked polyacrylamide gels [Cohen and Karger 1987] are used or solutions of polymers such as polydextran, polyethylene glycol or linear polyacrylamide [Widhalm et al. 1991; Ganzler et al. 1992; Guttman et al. 1993]. Polyacrylamide as sieving medium has the disadvantage of high UV-absorbance below 230 nm, thus other polymers as polydextran or polyethylene glycol, being UV-transparent also at 214 nm, are favored, allowing the detection of proteins with increased sensitivity. To suppress EOF and electrostatic interactions with the wall coated capillaries are used. A linear correlation between migration time and molecular mass is obtained allowing the determination of the molecular mass of unknown proteins. In Figure 14 the separation of a mixture of a monoclonal antibody and its deglycosylated form is shown. Both forms are clearly separated demonstrating the potential of CGE for e.g. monitoring deglycosylation.

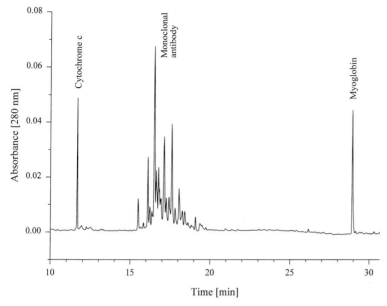

Figure 13. Separation of the different isoforms of a very basic monoclonal antibody by CIEF with cathodic mobilization. Focusing and chemical mobilization was performed in a coated capillary at 20 kV using Pharmalyte 3-10. As internal pI-markers cytochrome c and myoglobin were added [Schwer 1995].

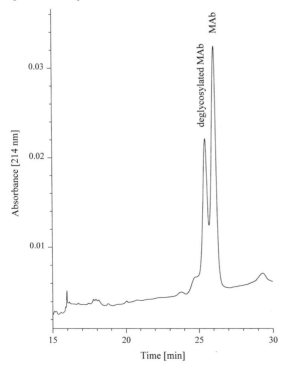

Figure 14. SDS-CGE separation of a monoclonal antibody and its deglycosylated form using a coated capillary and a polymer solution as sieving medium.

5 References

Altria, K.D., Dave, Y.K. (1993) *J. Chromatogr. 633*, 221.
Banke, N., Hansen, K., Diers, I. (1991). *J. Chromatogr. 559*, 325.
Bergman, T., Agerberth, B., Jörnvall, H. (1991) *FEBS 283*, 100.
Bruin, G.J.M., Chang, J.P., Kuhlman, R.H., Zegers, K., Kraak, J.C., Poppe, H. (1989) *J. Chromatogr. 471*, 429.
Bullock, J.A., Yuan, L.-C. (1991) *J. Microcol. Sep. 3*, 241.
Burgi, D.S. (1993) *Anal. Chem. 65*, 3726.
Burgi, D.S., Chien, R.-L. (1991) *Anal. Chem. 63*, 2042.
Burgi, D.S., Chien, R.-L. (1992) *Anal. Biochem. 2*, 306.
Bushey, M.M., Jorgenson, J.W. (1989) *J. Chromatogr. 480*, 301.
Camilleri, P., Okafo, G.N., Southan, C., Brown, R. (1991) *Anal. Biochem. 198*, 36.
Chen, F.A., Kelly, L., Palmieri, R., Biehler, R., Schwartz, H. (1992) *J. Liq. Chromatogr. 15*, 1143.
Cheng, Y.F., Dovichi, N.J. (1988) *Science 242*, 563.
Cheng, Y.-F., Fuchs, M., Andrews, D., Carson, W. (1992) *J. Chromatogr. 608*, 109.
Chien, R.-L., Burgi, D.S. (1991) *J. Chromatogr. 559*, 153.
Cobb, K.A., Dolník, V., Novotny, M. (1990) *Anal. Chem. 62*, 2478.
Cohen, A.S., Karger, B.L. (1987) *J. Chromatogr. 397*, 409.
Emmer, A., Jansson, M., Roeraade, J. (1991) *HRC 14*, 738.
Eriksson, K.-O., Palm, A., Hjertén, S. (1992) *Anal. Biochem. 201*, 211.
Foret, F., Fanali, S., Ossicini, L., Bocek, P. (1989) *J. Chromatogr. 470*, 299.
Foret, F., Sustácek, V., Bocek, P. (1990) *J. Microcol. Sep. 2*, 229.
Foret, F., Szoko, E., Karger, B.L. (1992) *J. Chromatogr. 608*, 3.
Fujimoto, C., Fujikawa, T., Jinno, K. (1992) *HRC 15*, 201.
Fujimoto, C., Muramatsu, Y., Suzuki, M., Jinno, K. (1991) *HRC 14*, 178.
Ganzler, K., Greve, K.S., Cohen, A.S., Karger, B.L., Guttman, A., Cooke, N.C. (1992) *Anal. Chem. 64*, 2665.
Giddings, J.C. (1969) *Sep. Sci. 4*, 181.
Giddings, J.C. (1989) *J. Chromatogr. 480*, 21.
Gilges, M., Husmann, H., Kleemiss, M.-H., Motsch, S.R., Schomburg, G. (1992) *HRC 15*, 452.
Grossman, P.D., Colburn, J.C., Lauer, H.H. (1989) *Anal. Biochem. 179*, 28.
Guttman, A., Cohen, A.S., Karger, B.L. (1990) *Anal. Chem. 62*, 137.
Guttman, A., Horváth, J., Cooke, N. (1993) *Anal. Chem. 65*, 199.
Guzman, N.A., Trebilcock, M.A., Advis, J.P. (1991) *J. Liq. Chromatogr. 14*, 997.
Haglund, H., Tiselius, A. (1950) *Acta Chem. Scand. 4*, 957.
Hjertén, S. (1985) *J. Chromatogr. 347*, 191.
Hjertén, S. (1990) *Electrophoresis 11*, 665.
Hjertén, S., Elenbring, K., Kilár, F., Liao, J.-L., Chen, A.J.C., Siebert, C.J., Zhu, M. (1987) *J. Chromatogr. 403*, 47.
Hjertén, S., Kiessling-Johansson, M. (1991) *J. Chromatogr. 550*, 811.
Hjertén, S., Kubo, K. (1993) *Electrophoresis 14*, 390.
Hjertén, S., Liao, J.-L., Yao, K. (1987) *J. Chromatogr. 387*, 127.
Hjertén, S., Valtcheva, L., Elenbring, K., Eaker, D. (1989) *J. Liq. Chromatogr. 12*, 2471.
Hjertén, S., Zhu, M. (1985) *J. Chromatogr. 346*, 265.
Huang, X., Zare, R.N. (1990) *Anal. Chem. 62*, 443.
Jandik, P., Jones, W.R. (1991) *J. Chromatogr. 546*, 431.
Jorgenson, J.W., Lukacs, K.D. (1981) *Anal. Chem. 53*, 1298.
Jorgenson, J.W., Lukacs, K.D. (1983) *Science 222*, 266.
Kaniansky, D., Marak, J. (1990) *J. Chromatogr. 498*, 191.
Kaniansky, D., Rajec, P., Švec, A., Havaši, P., Macášek, F. (1983) *J. Chromatogr. 258*, 238.
Kenndler, E., Schwer, Ch. (1991) *Anal. Chem. 63*, 2499.

Knox, J.H., Grant, I.H. (1987) *Chromatographia 24*, 135.

Kohlrausch, F. (1897) *Ann. Phys. Chem. 62*, 209.

Konse, T., Takahashi, T., Nagoshima, H., Iwaoka, T. (1993) *Anal. Biochem. 214*, 179.

Kuhr, W.G., Yeung, E.S. (1988) *Anal. Chem. 60*, 1832.

Lauer, H.H., McManigill, D. (1986) *Anal. Chem. 58*, 166.

Lindner, H., Helliger, W., Dirschlmayer, A., Jaquemar, M., Puschendorf, B. (1992) *Biochem. J. 283*, 467.

Mazzeo, J.R., Krull, I.S. (1991) *Anal. Chem. 63*, 2852.

Mazzeo, J.R., Martineau, J.A., Krull, I.S. (1993) *Anal. Biochem. 208*, 323.

McCormick, R.M. (1988) *Anal. Chem. 60*, 2322.

Mikkers, F.E.P., Everaerts, F.M., Verheggen, Th.P.E.M. (1979) *J. Chromatogr. 169*, 1.

Mikkers, F.E.P., Everaerts, F.M., Verheggen, Th.P.E.M. (1979) *J. Chromatogr. 169*, 11.

Moseley, M.A., Deterding, L.J., Tomer, K.B., Jorgenson, J.W. (1989) *J. Chromatogr. 480*, 197.

Offord, R.E. (1966) *Nature 211*, 591.

Olivares, J.A., Nguyen, N.T., Yonker, C.R., Smith, R.D. (1987) *Anal. Chem. 59*, 1230.

Pentoney, S.L., Zare, R.N., Quint, J.F. (1989) *Anal. Chem. 61*, 1642.

Rickard, E.C., Strohl, M.M., Nielsen, R.G. (1991) *Anal. Biochem. 197*, 197.

Schmalzing, D., Piggee, Ch.A., Foret, F., Carrilho, E., Karger, B.L. (1993) *J Chromatogr. 652*, 149.

Schwer, Ch. (1995) *Electrophoresis 16*, 2121.

Schwer, Ch., Gaš, B., Lottspeich, F., Kenndler, E. (1993) *Anal. Chem. 65*, 2108.

Schwer, Ch., Kenndler, E. (1991) *Anal. Chem. 63*, 1801.

Schwer, Ch., Kenndler, E. (1992) *Chromatographia 33*, 331.

Schwer, Ch., Lottspeich, F. (1992) *J. Chromatogr. 623*, 345.

Smith, R.D., Barinaga, C.J., Udseth, H.R. (1988) *Anal. Chem. 60*, 1948.

Stegehuis, D.S., Irth, H., Tjaden, U.R., Van der Greef, J. (1991) *J. Chromatogr. 538*, 393.

Swaile, D.F., Sepaniak, M.J. (1991) *J. Liq. Chromatogr. 14*, 869.

Swedberg, S.A. (1990) *Anal. Biochem. 185*, 51.

Towns, J.K., Regnier, F.E. (1991) *Anal. Chem. 63*, 1126.

Walbroehl, Y., Jorgenson, J.W. (1989) *J. Microcol. Sep. 1*, 41.

Wallingford, R.A., Ewing, A.G. (1987) *Anal. Chem. 59*, 1762.

Wallingford, R.A., Ewing, A.G. (1988) *Anal. Chem. 60*, 1972.

Widhalm, A., Schwer, Ch., Blaas, D., Kenndler, E. (1991) *J. Chromatogr. 546*, 446.

Wiktorowicz, J.E., Colburn, J.C. (1990) *Electrophoresis 11*, 769.

Zhu, M., Rodriguez, R., Wehr, T. (1991) *J. Chromatogr. 559*, 479.

Microseparation Techniques V: High Performance Liquid Chromatography

Maria Serwe, Martin Blüggel and Helmut E. Meyer

1 Introduction

Since the commercial introduction of high-performance liquid chromatography (HPLC) in the 1970s this versatile method has gained wide popularity, resulting in a phenomenal increase of successful applications described in literature [Henschen 1985]. HPLC is an efficacious method in peptide and protein analysis especially for the preparation of samples for further analysis by mass spectrometry or Edman sequencing. Due to the successful minimization of flow rate and column diameters during the last 10 years HPLC is today able to separate peptides and proteins also in those fields of biochemistry where only small amounts of sample are accessible.

The following paragraph is designed to give an overview of the theoretical aspects of HPLC. The main section will guide you through the practial aspects of HPLC. The last paragraph of this chapter includes some typical applications in peptide or protein analysis.

2 Principle of HPLC

The classical low-pressure chromatographic systems use soft-gel, while HPLC adopts micron-sized particles of high mechanical strength as supports for column packing materials, thus enabling liquid flow through the column at high pressure. Therefore HPLC is sometimes also termed as high pressure liquid chromatography. As HPLC is not so much a new technique but an advancement in the design of chromatographic systems, all common chromatographic modes by which peptides and proteins are fractionated on the analytical as well as preparative scale are also applicable.

In most of these chromatographic systems peptides and proteins are chromatographically resolved by a surface-mediated process. The differential adsorption of peptides and proteins on the surface of the packing material (stationary phase) depends on the sample and on the strength of the solvent (mobile phase). The various chromatographic modes differ according to the way the sample adsorpts. The most commonly applied modes in peptide and protein separation are size-exclusion, ion-exchange, affinity and partition chromatography [Chicz 1990; Mant 1991b].

Size-exclusion chromatography, also known as gel chromatography, discriminates between molecular species on the basis of size (or better hydrodynamic volume) due to differential permeation into matrices of controlled porosity. Large molecules which cannot enter the pores pass through the column unretained and are eluted first. Small molecules totally permeate the liquid volume within and between the particles and are eluted last. Mid-sized molecules will selectively permeate the pores depending on their relative size, and be eluted with a retention time between the two extremes. Size-exclusion chromatography can be used with a variety of mobile phases in near physiological conditions.

Ion-exchange chromatography resolves proteins according to accessible surface charges and their corresponding electrostatic interaction with surface-bound negatively charged (cation exchange) or positively charged (anion exchange) moieties. Displacement of the protein from the stationary phase is achieved at constant pH with increasing ionic strength of the mobile phase. Sodium or potassium chloride are the most commonly used displacement salts, their concentrations depend on the strength of interaction between the protein and the stationary phase.

Affinity chromatography is based on the bioaffinity of a protein for a specific ligand coupled to a solid support. The immobilized ligand interacts only with proteins that can selectively bind to it, whereas others are eluted unretained. The retained protein can later be released in a purified state.

Partition chromatography depends on the partitioning of the sample between the stationary phase on the solid support and the mobile phase that flows freely down the column.

This chromatographic mode is termed "normal phase" or hydrophilic interaction chromatography (HILIC) if the stationary phase is more polar than the starting mobile phase. The elution order is generally related to the increasing hydrophilic nature of the sample. The more hydrophilic a sample the slower it will be eluted and vice versa, the more hydrophobic the sample, the faster it will be eluted. The elution of sample molecules in order of increasing hydrophilicity is accomplished by decreasing the concentration of organic modifier in the mobile phase. The most popular support is microparticulate silica gel.

In contrast to this, partition chromatography is termed "reversed phase" if the starting mobile phase is more polar than the stationary phase. The support is also silica, but in contrast to "normal phase" the silanol groups are chemically derivatized with organosilanes, such as octadecyl (C18). In reversed phase chromatography (RPC) peptides or proteins are eluted with increasing organic solvent in order of increasing relative surface hydrophobicity. Compared to it, separations carried out with descending salt gradients at neutral pH are termed hydrophobic interaction chromatography (HIC). RPC and HIC both involve hydrophobic interactions between the solute and the stationary phase. Besides the eluting solvents these methods are distinguished by the ligand density of the stationary phase (in HIC approximately one-tenth of that used in RPC) and the fact that HIC is used for the separation of proteins in their native state, while harsher elution conditions in RPC tend to disrupt protein tertiary structure. This is especially true for complex, multiinteraction enzymes, whereas stabilized or crosslinked proteins as well as small peptides are less likely to loose biological activity. However, biological activity may be retained through proper chromatographic conditions or may be regained by post-chromatographic treatment.

Reversed phase chromatography, which has almost become synonymous with HPLC, is one of the most powerful separation techniques for peptides and proteins on the analytical as well as preparative scale. Besides speed and efficiency, the great variety in choice of mobile and stationary phase as well as column diameter account for this prevailing position.

In multidimensional chromatography RPC is widely used as the last step prior to micro-sequencing or mass spectrometry. Two-dimensional HPLC, for example, is common where a crude peptide extract or a complex protein digest is first fractionated on an ion-exchange column, followed by further purification of each fraction by gradient elution on a RP column (see 4 Applications). The successive use of two RP-columns with different packings (or the same packing with different solvents) is also possible in the separation of complex mixtures of peptides (e.g. enzymatic digests). RPC on micro-columns provides purified fractions in small volumes, which are mostly free of compounds which interfere with sequence analysis or mass spectrometry.

Availability of special equipment for capillary and even nano HPLC now results in enhanced sensitivity of limited amounts of samples.

3 Getting Started

Suggestions presented here derive from our own experience with HPLC and manufacturers' recommendations, as well as from several excellent articles dealing with practical and theoretical aspects of HPLC [Chicz 1990; Dolan 1991; Mant 1991a,b; Nugent 1991a,b; Chervet 1996; Grimm 1997].

3.1 Solvents

Mobile phase purity is a particulary important consideration in the HPLC of peptides and proteins: e.g. dust or other particulate matter plug columns, air-bubbles in the pump result in erratic low flow rates and pressure fluctuations; impurities and air-bubbles in the detector cell cause baseline noise or artificial peaks. Therefore, all solvents (water, organic modifier) and additives (e.g. buffer salts) should be of HPLC grade or of the highest quality available. All solutions used should be filtered through a 0.2 μm membrane filter and thoroughly deaerated prior to use. Continuous helium sparging is the most efficient way, but vaccum optionally combined with sonication is sufficient in most cases. An in-line solvent filter between the reservoir and the pump is mandatory to improve solvent reliability. The mobile phase is often a good microbial growth medium, especially when using acetate buffers or low organic solvent content; thus, use of fresh mobile phase each day is recommended the solvents have to be changed at least every week.

A major part of the excellent resolving power of HPLC is derived from the availability of ion-pairing reagents. Peptides are charged molecules at most pH values and the presence of different counterions will influence their chromatographic behavior. TFA is the anionic ion-pairing reagent in the most popular two-phase mobile system:

Solvent A: 0.1 % (v/v) TFA in water
Solvent B: 0.085 % (v/v) TFA, 84 % (v/v) acetonitrile in water.

The low pH (pH 2) of this unbuffered solution ensures protonation of carboxyl groups thereby increasing the interaction of peptides with the reversed phase sorbent. TFA, an excellent solvent for most peptides, is completely volatile and permits detection at wavelengths below 220 nm due to its high UV transparency. As the dielectric constant of the solvent changes with increasing organic modifier concentration, the absorption spectrum of TFA shifts, resulting in an upward baseline drift. Most of this difference can be compensated for by adding 15 % (v/v) more TFA to the water reservoir. If the presence of TFA is not sufficient to resolve a particular mixture of peptides efficiently, better results may be achieved through use of more hydrophilic (e.g. orthophosphoric acid) or a more hydrophobic (e.g. heptafluorobutyric acid) anionic ion-pairing reagent.

The use of TFA reduces the number of ions produced in the ionization process in online coupling of HPLC with an <u>E</u>lectro <u>S</u>pray <u>I</u>onization <u>M</u>ass Spectrometer (ESI MS). The best compromise between chromatographic resultion and quenching of the ion current is the reduction of the TFA content to 0.025 % and 0.02 % respectively.

The use of acetonitrile as organic modifier takes advantage of its low viscosity (resulting in low back pressure), high UV transparency at low wavelengths and high volatility (the acetonitrile can be easily removed from the peptides in solution). Alcohols, particularly isopropanol and methanol, are occasionally used for the separation of very hydrophobic or very hydrophilic proteins, respectively.

If gradient systems of higher pH are needed, the following buffered solutions are convenient:

Solvent A: 10 mM ammonium acetate, pH 6
Solvent B: 10 mM ammonium acetate, 84 % (v/v) acetonitrile

and

Solvent A: 0.2 % (v/v) hexafluoroacetone (HFA)/NH_3, pH 8.6
Solvent B: 0.03 % (v/v) HFA/NH_3, 84 % (v/v) acetonitrile.

The pH of these solvent systems is raised above the pK for the carboxyl groups; thus, peptides containing aspartate or glutamate residues elute earlier than in the TFA/acetonilrile system (Fig. 4). Besides, the ammoniumacetate/acetonitrile gradient system provides resolution of peptide species which differ in their phosphoylation state (Fig. 5).

3.2 Pump

Peptide separations on micro-columns or even nano LC require HPLC hardware that delivers reproducible gradients at low flow rates (20 μL/min or less). Without further equipment, such gradients can be achieved only with HPLC-systems that employ high-pressure micro-syringes. They deliver almost pulse-less flow and meet the dead-volume requirements for micro-columns. Solvent delivery must be pulse-free to limit the detector baseline noise and the damage to columns as well as to achieve optimal chromatographic resolution. Drawbacks of syringe pumps are the compression of liquids which disturbs flow and the need of refilling, which can cause problems with long equilibration time or slow gradients.

Reciprocating pistons have a small stroke making solvent changes rapid and accurate, and provide fast flow rate change. A major drawback, of course, is pump pulsation, which excludes generation of constant low flow rates needed for micro-columns (see also 3.4).

Wettable components of the pumping system must be chemically resistant to common mobile phases. Many of the older HPLC instruments have stainless steel pumping systems that will corrode in the presence of halide under acidic conditions. Titanium and very resistant plastics or ceramics have been used to fabricate pumping systems that are stable in both acid and base. These systems are generally more expensive than the stainless steel one, but give an added measure of confidence.

The valves and the pump seals are the parts of a pump most likely to give problems caused by bubbles, dirt and normal wear. The pump seal does not seal completely around the piston in many pumps, so the piston is damp behind the seal. To prevent abrasive damage of the pump seal and the piston resulting from crystalline buffer residues, it should be flushed behind the pump seal. Many HPLC systems have flushing ports at the rear of the pump head for this purpose allowing continuous flushing with water during the HPLC run.

3.3 Gradient

Convenient mixing of the solvents without disturbing the gradient is required at all flow rates. Dynamic mixers are the most versatile but have the disadvantage of introducing additional dead volume into the pumping system and are therefore not applicable for capillary or nano HPLC. Static mixers can be of lower volume than dynamic mixers but high-viscosity solutions are not mixed well. The volumetric flow rate and volume of the mixer and tubing connecting the mixer to the column determine the delay time between the start of a gradient program and actual solvent delivery to the head of the column. Replacing the mixer in a pump by a smaller one meets the needs for micro-HPLC, where long overall run-times caused by low flow rates are common.

Solvent gradients should begin with at least 3–5 % organic modifier and should not exceed 95 %. The extremes of solvent composition should be avoided if possible as mobile phase mixing and mobile phase delivery in syringe pumps is most difficult in this range. Besides, it is difficult to totally remove organic solvent from the stationary phase when equilibrating with water and this results in long equilibration times.

Gradient elution is virtually mandatory when chromatographing peptides and proteins. The change in eluent strength may be continuous (linear gradients) or stepwise according to the degree of resolution required. The majority of peptides are eluted from reversed phase columns prior to a 50 % acetonitrile concentration, optimum peptide resolution is generally obtained between 15 to 40 % concentration of acetonitrile in the mobile phase. A typical gradient for a peptide mixture (e.g. derived from a tryptic digest) is shown in Figure 1. After 30 min equilibration time the sample is injected and the percentage of solvent B is increased linearly from 5 % to 50 % in 0.5 % to 1 % per min. Finally the solvent is changed in 15 min to 95 % solvent B to ensure the eluation of hydrophobic peptides or not completely digested protein.

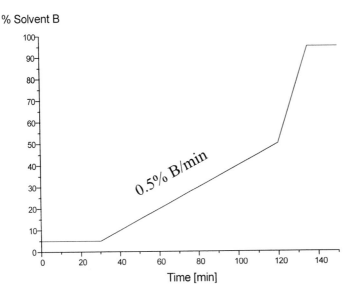

Figure 1. Typically used gradient for the separation of peptide on a C18 HPLC column.

3.4 Pre-Column Split

Capillary columns (100 to 500 μm internal diameter, I.D.) and nano columns (50 to 100 μm I.D.) require low flow rates (0.1–5 μL/min). Complementary equipment of conventional HPLC hardware with a pre-column split device is necessary to reduce the flow rate. Preinjection solvent split is possible for example by diverting most of the solvent, via a tee, through a balance column [Moritz and Simpson 1993].

3.5 Sample Preparation

All the material to be injected should be in solution and free from particulate matter. Sample preparation techniques removing as many interfering materials as is practical should be applied. When capillary columns or nano columns are used, centrifugation of the sample is mandatory. Surfactants are generally harmful to polypeptide separations and should be avoided. Even traces of SDS in a peptide sample reduce separation efficiency and peptide recovery during HPLC. Formic acid should be used cautiously because it shortens column lifetime and harms HPLC pumps. The sample solvent must be compatible with the mobile phase being used, so that the sample or buffer does not precipitate in the pores of the column packing. The sample solvent should not consist of too much organic content, otherwise the peptides or proteins are eluted during sample injection. To prevent further column problems, it is important to optimize the sample concentration for proper column loading. A sample volume of threefold the volume of the flow rate per minute is a reasonable volume. Sample components can also gradually coat the window of the detector cell thus decreasing the sensitivity and increasing the background noise. For optimal sample preparation see also chapter II.6.

3.6 Injector

Samples are most commonly introduced into the HPLC column via a manual injector. Autosamplers are not convenient for single injections of important samples, but work properly when injecting volumes of 5 µL or more. The loop should be of inert material like polyetheretherketone (PEEK) or titanium. The loop volume should be adapted to the sample volume and to prevent sample loss not be fully exploited. The injection syringe should be compatible with the injector septum. Loading of the loop can be monitored by the replaced solvent eluting from the injector.

3.7 Tubings

When biocompatible conditions are essential, connective tubings made of Teflon, PEEK, fused silica or titanium are employed instead of stainless steel. Conventional compression fittings are replaced by fittings that can be finger-tightened. To reduce extra-column dispersion, all tubing through which the sample passes should be of 0.1 mm I.D. or smaller and in as short length as is convenient. Larger I.D. (e.g. 0.25 mm) tubing is appropriate for other system connections (e.g. connecting the pump to the injector), but lengths should be kept as short as possible in order to minimize the gradient delay volume. Especially for nano HPLC the inner diameter of the tubing has to be optimized. Tubings with 100 µm I. D. cause long delay times, while tubings with an inner diameter of less than 25 µm result in high back pressure.

3.8 In-Line Filter, Guard Column

An in-line filter (0.5 µm porosity) and/or a guard column between the injector and the main HPLC column will help to prevent blockage of the column inlet frit with particulates from the sample, mobile phase or from instrument wear (e.g. pump seals and injector rotors). The use of a guard column is strongly recommended when working with complex samples, to remove materials that are irreversibly bound to the column packing or are so large that they block either the column frit or the packing bed. The best guard column to use is one that is prepacked with the exact same material (same support, particle size, pore size, and bonded phase) as the separation column.

In oder to reduce the volume in nano HPLC, the column is directly adjusted to the injector and an small inlet filter is placed inbetween.

3.9 Column

Alkylsilane-derivatized silica media, characterized by the type of hydrophobic ligand, are the dominant choice for HPLC of peptides and proteins. Typical ligands are n-octadecyl (C18), n-octyl (C8), n-butyl (C4), and phenyl. Several parameters have to be considered in the choice of packing:

- *Stationary phase:* While peptides are capable of intercalating the stationary phase of the packing, proteins have restricted access due to their large size and are assumed to interact at the outer surface. The alkyl chain length most effective for proteins and more hydrophobic samples is therefore in the shorter range (C4 to C8), while longer chains (C8 to C18) are normally used for small peptides as well as for more hydrophilic samples. In general, hydrophobicity and therefore the retention time increases with the chain length of the alky silanes. The C18 phase is especially recommended for mapping enzymatic digests (e.g. tryptic digest). The diphenyl phase (similar in hydrophobicity to C4) offers an alternate selectivity for polypeptides, particularly those containing aromatic side chains.
- *Pore diameter:* To ensure adequate mass transfer, it is recommended to use macroporous supports with a pore diameter between 300 and 1000 Å for protein separations. This allows better penetration into the pores, thus maximizing available sorbent surface area for protein-support interaction. For the separation of peptides, a pore size of 300 Å has become most popular. Nonporous media are acceptable for analytical separations, where speed is a major issue and small samples may be used.
- *Particle size:* As with small molecules, theory predicts that performance should go up as particle size goes down. To achieve maximum resolution in analytical separations, the use of particles of 5 μm or less is reasonable. 2–3 μm particles, however, perform higher column back pressure, shorten column life and perform greater susceptibility to plugging. For preparative applications, 10 to 20 μm media are often used because the resolution of 5 μm sorbents deteriorates quickly when they are overloaded.
- *pH stability:* When working with silica-based columns, most manufacturers recommend using a mobile phase pH between 2 and 7.5, as silica will slowly dissolve and greatly shorten column life at higher pH. Polymer-based columns or the newer stabilized silica columns offer a superior pH-stability.
- *Temperature:* Peptide or protein retention times generally decrease with increasing temperature, due to increasing solubility of the sample in the mobile phase. In addition, due to a more rapid transfer of the sample between the stationary and mobile phase, improved peptide resolution happens. Furthermore the viscosity of the mobile phase will be reduced with higher temperatures, resulting in lower back pressure. But in general, room temperature is adequate for peptide and protein separations. For good reproductivity of retention times stabilized column temperature is necessary. High temperature runs are important for the use of supports with smaller particle sizes (1 to 3 μm) for increased resolution and non-porous supports for ultrafast performance.
- *Special packings:* Flow-through or perfusive-particle chromatography is a powerful technique for the very rapid separation of peptides and proteins. Columns are

packed with derivatized porous polystyrene divinylbenzene and operate at very high flow rates (e.g. 5 to 10 mL/min for a 4.6 mm column; 40–100 µL/min for a 320 µm I.D. fused silica capillary). Like using conventional columns, the sample is transported to the surface of the particles by convective flow. The particles, however, contain two types of pores, so-called throughpores, which allow convective flow, and diffusive pores, which provide high adsorptive surface area. At high flow rates, convection dominates diffusion and allows direct access to the high surface area of the diffusive pores. Thus, rapid perfusive transport is made possible and column resolution and capacity maintained independent of flow rate [Kassel et al., 1993]. Use of non-porous RP-packings, made of monodisperse, spherical silica particles with a mean diameter of 2 µm, eliminates the intraparticulate diffusion that is responsible for band-broadening, losses in efficiency and resolution. As the solid core of these micropellicular packings is fluid-impenetrable, the columns are generally more stable at high pressures and elevated temperatures than conventional packings. Since the loading capacity of the column is a function of the surface area of the packing, this approach is not advantageous for preparative applications [Kitagawa 1989; Cox 1993].

Traditional 4.6 mm I.D. columns were quite adequate for most protein/peptide purification problems until the advent of microsequencing and mass spectrometry, which allow primary sequence information to be obtained from as few as 100 to 500 fmol of polypeptide. Thus, use of columns requiring less sample load but maintaining similar peak detection became necessary. Narrowbore (2 mm I.D.) and microbore (0.5–1 mm) columns as well as capillary HPLC (0.1–0.5 mm) are now commonly used for peptide separations. While most modern HPLC hardware is compatible with microbore columns, the use of fused silica columns for capillary (0.1–0.5 mm I.D.) or even nano (50–100 µm I.D.) HPLC requires special devices to reduce the flow-rate and to enhance the detectability of low sample amounts. Table 1 provides an overview of the different dimensions of HPLC columns suitable for the separation of peptides and proteins and for following microsequencing and for mass spectrometry analysis. Typical flow-rates and corresponding fraction volumes are stated for the respective column I.D.

The internal diameter is an important column parameter. The sensitivity or minimal detectable quantity is proportional to the square of the inner column diameter. The gain in sensitivity relative to conventional LC (set to 1) is about 200 for capillary

Table 1. Overview about HPLC columns and their parameters.

	Inner diameter [mm]	Relative gain in sensitivity	Flow rate [µL/min]	Fraction volume [µL]	Volume of flow cell [µL]
Conventional LC	4.6	$\equiv 1$	1000	500	15
Narrowbore LC	2.0	5	200	100	1.2
Microbore LC	0.5–1.0	35	50	25	1.2
Capillary LC	0.1–0.5	200	5	3	0.032
Nano LC	0.05–0.1	3500	0.2	0.1	0.003

LC, which became a standard method in highly sensitive peptide and protein analytics. Thus, compounds are eluted in smaller volumes, resulting in increase in peak concentration giving greater detector response and less unspecific adsorptions on surfaces. This enhances detection limits in sample limited situations. Lower flow rates used with smaller I.D. columns demand less solvent and reduce waste.

Flow rate and column length do not play such an important role in the separation of proteins because they adsorb to and are displaced from the reversed phase surface once and do not interact appreciable with the surface after displacement. In fact, long columns together with low flow rates often yield poor recovery of proteins due to the long residence time the protein spends on the column. Therefore short (20–50 mm) columns are recommended for the separation and purification of proteins. In contrast to proteins, peptides chromatograph by a mixture of adsorption and partitioning effects which lead to a positive effect of column length and flow rate on resolution. In general, the lowest flow rate consistent with near maximum resolution should be used. For peptide mixtures therefore 15 cm up to 25 cm columns should be applied.

As HPLC columns are expensive, precautions should be taken in order to extend their life. A proper application protocol (e.g. solvent used, pressure at the usual flow rate, notes about peak resolution) will help to quickly notice column deterioration. Besides it is useful to measure column aging by separation of a mixture of synthetic peptides under standard conditions and comparing records with special regards to peak retention time, peak shape and height. Sudden deterioration by blockage of the frit at the head of the column (resulting in significant rise in back pressure) and a void in the packing bed (peak broadening or splitting) can be quickly noticed. When strongly retained compounds are acccumulated on the column (rise in back pressure, deterioration of sample resolution) an effective cleaning procedure (gradient elution from 0.1 % (v/v) aqueous TFA to 0.1 % (v/v) TFA in isopropanol with repetitive gradients) should be applied. If deteriorations are permanent and cannot be cured, the column must be replaced. Storage after every day's use should be in a high concentration of organic solvent (e.g. 90 % acetonitril), so that the columns are cleaned. For long term storage, 100 % methanol is recommended by many manufacturers. Prior to the first run performed on a previously stored column, a gradient run up to 100 % B in the absence of sample (e.g. with injection of solvent B) should be carried out. This will serve to remove any impurities from the column that may have accumulated during storage. Because of air-bubbles and the low flow rate it is recommend for capillary or nano LC to equilibrate the HPLC system with several blank runs over night.

3.10 Detection

Detection of peptides during HPLC is generally based on peptide bond absorbance at low ultraviolet (UV) wavelengths. The absorption maximum for the peptide bond is about 187 nm but detection below 210 nm can suffer from interference due to impurities present in buffers, solvents or the sample itself. Thus, the common use of 214 nm as detection wavelength is a good compromise between detection sensitivity and potential detection interference. In addition to peptide bond absorbance, the aromatic side chains of tyrosine, phenylalanine and tryptophane absorb light in the 250 to

290 nm UV range. Thus, dual wavelength detectors can discriminate peptides containing aromatic amino acids.

For the above reasons, UV detectors are the most widely used LC detectors with protein and peptide samples. For certain components, including tryptophan-containing proteins, which fluoresce under UV-illumination, and for fluorescence labeled peptides fluorescence detectors can provide higher sensitivity and selectivity than absorbance detection, particularly in the analysis of complex samples.

The detector cell volume (see also Table 1) must be minimized to prevent postcolumn mixing, band broadening and collection errors due to a delay of several seconds before the peak leaves the detector. To prevent loss in sensitivity caused by the small cell volume (e.g. short optical path length), the use of ultrasensitive UV flow cells, e.g. Z- or U-shaped flow cells with longer optical path, is necessary [Chervet et al. 1989]. For capillary LC and nano LC the typical detector cell is Z-shaped and contains a cell volume of 35 nL and 3 nL respectively. The optical path length is 8 mm.

Due to the rapid developement of mass spectrometry, Electro Spray Ionization mass spectrometers coupled online to a HPLC column are also used as detectors in analytical HPLC nowadays. The total ion current is monitored and the resulting chromatogram is quite similar to the UV chromatogram. Additionally, the peptide or protein mass is recorded, which can be very useful in peptide or protein identification. Mass spectrometry as a detection method in HPLC is also highly sensitive. Because of the mass selectivity of these detectors they are in some cases even more sensitive than UV detectors; e.g. a compound with a known molecular weight can be resolved with the help of a mass trace, although it is not resolved by HPLC itself.

3.11 Fractionation

To assure optimal peak pooling the different fractions should be hand-collected in small plastic cups. Automatic peak detectors provide problems at low flow rates. If more than one peak elutes during the time the peak detector is counting down the transit time from the UV detector to the fraction collector, and hence the peaks are mixed together [Stone and Williams 1993].

The use of columns from 1 mm down to 300 μm inner diameter permits collection of peptide peaks in a volume of several microliters. This minimizes the losses from adsorption to the collection tube observed when a low amount of sample is collected and stored in a larger volume typical for conventional columns. After collecting, if the fractions are not directly applied to further analysis, rapid freezing (–80 °C) is strongly recommended.

For capillary columns of 180 μm inner diameter with a typical fraction volume of as few as 0.5 to 2.0 μL collection of samples in cups is not practicable. HPLC fractions has to be fractionated directly on targets suitable for further analysis. This can be achieved by a computer controlled robot (Fig. 2). For Edman degradation the samples are transferred from the HPLC capillary directly to polyvinylidene difluoride (PVDF) membrane fixed on the moveable table of the robot. After evaporation of the solvent an area of approximately 2 mm^2 of the membrane is excised and subjected to Edman degradation. To analyze the samples with MALDI mass spectrometry they are fractionated directly on the MALDI target. The necessary UV absorbing MALDI

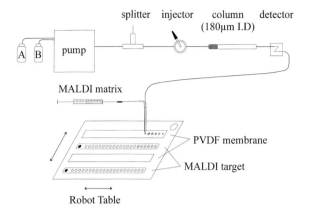

Figure 2. Off-line coupling of capillary HPLC with Edman sequencing and MALDI mass spectrometry by a robot. Fractions of as few as 0.5 μL can be directly subjected for further analysis by one transfer step only.

matrix can be added continuously. The HPLC fraction can be split with this robot system for MALDI mass spectrometry and Edman degradation in any requested ratio, so that both analyses can be performed simultaneously.

4 Applications

HPLC is a powerful tool in peptide and protein separation. Although the right choice of column, solvents and gradient is necessarry to succeed in a specific question. Several typical applications in modern analysis of peptides or proteins are given in this chapter.

If not stated otherwise, peptides were detected by their absorbance at 214 nm, setting the sensitivity to absorption units at full scale (AUFS) as indicated. Solvent data in percent mean v/v. For further data on instrumentation see references.

If peptides from enzymatic digests are not separated well in the first chromatography, a second, tandem-arranged HPLC can achieve optimum resolution. Figure 3 [Serwe et al. 1993] depicts the elution of peptides gained from Lys-C digestion of the regulatory light chain (LC25) of obliquely striated muscle myosin from earthworm, *Lumbricus terrestris*. Fraction 32 was subjected to a second HPLC (Fig. 4) carried out in a different solvent system (HFA/NH₃/acetonitrile) on the same column, succeeding in the separation of two peptides further analyzed by sequence analysis. As the pH of the solvent system (pH 6) is raised above the pK for carboxyl groups (e.g. pH 4.5), peptides containing aspartate or glutamate residues elute earlier according to their acidic character.

Fraction 33 in Figure 3 was injected onto a Vydac-phenyl column (2.1 x 50 mm). Separation was achieved with a gradient of TFA/acetonitrile to receive better resolution for a peptide containing tryptophane (data not shown). Diphenyl offers an alternate selectivity to aliphatic reversed phases, particulary for those containing aromatic

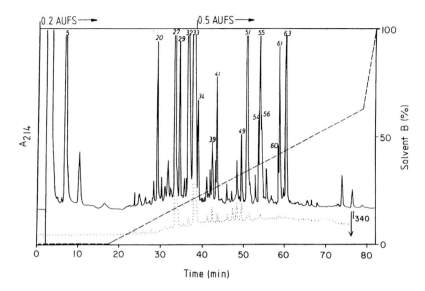

Figure 3. Separation of endoproteinase Lys-C peptides from LC25 [Serwe et al., 1993]. 10 nmol of the digested protein were applied onto a 5-μm Vydac C18 RPC column (4.6 x 150 mm). Separation was performed using a gradient of 0.1 % TFA (solvent A) vs. 84 % acetonitrile/0.08 % TFA as indicated (broken line) at a flow rate of 1 mL/min. Fluorescence of tryptophan-containing peptides was measured at 285 nm excitation and 340 nm emission (punctuated line).

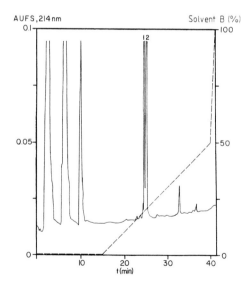

Figure 4. Rechromatography of fraction 32 from the separation of Lys-C peptides from LC25 (Figure 3).

800 pmol peptid were injected into a 5-μm Vydac C18 RPC column (4.6 x 150 mm). Elution was performed applying a gradient system consisting of 0.2 % HFA/NH$_3$, pH 6 vs. 0.03 % HFA/NH$_3$, 84 % acetonitrile as indicated (broken line), the flow rate was 1 mL/min. The designated fractions were subjected to sequence analysis.

side chains. Rechromatography on a smaller I.D. column is generally useful to minimize sample loss and increase sample concentration.

A convenient solvent system of higher pH is the ammoniumacetate/acetonitrile gradient system at pH 6. As shown in Figure 5 it provides resolution of peptides only

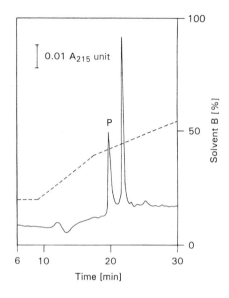

Figure 5. Separation of a synthetic peptide from its phosphorylated form.
An aliquot of a phosphorylation mixture containing 1 nmol peptide was applied onto a 5 µm SGE C18 RPC column (2 x 250 mm). Separation was performed with a gradient system consisting of 10 mM ammoniumacetate, pH 6 (solvent A) and 10 mM ammoniumacetate/84 % acetonitrile (solvent B) as indicated (broken line). The flow rate was 80 µL/min. P designates the phosphorylated peptide.

varying in the phosphorylation state of only one serine residue. A synthetic peptide of 19 amino acid residues derived from the rabbit skeletal muscle phosphorylase kinase α subunit sequence (provided by Dr. Chen, Research Institute of Occupational Medicine at the Ruhr-University of Bochum), was phosphorylated in vitro by a recombinant mitogen-activated protein kinase (provided by Dr. E. Mandelkow, Research Unit for Structural Molecular Biology, Max-Planck Gesellschaft, Hamburg). The more hydrophilic phosphorylated peptide eluted about 2 min earlier than the unphosphorylated one.

To analyze the sites of phosphorylation, proteins can be labeled with [^{32}P] and the phosphorylated peptides generated by an enzymatic digest can be monitored by their radioactivity. Figure 6 [Lehr et al. 1998] shows a rechromatography of phosphorylated peptides from a 107 kD protein. The human Gab-1 protein was phosphorylated with [^{32}P]-γ-ATP using a tyrosine kinase. The protein was digested with trypsin and the peptides were separarted by ion exchange chromatography (4.6 mm x 50 mm, A: 20mM NH$_4$Ac pH 7.0; B: 1 M K$_2$HPO$_4$/KH$_2$PO$_4$ pH 4,0). Each fraction containing radioactive phosphopeptides was further separated by reversed phase HPLC using a 800 µm x 250 mm C18 column at a flow rate of 17 µL/min. The fractions were analyzed for radioactivity (Fig. 6) and 1 µL of all positive fractions was taken for nanospray ESI-MS/MS experiments the remaining for Edman degradation. With both analyses three of the four radioactive peptides (Fig. 6) have been identified and the phosphorylation site was determined. All in all 8 out of 18 tyrosine were found to be at least partially phosphorylated.

Elution conditions of HPLC (e.g. low pH, high organic solvent content) are not convenient for the isolation of proteins in their native state. This is especially true for complex, multiinteracting enzymes. A positive effect of the tertiary structure disrupting conditions of HPLC is shown in Figure 7 [Weber 1989]. Native phosphorylase kinase from rabbit skeletal muscle is rapidly resolved into individual subunits by HPLC in TFA/acetonitrile.

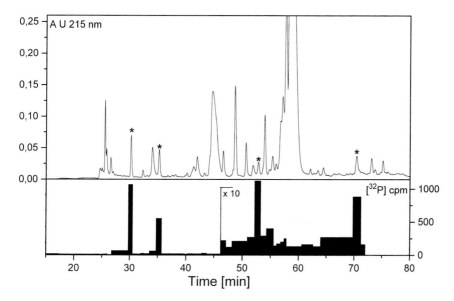

Figure 6. Chromatogram of a tryptic digest of phosphorylated recombinant GST-Gab-1 protein. The upper curve shows the UV absorption at 215 nm. The peptides were hand-collected and the radioactivity of each fraction was measured (lower trace, [^{32}P] cpm = counts per minute). Four peptides (*) contained [^{32}P]. The peptides were identified and the phosphorylation sites were localized by mass spectrometry and Edman sequencing.

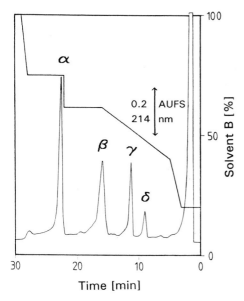

Figure 7. HPLC of holophosphorylase kinase [Weber 1989].

1 mg of the protein was applied onto a 10-μm Vydac C4 RPC column (20 x 50 mm). Elution was carried out using a gradient system consisting of 0.1 % TFA vs. 0.085 % TFA/84 % acetonitrile as indicated (solid line). The flow rate was 7.8 mL/min.

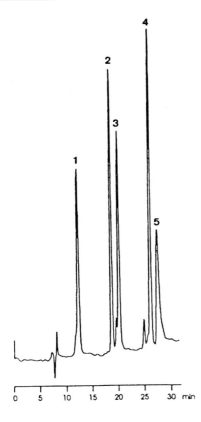

Figure 8. Separation of proteins by using a micro gradient [Chervet, 1989].
A mixture of 5 proteins (1 = ribonuclease; 2 = insulin; 3 = cytochrome C; 4 = lysozyme; 5 = bovine serum albumine; 60 ng abs./compound) was injected into a 5-µm fused silica capillary column (320 µm i.d. x 30 cm, C18, 300 Å; LC Packings). Separation was carried out applying a gradient with the solvents A: 0.1 % TFA and B: 0.1 % TFA/95 % acetonitrile (49 % to 70 % B in 30 min) at a total flow rate of 4µL/min. Temperature was set to 20 °C. Detection was performed at UV 220 nm with a 20 mm Z-shaped flow cell (LC Packings), range 0.2 AUFS.

Figure 8 [Chervet 1989] depicts an application of 320 µm I.D. fused silica columns to separate proteins or peptides, respectively. Microgradients were produced by a microflow processor; detection was performed with a 20 mm Z-shaped flow cell.

Figure 9 shows the chromatogram of a sample containing 10 peptides in the range of 0.5 to 3 pmol. This sample was used for a LC contest on the 4th meeting "Micromethods in Protein Chemistry" at Martinsried, Germany (4. Arbeitstagung "Mikromethoden in der Proteinchemie", 1997). The task was to resolve these peptides and to quantitate the UV absorption of each peptide. The chromatogram shows more than 10 signals due to the unstability of peptides in low amounts. The signal 10* (Fig. 9) for example was identified by a LC-MS analysis as oxidized peptide no. 10. All eight participating companies used C18 columns (mainly I.D. 180 µm) combined with an acetonitril/TFA gradient and were able to resolve at least 10 of the predominant 12 peptides.

Figure 10 shows the separation of tryptic peptides of lactate dehydrogenase *(rabbit)*. 650 fmol of the digested protein was applied onto a nano column with an inner diameter of 75 µm (C18, 75 µm x 150 mm). The gradient of Figure 1 was used. The flow rate was 200 µL/min and was achieved by splitting the pump flow of 12 µL/min with a dead volume free T.

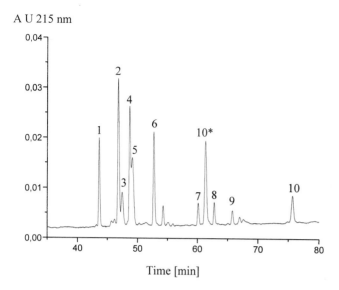

Figure 9. Separation of a contest sample which contains 10 peptides in the range of 0.5 to 3 pmol. A C18 column (320 μm x 25 cm) with an acetonitril/TFA gradient (Figure 1) was used. Additional signals were detected due to the unstability of peptides in samall amounts. (e.g. peak 10* is due to oxidation of peptide 10).

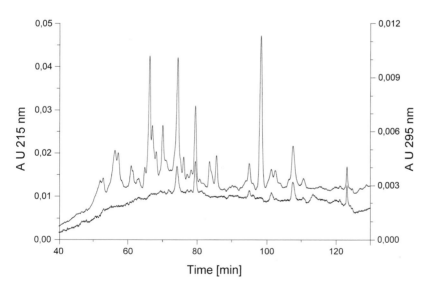

Figure 10. Separation of a tryptic digest of lactate dehydrogenase from *rabbit*. 650 fmol of the digested protein were injected into a nano comun (75 μm I.D. x 15 cm, C18, LC Packings) and separated with a solvent system containing TFA/acetonitrile (see Fig.1). The flow rate was 200 nL/min. Detection was performed at 215 nm (upper curve) and 295 nm with a Z-shaped flow cell (35 nL, LC Packings).

5 References

Chervet, J.P., Ursem, M., Salzmann, J.P. (1996), *Anal. Chemi.* 68 (9), 1507–1512, Instrumental Requirements for Nanoscale Liquid Chromatography.

Chervet, J.P., Ursem, M., Salzmann, J.P. and Vannoort, R.W. (1989), *J. High Resolution Chromatography* 12 (5), 278–281, Ultra-sensitive UV detection in micro separation.

Chicz, R.M. and Regnier, F.E. (1990), *Methods Enzymol.* 182, 392–421, High-performance liquid chromatography: effective protein purification by various chromatographic methods.

Cox, G.B. (1993), Preparative reversed phase chromatography of proteins, in: *Chromatography in Biotechnology* (Horvath, C. and Ettre, L.S., eds.), pp 165–182, American Chemical Society, Washington, DC.

Dolan, J.W. (1991), Preventive maintenance and troubleshooting LC instrumentation, in: *High-Performance Liquid Chromatography of Peptides and Proteins: Separation, Analysis, and Conformation* (Mant, C.T. and Hodges, R.S., eds.), pp 23–29, CRC Press, Boca Raton.

Grimm, R., Serwe, M., Chervet, J.P. (1997), LC/GC 15, 960–968, *Capillary LC Using Conventional HPLC Instrumentation.*

Henschen, A., Hupe, K.-P., Lottspeich, F., Voelter, W. (1985), *High Performance Liquid Chromatography in Biochemistry*. VCH, Weinheim.

Kassel, D.B., Luther, M.A., Willard, D.H., Fulton, S.P. and Salzmann, J.P. (1993), Rapid purification, separation and identification of proteins and enzyme digests using packed capillary perfusion column LC and LC/MS, in: *Techniques in Protein Chemistry IV* (Hogue Angeletti, R., ed.), pp 55–64, Academic Press, Inc., San Diego.

Kitagawa, N. (1989), New hydrophilic polymer for protein separations by HPLC, in: *Techniques in Protein Chemistry* (Hugli, T.E. ed.), pp 348–356, Academic Press, Inc., San Diego.

Lehr, S., Kotzka, J., Klein, E., Siethoff, C., Knebel B., Tennagels N., Affüpper, M.,Klein, H. W., Brüning, J. C., Meyer, H. E., Krone, W., Müller-Wieland, D. (1998), Human Gab-1 Protein: Identification of Tyrosine Phosphorylation Sites by EGF Receptor Kinase in Vitro, (Submitted to Biochemistry).

Mant, C.T. and Hodges, R.S. (1991a), Mobile phase preparation and column maintenance, in: *High-Performance Liquid Chromatography of Peptides and Proteins: Separation, Analysis, and Conformation* (Mant, C.T. and Hodges, R.S., eds.), pp 37–45, CRC Press, Boca Raton.

Mant, C.T. and Hodges, R.S. (1991b), The effects of anionic ion-pairing reagents on peptide retention in reversed phase chromatography, in: *High-Performance Liquid Chromatography of Peptides and Proteins: Separation, Analysis, and Conformation* (Mant, C.T. and Hodges, R.S., eds.), pp 327–341, CRC Press, Boca Raton.

Meyer, H.E. and Mayr, G.W. (1987), *Biol. Chem. Hoppe-Seyler* 368, 1607–1611, Nπ-methylhistidine in myosin-light-chain kinase.

Moritz, R.L and Simpson, R.J. (1993), Capillary liquid chromatography: a tool for protein structural analysis, in: *Methods in Protein Sequence Analysis* (Imahori, K. and Sakiyama, F., eds.), pp 3–10, Plenum Press, New York.

Nugent, K.D. (1991a), Commercially available columns and packings for reversed phase HPLC of peptides and proteins, in: *High-Performance Liquid Chromatography of Peptides and Proteins: Separation, Analysis, and Conformation* (Mant, C.T. and Hodges, R.S., eds.), pp 279–287, CRC Press, Boca Raton.

Nugent, K.D. and Dolan, J.W. (1991b), Tools and techniques to extend LC column lifetimes, in: *High-Performance Liquid Chromatography of Peptides and Proteins: Separation, Analysis, and Conformation* (Mant, C.T. and Hodges, R.S., eds.), pp 31–35, CRC Press, Boca Raton.

Serwe, M., Meyer, H.E., Craig, A.G., Carlhoff, D. and D'Haese, J. (1993), *Eur. J. Biochem.* 211, 341–346, Complete amino acid sequence of the regulatory light chain of obliquely striated muscle myosin from earthworm, Lumbricus terrestris.

Stone, K.L. and Williams, K.R. (1993), Enzymatic digestion of proteins and high-performance liquid chromatography peptide isolation, in: *A Practical Guide to Protein and Peptide Purification for Microsequencing* (Matsudaira, P., ed.), pp 43–69, Academic Press, Inc., San Diego.

Weber, C. (1989), *PhD Thesis*, Institute of Physiological Chemistry, Ruhr-University of Bochum, Entwicklung einer spezifischen Methode zur Lokalisierung von Phosphoserin.

Sample Preparation I:
Removal of Salts and Detergents

Dorian Immler, Ehrenfried Mehl,
Friedrich Lottspeich and Helmut E. Meyer

1 Introduction

Most analytical techniques used in protein structure analysis like amino acid sequence analysis, amino acid composition analysis or mass spectrometry require pure samples carrier free and devoid of contaminants as salts, detergents or dyes. Therefore, the last step in a purification protocol has to be planned carefully and with foresight to the requirements of the following analytical technique. Prior to analysis the protein has to be precipitated, or to be transferred in a completely volatile solvent or immobilized onto a chemically inert solid support, where analytical reactions can be performed directly.

To remove salts or polar contaminants from milligram or large microgram quantities of soluble proteins or peptides causes usually no major problems. Samples are easily kept in solution and with some care no major losses will occur. Thus, high and often almost complete recoveries are commonly obtained with all the methods in use, i.e. precipitation, ultrafiltration, dialysis or gel chromatography.

However, when working on a microscale, i.e. low microgram or nanogram amounts of protein, severe problems arise due to the severe loss of sample material. This is caused by the inevitable contact of the sample with surfaces by either hydrophobic or hydrophilic interactions of the amphiphilic protein sample; and at least the equivalent of a monolayer of sample protein is usually irreversibly adsorbed to any surface. Almost complete loss of nanogram or even microgram amounts of protein may happen in few minutes. The know how to recover the adsorbed protein, or better, how to avoid or to minimize adsorptive losses on a microscale is one of the prerequisites for successful work in protein chemistry.

Hydrophobic membrane proteins are difficult to handle not only on the microscale, since almost always these proteins have to be isolated and purified in the presence of detergent and they become insoluble when these amphiphilic substances are removed. Major adsorptive losses due to aggregation, precipitation or hydrophobic interaction are common problems. Thus, handling of membrane protein requires great skill, experience and usually large quantities of starting material.

In the following, a brief strategic outline is given how removal of salts and detergents may be approached. Since problems are completely different if the substance to be removed is hydrophilic or hydrophobic they will be discussed separately. Some rules which a scientist working in microanalysis should always follow are summarized as the "Golden Rules for Handling Proteins" at the end of the chapter.

2 Removal of Salts or Polar Components

The most common situation where salts have to be removed is after an enzymatic cleavage of a protein. Here, the strategic demand to avoid an additional desalting step led to the big success of reversed-phase HPLC and the volatile 0.1 % trifluoro-acetic acid/acetonitrile solvent system in protein chemistry [Yang and Kratzin 1981]. Sample concentration, separation of the cleavage products and the desalting is done efficiently and simultaneously in one step.

Often, the last purification step cannot be performed in a salt free solution. One is faced with the situation that a protein is purified to homogeneity but the last step was done in buffer solution (e.g. ion exchange chromatography, gel chromatography or hydrophobic interaction chromatography). All salt has to be removed prior to further protein analysis. The same is true when proteins are subjected to chemical modifications (e.g. reduction and alkylation) where salts, reagents and reaction by-products are present in large quantities and have to be removed. This problem can be greatly minimized by the substitution of buffer salts wherever possible, e.g. Tris buffers and sodium hydrogen carbonate or acetate can often be replaced by ammonium hydrogen carbonate or ammonium acetate. These volatile buffers are removed during drying and pose no problem to mass spectrometry.

However, these kinds of contaminants usually do not cause major problems in removal.

2.1 Protein-Binding Membranes

If the primary goal is to sequence an already pure protein, one of the best ways is to desalt it directly on a hydrophobic sequencing support like a PVDF (polyvinylidene difluoride) [Matsudeira 1987] or a siliconized glass fiber membrane [Eckerskorn et al., 1988]. The procedure is very simple and fast; small devices are commercially available, e.g. Prospin from PE-Applied Biosystems. These tools are in principle pipette tips where at the bottom a hydrophobic membrane is mounted. Prior to use the membrane has to be prewetted with an organic solvent like acetonitrile. Then, the protein-containing solution will be applied by centrifuging through the membrane. Proteins strongly adsorb to the hydrophobic surface and can be washed extensively with water without any risk of loss. The big advantage of this desalting procedure is that even large volumes can be continuously applied and the sample is concentrated and desalted at the same time. In our experience the commercial devices give excellent results down to 50 pmol protein. With lower amounts losses are remarkable and sometimes no sequence information can be obtained even from a protein with free N-terminus applying 20–30 pmol. For microsequencing it is a command to use smaller devices which can be self-made with a smaller surface area to avoid such complete losses.

If the protein is to be sequenced but is not perfectly pure, the easiest way to prepare such a sample is to perform a SDS-PAGE separation followed by electroblotting of the proteins onto a hydrophobic membrane (see above), staining with Coomassie Blue, destaining and extensive washing with water. The recoveries are usually between 80 and 100 %, and while at the same time contaminants are removed an additional purification step was performed. Again, it is very important to keep the separation device small and the protein concentration high if lower picomole amounts of proteins are handled.

To desalt especially a solution containing a larger protein and/or to reduce it in volume a possibility is the use of microconcentrators (e.g. Centricon) or precipitation. However, this procedure has the risk to lose hydrophobic proteins and is therefore only recommended for hydrophilic proteins.

2.2 High Performance Liquid Chromatography

If a protein in solution is desired which is advantageous if the protein of interest requires additional handling after desalting, like cleavage or chemical modification, or if electrospray mass spectrometry has to be performed almost exclusively reversed-phase HPLC is the way of choice. Here again, large volumes of diluted protein solutions can be applied and at the same time concentrated onto the column using the aqueous phase of the solvent system. Commonly, 0.1 % trifluoroacetic acid in water is used due to its transparency in the UV range and its volatility. When the protein is applied a gradient with an organic modifier, usually acetonitrile, is developed. The shape of the gradient and the size and the packing of the reversed-phase column should be carefully adapted to the individual protein sample. If there is only one component to be desalted a rather steep gradient should be applied (i.e. 2–5 %/min) and a small short column (10 mm x 1 mm or less) with a rather hydrophilic reversed-phase packing (C4 and large pore size material) should be used, parameters maximizing the recovery. If the protein of interest is not very pure a flatter gradient can be used to remove contaminating substances at the same time. With small and hydrophilic proteins or with peptides reversed-phase HPLC can be used without too much risk and it is the almost ideal sample preparation for on-line ESI-MS.

Desalting large proteins with reversed-phase columns always risks low or no recovery. Sometimes, a second gradient system consisting of 1 % formic acid as starting solvent and 70 % formic acid/20 % isopropanol/10 % water can be successfully applied allowing the recovery of the desired protein from a reversed-phase column [Wildner 1987]. With this gradient system the UV detection can only be done at a higher wavelength, i.e. 280 nm. Special care is needed after the run to remove the residual formic acid to protect the HPLC from corrosion. Alternatively, the injection of 70 % formic acid in isopropanol may elute proteins adsorbed to the column material.

2.3 Desalting on Microcolumns

By using microcolumns (<0.3 mm innes diameter) with reversed-phase material even samples in the low picomole or high femtomole range can be successfully desalted for a subsequent analysis by mass spectrometry [Corchesne 1997]. No separation from other protein contaminants is achieved but the procedure removes salts good enough to obtain a sample suitable for a mass fingerprint- or MS/MS-analysis.

A procedure for desalting of samples in even smaller volumes using nanospray needles filled with Poros-Reversed Phase material has also been described but it is more specialized and also more difficult to perform than the method mentioned above [Wilm 1996].

3 Removal of Detergents and Apolar Contaminants

Frequently, the protein chemist is faced with a situation where hydrophobic substances (e.g. detergents) are present in a protein sample. It is much more difficult to get rid of these types of contaminants compared to the removal of hydrophilic substances. Detergents are usually only used in a purification scheme if a protein cannot be kept in solution without the aid of such substances. Their removal often causes solubility problems accompanied by severe precipitation or aggregation of the protein. Therefore, procedures for detergent removal have to be assisted by other types of solubilizing agents like strong acids or chaotropes to keep the protein in solution. Thus, one possibility is the removal of detergents by gel chromatography in the presence of organic solvents (e.g. formic acid) or urea. A procedure used with a larger amount of proteins, the dilution of the detergent containing solution below the critical micelle concentration (i.e. the maximum concentration where detergent monomers can exist) and subsequent ultrafiltration is seldom successful with micro amounts due to adsorptive losses. A procedure efficient with rather concentrated protein solutions (>50 µg/mL) based on the Pinkerton principle is the binding of the small detergent molecules to the hydrophobic interior of a chromatographic support (e.g. Extracti-Gel, Pierce). These stationary phase particles exhibit a hydrophilic and low-adsorptive outer surface and a hydrophobic inner core with small pores where the small detergent molecules can penetrate and bind, the large proteins are excluded. With more diluted protein solutions adsorptive losses are remarkable.

Also ionic retardation of the detergent molecules on specially designed column materials (e.g. AG11A8, Biorad) is a possibility to remove the detergents from a protein solution; here again, losses with micro amounts are prohibitively high.

3.1 Detergents

However, there are many different types of detergents available with quite distinct physicochemical properties [Hjelmeland 1990a, 1990b; Neugebauer 1990]. Therefore, the right choice of a detergent has to take into consideration not only the preservation of the functional integrity of the protein to be purified but also which kind of analysis should be performed afterwards.

Very basically, detergents can be divided into polymers like Triton or NP 40 and chemically uniform substances like CHAPS. Polymers are in general more problematic with mass spectrometry because they result in a large number of peaks in mass spectra. During an HPLC separation of peptides on reversed phase material they tend to elute in a number of peaks in the same part of the gradient as peptides. This makes analysis of the fractions by mass spectrometry more difficult or impossible although Edman degradation is normally not affected.

Some detergents are removed very easily others are almost impossible to get rid of. Generally, detergents with low critical micellar concentrations (i.e. Triton, NP40) are hard to remove due to the formation of large micelles. They are therefore not separated from proteins by centricons or similar devices. Over that, such detergents contain polyethylene glycol moieties disturbing MS measurements. For MS purposes detergents with a high critical micellar concentration are usually better suited; on the one hand most of them have a distinct molecular weight (like octyl-glucose, octyl-mannose) and on the other hand they can be removed much easier.

For microgram amounts of protein the above mentioned methods are not well suited and other alternatives have to be applied in this case. Which one of the following procedures will be chosen depends on the behavior of the individual protein and sometimes it has to be found out by try and error.

3.2 Protein-binding Membranes

As described above in the section on salt removal this kind of device can also be used for hydrophobic proteins. However, due to the strong adsorption to the PVDF membrane a large aliquot of the sample can become resistant against Edman degradation and as much as 50 pmol or more may become inaccessible due to this phenomenon. The amount of protein lost during this step depends on the size of the membrane. Therefore this type of sample preparation is only second choice. Small HPLC columns containing short chain reversed-phase materials may be used for the removal of detergents and other hydrophobic contaminants (see 2.2).

3.3 Precipitation

Numerous precipitation protocols exist, which use the different solubility of detergent bound and detergent-free protein in certain solvent systems and temperatures [Henderson 1978; Sandri 1993]. Some examples of protocols are listed below. Since the solubility of the protein in the presence of detergent depends on many parameters, which are difficult to predict exactly (detergent concentration, critical micellar concentration under actual salt and temperature conditions, strength of protein-detergent interaction under certain solvent conditions) protein recoveries can hardly be foreseen. A common problem is that all the precipitation protocols in the literature are not generally applicable, that each one is established and proven only with a very limited number of proteins, and usually success with an unknown protein cannot be guaranteed.

In some cases the precipitated protein cannot be resolubilized. Especially hydrophobic proteins tend to stay precipitated even in the presence or detergents like SDS. However, in this state the protein is not accessible to proteases and is therefore lost for the purpose of enzymatic cleavage.

Precipitation with 80 % acetone
- Add precooled (–80 °C) acetone to the protein solution, end-concentration of acetone is 80 % and keep the mixture cool at –30 °C for some hours
- Centrifuge for 10 min at 4 °C at 14.000 rpm
- Decant the supernatant and resuspend the pellet in a small volume of cold (–80 °C) acetone
- Keep it at –30 °C for the next hour
- Repeat step 2
- Take the pellet for further analyses.

Precipitation with 80 % acetone under acidic conditions
- Add precooled (–80 °C) acetone/0.2 % (v/v) HCl to the protein solution, end-concentration of acetone is 80 % and keep the mixture cool at –30 °C for some hours
- Centrifuge for 10 min at 4 °C at 14.000 rpm
- Decant the supernatant and resuspend the pellet in a small volume of cold (–80 °C) acetone/0.2 % HCl
- Repeat step 2
- Take the pellet for further analyses.

Precipitation with trichloroacetic acid /desoxycholat
- Add to the protein solution a solution containing sodium desoxycholat (20 mg/mL, pH>8) end-concentration of sodium desoxycholat 80–200 µg/mL
- After 30 min at 4 °C add 20 % trichloroacetic acid, end concentration 6 % and place this mixture on ice for 1 h
- Centrifuge for 10 min at 4 °C at 14.000 rpm
- Decant the supernatant and resuspend the pellet in a small volume of cold (–80 °C) acetone
- Keep it at –30 °C for the next hour
- Repeat step 2
- Take the pellet for further analyses.

Precipitation of a protein from a detergent containing solution
- Sample: 3µg of a 50 kDa proteins containing in 1.2 mL of the following solution: 20 mM HEPES/KOH pH 7.8, 12 mM MgCl2, 1 mM EDTA, 1 mM DTT, 1 M KCl, 0.1 % NP40, 50 % glycerol
- Add to this sample solution an equal volume of 20 % trichloroacetic acid and keep it on ice for 30 min
- Centrifuge for 10 min at 4 °C at 14.000 rpm
- Decant the supernatant and resuspend the pellet in a small volume of cold (–80 °C) acetone
- Keep it at –30 °C for the next hour

- Repeat step 2
- Take the salt and detergent free pellet for further analyses.

Removal of SDS by ion-pair extraction
- Prepare a extraction solution containing acetone:triethylamine:acetic acid:water (85:5:5:5, v/v/v/v)
- Add a 20-fold volume of the above solution to the protein and SDS containing sample
- Keep it on ice for 1 h
- Centrifuge for 10 min at 4 °C at 14.000 rpm
- Wash the pellet twice with fresh solution
- Wash the pellet with anhydrous acetone to remove traces of the extraction solution.

3.4 SDS-PAGE

On the microscale, the only generally applicable and recommended method to remove detergents and other hydrophobic contaminants is a polyacrylamide gel electrophoresis step. The protein is applied onto a polyacrylamide gel (eventually simultaneously separated from contaminating proteins) and either transferred onto a chemical inert membrane via electroblotting for direct amino acid sequence analysis or cleaved directly in the gel if internal sequences have to be established.

Although SDS-PAGE is widely used as a last step in protein purification one should be aware of its drawbacks. When membrane proteins are purified in the presence of ether moiety-containing detergents like Triton X-100 or NP-40 these detergents are not completely removed during electrophoresis. Most molecules are replaced by SDS during sample preparation but very often the remaining detergent can interfere with mass spectrometry, deteriorating results or making it impossible.

The staining method for a gel should also be carefully chosen. Some staining protocols include covalent modifications of proteins, e.g. the sensitization step in many silver-staining protocols. Glutardialdehyde does not only increase sensitivity but also crosslinks proteins leading to decreased recoveries and artificial mass changes hindering protein identification by a mass fingerprint. Protocols without the use of such modifiers [Shevchenko 1996] do not interfere with mass analysis but their sensitivity is not remarkably higher than a standard Coomassie staining.

Coomassie on the other hand binds to proteins very tightly. Even after destaining a certain portion of the protein will be loaded with Coomassie. This complex may be inaccessible to proteases. Therefore, staining should be done no longer than absolutely necessary and destaining should be done thoroughly before applying a protease to the gel piece.

In many cases negative staining using zinc and imidazole is a good choice [Fernandez-Patron 1995]. It is done in about 20 min, almost as sensitive as silver staining procedures and completely reversible within a few minutes. Gels can be destained and then electroblotted or used for in-gel digestions. A staining kit is commercially available (BioRad) although all solutions can be easily prepared in-house from chemicals of p.a. grade.

3.5 Concentrating Gels

A critical parameter when performing in-gel digestions is the protein-to-matrix ratio as in solution losses due to adhesion and the background noise due to contaminants increase with the gel volume. Therefore, it should be kept as low as possible.

A fast and easy protocol for a concentrating gel-system has been described [Gevaert 1996]. The method uses low-percentage acrylamide gels cast in a Pasteur pipette. It takes about 4–5 h and is applicable for samples down to less than 1 pmol. The resulting gel piece has a volume of 3–4 mL. Using such a gel piece for an in-gel digestion has the advantage of low extraction volumes. This helps keeping the peptide concentration in the extraction solutions high preventing sample loss due to adhesion to vessel walls. Additionally, small volumes of protease solution help preventing autodigestion. The authors describe concentrating gels as a sample preparation for MALDI-MS but with some modifications to the digestion procedure digests resulting from a concentrating gel can be directly used for nanospray-ESI-MS/MS analysis without additional desalting steps [Immler 1998].

4 "Golden Rules" for Protein and Peptide Handling

1. Take your unprotected hands away from microsamples or you will find keratin, glycine, alanine ...
2. Take your unprotected hands away from microsamples or you will find keratin ...
3. Take your unprotected hands away from microsamples ...
4. Avoid any unnecessary sample transfer.
5. Minimize the number of purification steps.
6. Avoid purification steps which increase the sample volume, especially during the late stages of a purification scheme.
7. Take the "right" detergents if needed to purify your protein (see 3.1).
8. Adjust column dimensions to your sample size.
9. Be aware: an individual peptide has a concentration limit, usually between 100 and 500 fmol/µL, below which it is irreversibly adsorbed onto surfaces.
10. Therefore, keep the sample volume as small as possible and the peptide concentration as high as possible, respectively.
11. Work fast, because even at −80 °C peptide samples are not completely stable.
12. Keep your protein protected from oxygen (especially important for mass spectrometry).
13. Use quartz glassware for water and all buffer solutions to minimize contamination of mass spectrometry samples with sodium and potassium.
14. Excise a PVDF-blotted or -spotted protein band precisely, it improves the initial and repetitive yield during sequence analysis.

5 References

Corchesne, P.L., Patterson, S.D.: Manual Microcolumn Chromatography for Sample Cleanup Before Mass Spectrometry, *Biotechniques* 22 (1997) pp. 244–250

Eckerskorn, C., Mewes, W., Goretzki, H.W., Lottspeich, F.: A new siliconized-glass fiber as support for protein chemical analysis of electroblotted proteins, *Eur. J. Biochem.* 176 (1988) 509–519

Fernandez-Patron, C., Calero, M., Collazo, P.R., Garcia, J.R., Madrazo, J., Musacchio, A., Soriano, F., Estrada, R., Frank, R., Castellanos-Serra, L.R., Mendez, E.: Protein Reverse Staining: High-efficiency Microanalysis of Unmodified Proteins detected on Electrophoresis Gels, *Anal. Biochem.* 224 (1995) 203–211

Gevaert, K., Verschelde, J.-L., Puype, M., Van Damme, J., Goethals, M., De Boeck, S., Vandekerckhove, J.: Structural Analysis and Identification of Gel-Purified Proteins, available in the Femtomole Range, using a novel Computer Program for Peptide Sequence Assignment, by Matrix-assisted Laser Desorption Ionization –Reflectron Time-of-Flight – Mass Spectrometry, *Electrophoresis* 17 (5) (1996) pp. 918–924

Henderson, L.E., Oroszlan, S., Konigsberg, W.: A Micromethod for Complete Removal of Dodecyl Sulfate from Proteins by Ion-Pair Extraction, *Anal. Biochem.* 93 (1979), pp. 153–157

Hjelmeland, L.M.: Solubilization of Native Membrane Proteins, *Methods Enzymol.* 182 (1990) pp. 253–264

Hjelmeland, L.M.: Removal of Detergents from Membrane Proteins, *Methods Enzymol.* 182 (1990) pp. 277–282

Immler, D., Gremm, D., Kirsch, D., Spengler, B., Presek, P., Meyer, H.E.: Identification of Phosphorylated Proteins from Thrombin-activated Human Platelets isolated by 2-D-Gel Electrophoresis by Nanospray-ESI-MS/MS and LC-ESI-MS, *Electrophoresis* (1998) 19, 1015–23

Matsudaira, P.: Sequence from Picomole Quantities of Proteins electroblotted onto Polyvinylidene Difluoride Membranes, *J. Biol. Chem.* 262 (1987) pp. 10035–10038

Neugebauer, J.M.: Detergents: An Overview, *Methods Enzymol.* 182 (1990) pp. 239–253

Sandri, M., Rizzi, C., Catani, C., Carraro, U.: Selective Removal of Free Dodecyl Sulfate from 2-Mercaptoethanol-SDS-Solubilized Proteins before KDS-Protein Precipitation, *Anal. Biochem.* 213 (1993), pp. 34–39

Shevchenko, A., Wilm, M., Vorm, O., Mann, M.: Mass Spectrometric Sequencing of Proteins from Silver Stained Polyacrylamide Gels, *Anal. Chem.* 66 (1996), pp. 850–858

Wildner, G., Fiebig, C., Meyer, H.E., Dedner, N.: The Use of HPLC for the Purification of the QB-Protein, *Z. Naturforsch.* 42c (1987), pp. 7849–741

Wilm., M., Shevchenko, A., Houthaeve, T., Breit, S., Schweigerer, L., Fotsis, T., Mann, M.: Femtomole Sequencing of Proteins from Polyacrylamide Gels by Nano-Electrospray Mass Spectrometry, *Nature* 379 (1996), pp. 466–469

Yang, C.-Y., Kratzin, H.: Chromatography and Rechromatography of Peptide Mixtures by Reversed Phase HPLC, in: *High Performance Liquid Chromatography in Protein and Peptide Chemistry,* Lottspeich, Henschen, Hupe (editors), Walter de Gruyter, 1981, pp. 283–292

Sample Preparation II: Chemical and Enzymatic Fragmentation of Proteins

Roland Kellner and Tony Houthaeve

1 Strategy

Chemical or enzymatic fragmentation is required to study the covalent structure of a protein. Proteins under investigation are fragmented either in solution, or in gel matrix, or in situ attached onto a membrane. Resulting peptide fragments are then isolated and analyzed, and the information gained can be used to clone the structural gene, to identify an N-terminally blocked protein, or to study structure and function.

Protein amounts less than 100 pmol (= 3 µg for a 30 kDa protein) are difficult to handle. Any desalting step, precipitation or lyophilization, etc., carries the risk of sample loss due to unspecific adsorption onto surfaces or due to insolubility after a drying step. Therefore, on the one hand the number of purification steps must be minimized. On the other hand, extremely high purity is required for microcharacterization techniques like Edman sequencing or mass spectrometry. Additionally, conflicts may arise because of incompatibility of buffers or detergents with a forthcoming fragmentation or separation step. For these reasons only a limited number of isolation techniques are useful for microscale sample preparation and fragmentation. Reversed phase HPLC and gel electrophoresis combined with electroblotting are typically used in most applications (see also Chapter II.6).

The quantitation of proteins is difficult but important for project planning or interpretation. Classical techniques like Lowry or Bradford require microgram amounts [Stoscheck 1990; Fountoulakis 1992] and they are not applicable on the microscale. Generally, the quantities are estimated by the intensity of a Coomassie Blue stained gel spot. But proteins are stained very differently by Coomassie Blue and over- or underestimation can easily give an error of factor ten. Amino acid analysis is the method of choice to achieve quantitative information on protein samples (see Chapter III.1).

The chemical or enzymatic fragmentation of a native protein may be hindered due to its secondary and tertiary structure caused by disulfide bridges. The reduction of a cystine to cysteines elongates the protein backbone and allows fragmentation to occur better. Subsequently alkylation of thiol groups yields stable cysteine derivatives and these can also be detected.

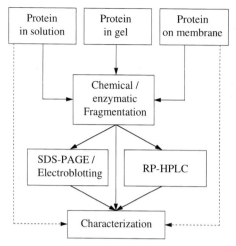

Figure 1. Strategy for characterizing proteins.

In general there are three ways to apply a protein for a cleavage reaction: (1) in solution; (2) bound onto a membrane; (3) in the polyacrylamide gel matrix (Figure 1).

Proteins in solution are supposed to be the simplest form to start a digest – however, it is often difficult to get the protein into solution. A solubilization buffer may contain denaturing detergents and chaotropic salts which might affect protease activity or interfere with a subsequent chromatographic separation. For these samples an extra purification step is required.

Electrophoretic separation and electroblotting of a protein onto a membrane is a very flexible strategy for protein microisolation. Membrane supports in use for protein characterizations are either of polyvinylidene difluoride (PVDF) [Pluskal 1986; Matsudaira 1987], siliconized glass fiber [Vandekerckhove 1985; Eckerskorn 1988] or nitrocellulose [Aebersold 1987]. The blotted protein band is visualized by staining, e.g. by Coomassie Blue, Amido Black or Ponceau S, and then excised in order to be applied for N-terminal sequencing, fragmentation, or amino acid analysis. PVDF membranes are inert and withstand the harsh conditions of Edman chemistry or acid hydrolysis. The protein bound onto PVDF can be applied for an N-terminal sequencing approach and if it is shown to be blocked, that piece of PVDF can still be used to check the quantity or determine the composition by amino acid analysis. Glass fiber membranes are routinely used as a support in protein sequencers but they are not recommended for amino acid analysis. Nitrocellulose binds proteins less strongly than PVDF does and therefore more hydrophobic peptides are recovered from this membrane. However, the disadvantage is that it is neither compatible with Edman chemistry nor with acid hydrolysis; it decomposes e.g. at higher acetonitrile concentrations. Nitrocellulose is therefore only chosen when the protein will be identified via internal fragmentation.

Proteins can also be cleaved, either chemically or enzymatically, within the polyacrylamide gel matrix. After separation on a 1D- or 2D-SDS PAGE the protein is visualized by Coomassie Blue staining and the band is cut out by a scalpel. The gel piece is washed, dried and then the protease and buffer are added. While reswelling the matrix takes up the enzyme and digestion starts. Alternatively, the gel slice can be placed in the sample well of a second gel and then overlaid with protease. Then the digestion proceeds in the stacking gel, and proteolytic fragments will be separated immediately by the ongoing electrophoretic separation [Cleveland et al. 1977]. Recovery of the resulting peptides is achieved either by electroblotting [Kennedy 1988] or by elution from the gel matrix with organic solvents and separation by RP-HPLC [Eckerskorn 1989]. The peptide mixture can also be directly analyzed using MALDI-MS [Schaer 1991] or nano-ES-MS [Wilm 1996].

The in-matrix protocol uses only two steps for protein isolation and digestion (1. gel electrophoresis, 2. fragmentation of the gel-embedded protein) while the blotting steps strategy needs three (1. gel electrophoresis, 2. electroblotting, 3. fragmentation of the membrane-bound protein). Reduced handling of the sample helps to improve the yield of the peptide fragments which need to be determined and therefore the in-matrix digestion is advantageous for small sample amounts (<50 pmol) [Rojo 1997] and for proteins that show poor blotting efficiency.

2 Denaturation, Reduction and Alkylation

Fragmentation reactions are facilitated by denaturing the investigated protein as the cleavage sites are not equally accessible due to the secondary and tertiary structure of a native protein. Denaturation disrupts the structure of the protein chain and its conformation approaches that of a random coil. Detergents, urea or guanidine hydrochloride are used as denaturants. Detergents are mainly used for solubilization and disaggregation of membrane proteins [Neugebauer 1990]. Because urea can be contaminated by cyanate ions which cause carbamylation of the amino groups of a protein [Cole 1966], thereby blocking the N-terminus. Guanidine hydrochloride is the preferred denaturing agent. For denaturation a 6 M guanidine solution is applied and subsequently diluted to give a 1–2 M concentration which is compatible with the activity of many proteases [Riviere 1991].

Disulfide bonds can also hinder proteolysis. Furthermore, cleavage fragments are difficult to identify if they comprise two peptide chains still linked by an S–S bond. Therefore, the reduction and alkylation of cystines facilitates cleavage reactions [Means 1995].

Firstly, the S–S bridges are cleaved by reduction, yielding two cysteines (Figure 2). This is achieved by dithiothreitol (DTT), 2-mercaptoethanol or tributylphosphine as reducing agents. DTT (Cleland' s reagent) [Cleland 1964] has some advantages, e.g. a low redox potential, and at pH 8 the reduction of a cystine is completed in a few minutes; it is resistant to air oxidation; its odor is not as unpleasant as that of 2-mercaptoethanol. Tributylphosphine [Rüegg 1977] provides the advantages of a vapor-phase reaction [Amons 1987].

Secondly, the thiol groups are modified (Figure 3). Cysteine residues in a protein are modified either before gel electrophoresis or after the proteolytic digest prior to HPLC [Tempst 1990]. Modification is necessary to stabilize the thiol groups. Cysteine is destroyed during Edman degradation and cannot be identified anymore. The two most frequently used agents for alkylation are 4-vinylpyridine [Raftery 1966], yielding 4-(pyridylethyl)cysteine, and iodoacetic acid (or iodoacetamide), yielding S-car-

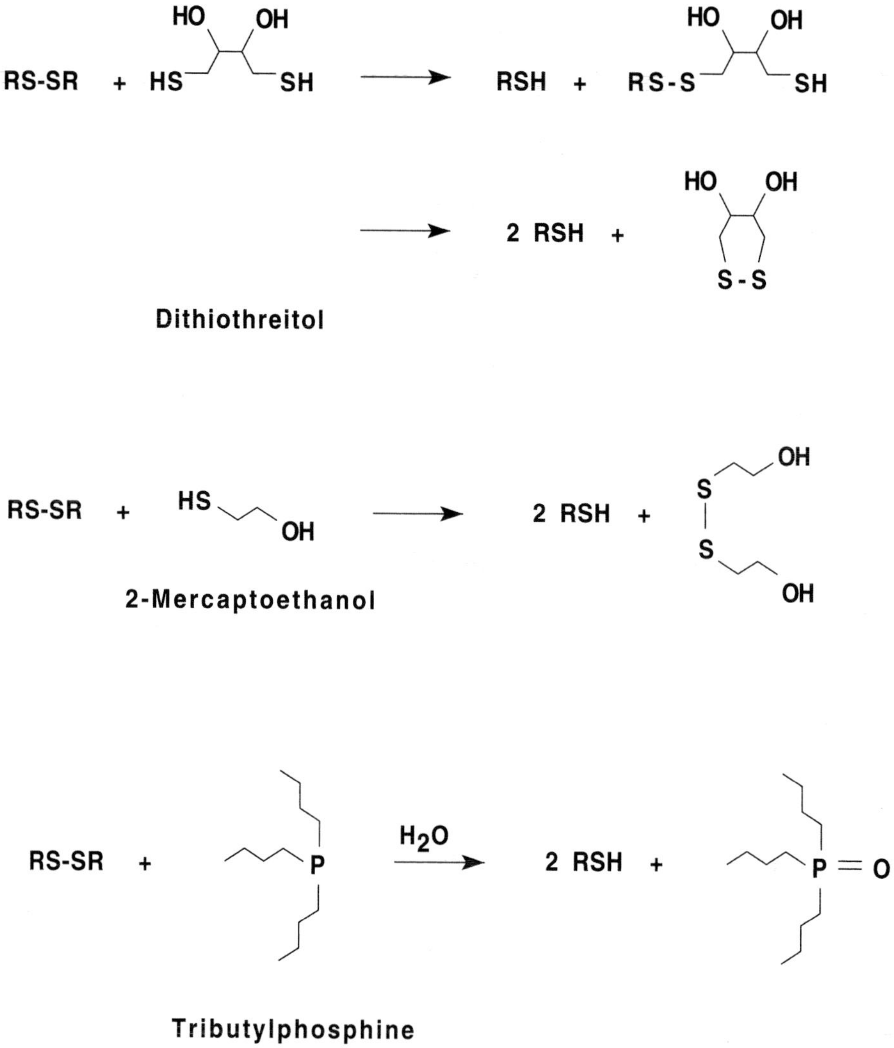

Figure 2. Reduction of disulfide bonds.

boxymethyl (or carboxamido methyl) cysteine [Crestfield 1963]. The pyridylethylation needs to be followed immediately by separation of the reaction mixture because prolonged incubation causes several side reactions of the excess reagent (i.e. His, Trp and Met may be modified). The reduction/alkylation should be fitted into the overall protocol in such a way that no extra purification step is required, e.g. prior to the last electrophoretic separation.

It should be noted that residual acrylamide monomers in polyacrylamide gels are known to couple to thiol groups. This side reaction has also been applied for use as a stable S-alkylation reagent [Brune 1992].

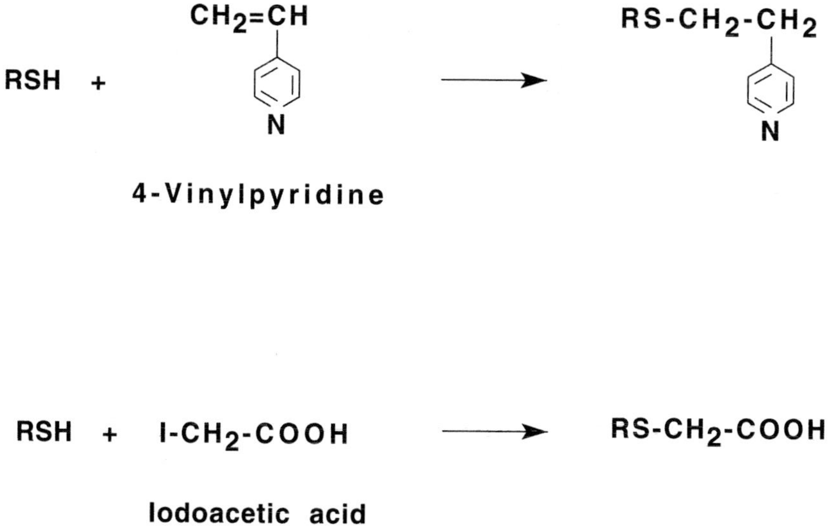

CH₂=CH .... RS-CH₂-CH₂

RSH +

4-Vinylpyridine

RSH + I-CH₂-COOH ⟶ RS-CH₂-COOH

Iodoacetic acid

Figure 3. Alkylation of cysteine residues in proteins.

A variety of other protein modifications is described in the literature [Glazer 1975; Darbre 1986; Allen 1989] but they play no significant role in protein microcharacterization. A nanomole amount of protein is generally needed for a modification reaction because of inherent difficulties like multiple product formation or side reactions. Applications in the microscale range are therefore nearly impossible. The reduction/ alkylation of cysteine is a rare exception.

3 Enzymatic Fragmentation

The enzymatic fragmentation of proteins has several advantages:

- Only a catalytic amount of the protease is needed
- High specificity
- High cleavage yield
- Side reactions are negligible
- Proteases are commercially available in high quality.

The choice of an enzyme is according to given information and the purpose of the fragmentation. Knowledge of the amino acid composition or even the primary sequence helps in selecting an enzyme that will give either an excessive or a more limited fragmentation – so many small or a few large peptide fragments will occur. Specific proteases can be used to cleave the peptide bond either at basic, acidic or hydrophobic side chains.

A typical protein of 300 amino acid residues can be calculated from a protein database (Swiss-Prot or PIR) by the total number for each amino acid residue versus the

Table 1. Calculated number and length of peptide fragments for a hypothetical 300 residue protein.

Amino acid	Specific cleavage	No. of fragments	Average length
Phe/Trp/Tyr/Leu	Chymotrypsin	54	6
Lys/Arg	Trypsin	35	9
Glu	Glu-C	20	15
Lys	Lys-C	19	16
Arg	Arg-C	17	18
Asp	Asp-N	17	18
Met	CNBr	8	38
Trp	BNPS	5	60

number of database entries. This example helps to estimate the number of specific fragments and their average length (Table 1).

An enzymatic fragmentation generates a peptide mixture which is a characteristic of the substrate; it is called a "peptide map", a "fingerprint" or a "mass map" of a protein. The HPLC chromatogram, the gel pattern or a mass spectra of the mixture can be used for comparative studies.

Limited enzymatic fragmentation is used for protein chemical studies. The identification of post-translational modification sites is often achieved by combining two different fragmentations. A first cleavage may separate a discrete region of a protein. Subsequently, a second digest generates smaller fragments, one containing the modified amino acid residue. This strategy helps to isolate a target peptide of reasonable size in order to be analyzed by Edman degradation and/or mass spectrometry. Edman degradation requires peptide fragments where the modification is located close to the N-terminus, otherwise an unequivocal identification becomes even more difficult.

For internal sequence analysis ca. 5–10 times more starting material is needed than for a direct N-terminal approach. This is because both the fragmentation of a protein and the isolation of peptides causes sample loss. The yield of individual peptide bond cleavages varies according to the influence of neighboring amino acid side chains on the protease activity. In particular, a proline residue adjacent to the cleavage site hinders proteolysis because of structural reasons; and, for example, trypsin is restricted if Lys/Arg is followed by an acidic side chain like Glu. Side chain modifications like glycosylation can also prevent proteolytic attacks. The recovery of fragments from a membrane or a polyacrylamide gel by solvent elution gives various yields according to the solubility of the peptides and their hydrophobicity. The same is true for the subsequent separation by RP-HPLC.

3.1 Enzymes

Proteases important for protein characterization studies can be divided into endo- and exoproteases (Table 2) and their ability to hydrolyze specific peptide bonds.

Endoproteases attack the protein backbone, generating peptide fragments. They are mainly used for protein studies. There is a very high specificity available to cleave

Table 2. List of the enzymes most commonly used for protein structure analysis.

Enzyme	EC. No.	pH	Accession No.
Endopeptidases			
Trypsin	3.4.21.4	8.0-9.0	P00760
Chymotrypsin	3.4.21.1	7.5-8.5	P00766
Endoproteinase Asp-N		6.0-8.0	[1]
Endoproteinase Arg-C	3.4.22.8	8.0-8.5	X63673
Endoproteinase Glu-C	3.4.21.19	8.0	P04188[2]
Endoproteinase Lys-C	3.4.21.50	8.0-8.5	[1]
Pepsin	3.4.23.1	2.0-4.0	P00791
Thermolysin	3.4.24.4	7.0-9.0	A24924
Exopeptidases			
Carboxypeptidase A	3.4.17.1	7.0-8.0	P00730
Carboxypeptidase B	3.4.17.2	7.0-9.0	P00732
Carboxypeptidase P	3.4.16.1	4.0-5.0	[1]
Carboxypeptidase Y	3.4.16.1	5.5-6.5	P00729
Cathepsin C	3.4.14.1	5.5	P80067[1]
Acylamino acid releasing enzyme	3.4.19.1	7.5	P19205[3]
Pyroglutamate aminopeptidase	3.4.19.3	7.0-9.0	P28618

[1] Primary structure not yet defined.
[2] Specificity depends on incubation buffer; Asp-Xaa bonds are additionally cleaved in phosphate buffer.
[3] P19205 refers to pig, the commercial enzyme is from horse.

exclusively adjacent to charged residues. Proteases like chymotrypsin, pepsin or thermolysin cleave adjacent to several amino acid residues and yield a more extended fragmentation.

Exoproteases attack the N- or C-terminal amino acids. They are of special importance for the removal of N-terminal blocking groups like a pyroglutamate residue [Jansenss 1994] or a N-acetyl group [Farries 1991]. Fragmentations vary considerably with the neighboring amino acid residues. Additionally, applications are restricted for peptides rather than for proteins. Acylamino acid releasing enzyme (AARE) can be used to remove N-acetylated termini. However, this enzyme cleaves only peptides up to a maximum length of ca. 40 residues [Mitta 1989]. Therefore the protein must be fragmented first and the acetylated N-terminal peptide needs to be isolated (!) in order to apply an AARE digestion. Cathepsin C repetitively cleaves dipeptides from the amino terminus of a protein and can be of interest for certain applications.

The C-terminal degradation using carboxypeptidases provides the complementary information to the Edman degradation. The released amino acids have to be identified which might be difficult for small amounts and also for repetitive amino acids. The combination with MS has given some new perspectives to this application by identifying the shortened peptide instead of the released amino acids [Patterson 1995].

Enzymes are identified by EC code numbers which are based on the classification worked out by the Enzyme Commission. The code number contains four elements with the following meanings: the first figure defines one of six main divisions (e.g. EC 3. are hydrolases); the second figure indicates the nature of the bond hydrolyzed

(EC 3.4. are peptide hydrolases); the third figure specifies the catalytic mechanism of the active center (EC 3.4.21. are serine proteinases); the fourth figure is the serial number of the enzyme in its subclass (EC 3.4.21.4 designates trypsin).

A list of the enzymes most frequently used is given in Table 2. These proteases are commercially available from several suppliers as high-quality products.

3.2 Practical Considerations

The experimental parameters for enzymatic fragmentations are: buffer, pH, temperature, time, enzyme/substrate ratio. The buffer is chosen according to the required pH range; volatile buffer systems are preferred. For most digestions 0.1 M bicarbonate or N-ethylmorpholine buffers support an optimal milieu of about pH 8 and they can be removed by lyophilization. Incubation at 37 °C may last for 4 h or overnight. For digestions in solution an enzyme/substrate ratio (w/w) of 1:10 to 1:100 is recommended. For digestions where the protein is either bound onto a membrane or it is placed in the polyacrylamide matrix a ratio as high as 1:1 may be used. The proteases are dissolved in concentrations of 0.1–1 µg/µL immediately prior to use. Trypsin requires calcium ions (2 mM) to stabilize the active center; pyroglutamate aminopeptidase requires a thiol compound for activation and bivalent metal ions are complexed by EDTA. The other proteases need no special additives although their activity can be influenced by the addition of chaotropic reagents [Riviere 1991].

Autohydrolysis of the proteases is sometimes observed and care must be taken that the sequence interpretation will notify those peptides. A database search should be routinely performed for the identified amino acid sequences to overcome this pitfall. Attempts have been made at increasing resistance against autohydrolysis, e.g. by the modification of trypsin. However, autolytic fragments were still observed. Together with every digestion, a blank sample must be processed simultaneously – a piece of matrix from the same origin and with identical history as the investigated protein. Either the solution of a dissolved protein, or a piece of membrane from the same blot, or a gel piece from the same gel without containing any protein should be processed identically. This helps to mark artefact peaks in the HPLC trace resulting from solvents used for extractions, from detergents or dyes, or to detect autolytic fragments from the protease used for the digest. A dual wavelength detection system recording the peptide bonds at 214 nm and the aromatic groups at 280 nm can help to identify and exclude artificial peaks.

3.2.1 Practical Considerations for On-Membrane Digestions

For digestions of proteins bound onto PVDF or nitrocellulose a pretreatment is necessary to prevent nonspecific binding of the protease during digestion. The free surface of the membrane is masked by the adsorption of polyvinylpyrrolidone (mw: 40 kD; PVP-40) before the enzyme is added (0.5 % PVP-40 in methanol, 30 min, 37 °C). Unfortunately, the PVP quenching can cause a huge artefact peak in the subsequent HPLC separation. Extensive washing with water (5–10 changes) is therefore necessary to wash off as much PVP-40 as possible. However, PVP-40 can never be removed completely from the membrane and this is an inherent drawback to this procedure.

To overcome the disadvantage of the PVP-40 procedure the use of hydrogenated Triton X-100 (RTX-100) has been described [Fernandez 1994]. The membrane-bound protein is digested in a buffer containing 1 % RTX-100 which serves as both an elution agent for peptides from the membrane and a block for prevention of enzyme adsorption to the membrane. The detergent Tween-20 can also be used to recover hydrophobic peptide fragments by adding 0.02 % to the extraction solvents [Ward 1990; Kurzchalia 1992]. Of major importance is the fact that Tween-20 does not interfere with the subsequent chromatographic separation.

The membrane pieces are then covered with digest buffer (<100 µL) and the digest is performed overnight at 37 °C. The reaction is stopped by acidifying the solution. The supernatant is then taken off and peptides are extracted from the membrane with either 80 % formic acid [Bauw et al. 1989] or acetic acid / acetonitrile mixtures [Fisher 1991]. The extraction can benefit from elevated temperatures and ultrasonication steps. Finally all solutions are pooled (<1mL) and lyophylized in a vacuum centrifuge to remove all volatile chemicals. The resulting dried pellet can be further analyzed after reconstitution.

The on membrane digest procedure has been succesfully applied with sample amounts even below 50 pmol.

3.2.2 Practical Considerations for In-Gel Digestions

For protein characterization on minute amounts it is straightforward to digest the sample directly within the (polyacrylamide) matrix. Resulting peptide fragments can be eluted, fractionated by microbore HPLC and applied on an Edman sequencer or they can be determined as a mixture e.g. by MALDI-MS. The in-gel approach is advantageous compared to alternative methods like (1) electroblotting onto an appropriate membrane; (2) electroelution; or (3) in matrix digestion, electrophoretic separation and subsequent electroblotting. To handle the protein directly in the gel slice circumvents one step from the overall protocol and therefore avoids a sample loss for that additional step.

Coomassie stained proteins in gel slices may result from one or several, dried or fresh 1D- or 2D-spots, or from concentrating gels. For a concentrating gel several gel spots, up to 30, can be pooled and reconcentrated in a gel that is either out of poly-acrylamide [Rasmussen 1991] or out of agarose [Rider 1995].

Polyacrylamide gels are very inhomogeneous sample matrices. The purity of chemicals and the protocols used for preparing the gel can vary tremendously. For example, storage and prerunning of gels have an influence on the residual monomer concentration. Loss of 5–10 % protein comparing freshly prepared gels and gels stored overnight is described [Liang 1991]. A possible interaction occurring during gel electrophoresis is the modification of free thiol groups of cysteine by acrylamide monomers [Brune 1992] or by β-mercaptoethanol (unpublished results).

The diffusion of the enzyme into the matrix and from the peptide fragments out of the matrix depends on the pore size and the thickness of the polyacrylamide gel. A large pore size (low percentage gel) and a thin gel are advantageous. Of course, a compromise has to be found in respect to the separation performance of the given proteins.

After gel electrophoresis and subsequent Coomassie Blue staining a clearly visible protein spot is necessary to run an in-gel digest. The time for staining and destaining the gel should be short, e.g. a 0.5 mm gel is stained for 1 min and destained within 10 min. It is sufficient if Coomassie Blue stains just the molecules at the gel surfaces. Destaining acidifies the gel which is in conflict to the pH requirements for the protease afterwards. Therefore the acetic acid concentration in the staining and destaining solution should be only 1 % instead of the usual 10 % [Ferrara 1993]. The protein band has to be cut out precisely along the edge in order to avoid needless polyacrylamide, as the ratio of protein to gel is a critical factor! Not more than 3–4 average sized gel pieces of one protein should be combined.

Once excised, the gel slices are washed with ammonium bicarbonate and 50 % acetonitrile, in order to remove excessive Coomassie Blue dye [Rosenfeld 1992] and to bring up the proper pH. Gel shrinkage is achieved by lyophilization in a vacuum centrifuge or, alternatively, by submerging in pure acetonitrile. The enzyme is added in a volume just enough to re-swell the gel pieces. When re-swelling is completed, ca. 15 min, some additional buffer is added to keep the gel pieces covered with solution. All the proteases applied routinely for peptide mapping can be used for the in-gel digest procedure. For example, applications are described using chymotrypsin, endoproteinase Glu-C and papain [Cleveland et al. 1977], endoproteinase Glu-C [Kennedy 1988] and trypsin [Eckerskorn 1989]. A modified trypsin is commercially available which has some advantages regarding its activity in the presence of SDS and organic solvent, and it is more stabile against autodigestion (however, autodigestion does occur). Interesting is that the digest buffer can be supplemented with chaotropes and solubility increasing additives [Riviere 1991], resulting in sometimes elevated reactions.

The cleavage fragments are extracted out of the gel slice with acidic and organic solvents by passive diffusion. One application describes the elution of the peptides by shaking with 75 % (v/v) trifluoroacetic acid in water for 4 h followed by 50 % (v/v) trifluoroacetic acid in acetonitrile for an additional 4 h and this was repeated twice [Eckerskorn 1989]. Although 75 % trifluoroacetic acid in water is a strong solvent, extraction can as well be done with a 5 % acidic solution [Wilm 1996].

0.02 % Tween 20 was added as a detergent to the solvents in order to dissolve the more hydrophobic peptide fragments [Ward 1990; Kurzchalia 1992]. Thereby it is important that the detergent is compatible and not interfering with the subsequent chromatographic separation. Mass spectrometic techniques are very sensitive regarding detergents (<0.0005 % of SDS) and one should try to completely avoid them. In MALDI-MS the analytes must be included in the matrix crystals, a process that is disturbed by detergents and salts. An on-target clean-up procedure has been described that can partially cope with these difficulties using beads from RP chromatography [Gevaert 1997].

A basic extraction using 0.1 M ammonia was reported to be slightly less effective [Tetaz 1993]. A tryptic digest of bovine serum albumin was analyzed and three extraction procedures were compared: (1) the standard TFA/acetonitrile extraction; (2) an ammonia extraction; (3) a HPLC method (briefly: the gel pieces were digested while packed into an empty HPLC precolumn; peptides were extracted at high pressure by 0.1 % TFA and recovered on the downstream RP-HPLC column). The procedures worked equally well, although yields were somewhat lower for the ammonia-extracted peptides. After extraction, the combined peptide mixture is lyophilized almost to complete dryness (overdrying causes sample loss!) and is kept frozen or used immediately for analysis.

Characterizations of unknown proteins using the in gel digest procedure have been reported with starting amounts (on the gel) of down to 10 pmol. Below this amount special precautions have to be taken, such as clean room conditions to avoid dust and keratine contamination.

3.2.3 Automation of Digest Procedures

An indispensable requirement for proteome studies is the automatization of protein digestion. The need to handle hundreds of samples can only be addressed by robotics and automated procedures. The chemical or enzymatic fragmentation of proteins will significantly be improved by an automated digest procedure; automated handling will minimize background contamination and enlarge sample throughput. The idea is to automate the excission of 2D-spots, robotic handling of the enzymatic fragmentation and application on a mass spectrometer target. Data acquisition and interpretation is software-driven and known proteins are identified immediately.

A system for automating the protein manipulation steps was introduced [Miller 1994]: in a static workstation the protein sample is immobilized on a biphasic column and various manipulations are enabled. Briefly, sample washing and desalting, reduction and alkylation, subsequent chemical or enzymatic fragmentation are possible under temperature control and inert gas conditions. By semi-automated operation peptide fragments can be eluted after in-gel digestions; a RP-HPLC column might be connected on-line [Eckerskorn 1996]. The workstation is a useful tool for single protein manipulation but lacks the availability for processing several samples simultaneously (e.g. processing a blank sample).

Multiple samples are processed in solution using a similar principle as described above [Hsieh 1996]. An autosampler injects the samples into a reactor chamber where a chemical modification (reduction/alkylation), a proteolytic digestion using an immobilized enzyme (Trypsin or Glu-C), RP-HPLC separation and the possible LC-MS connection are performed. This system is an interesting device e.g. for recombinant proteins in solution when efficiency and sample throughput of the mapping procedure and exclusion of manual intervention/contamination is the major focus.

The implementation of the digest procedure onto a modified sample robot permitted to handle up to 32 (96) protein samples in parallel [Houthaeve 1995]. The proteins can be digested either in solution, in gel or on blot. They are loaded in flowthrough reactors and placed in a thermocontrolled reactor block. The robot platform does allow the use of different enzymes, different reduction/alkylation conditions and different extraction conditions. The digest robot is compatible with RP-HPLC peptide separation and Edman sequencing, MALDI-MS and ESI-MS.

Proteomics is a very busy field and new achievements with the automation of digest procedures can be expected in the near future.

4 Chemical Fragmentation

Chemical fragmentation of proteins is complementary to enzymatic digestions. Cleavages can be performed at amino acid residues where no enzyme is available. Reagents are nearly insensitive to salts or detergents, and fragmentations are possible in cases when the use of proteases is restricted.

A variety of chemical methods for cleavage have been reviewed [Kasper 1975; Fontana 1986]. However, as early as 1975 it was stated that " ... these methods have not been developed to the stage where they may be routinely applied in a predictable fashion to the chemical fragmentation of proteins ...". The disadvantages are: "(a) low cleavage yields, (b) lack of specificity, (c) undesirable side reactions, and (d) wide variability in the reactivity of the sensitive bond" [cited from Kasper 1975]. Since then, no major improvement has been achieved.

Nevertheless, a few chemical fragmentation procedures are important alternatives to enzymatic cleavages and are valuable tools for the elucidation of protein structures.

4.1 Cyanogen Bromide Cleavage

Cyanogen bromide (CNBr) cleavage has become the most widely used chemical fragmentation method for proteins since its introduction by Gross and Witkop in 1961. CNBr provides the following advantages:

- Met-Xaa bonds are cleaved specifically and nearly quantitatively.
- Methionine residues are usually rare in proteins, and generally a few, large peptide fragments are obtained.
- The reaction is generally performed in 70 % formic acid, which is also a strong solvent.
- Reagent and by-products are volatile.
- No appropriate enzyme is available to cleave at methionine positions.

CNBr is a highly toxic reagent and proper care has to be taken! It must be handled in a fume hood and has to be stored under nitrogen. CNBr may decompose during storage and then change color, becoming yellow – only colorless stocks should be used.

A Met-Xaa bond is cleaved, converting methionine into a C-terminal homoserine residue and creating a new amino terminal NH_2-Xaa. The selectivity of CNBr can be explained by the reaction mechanism (Figure 4) [Inglis 1970]. The sulfur from the methionine side chain electrophilically reacts with CNBr (1) forming a sulfonium ion (2). Methyl thiocyanate is released while an intermediate imino ring is formed involving the carbonyl group from methionine (3). Then the iminolactone is hydrolyzed and the Met-Xaa bond is cleaved (4). Finally, a peptide fragment with a new amino terminus is released (5) and homoserine results in the C-terminal position. Homoserine and homoserine lactone are interconvertible and form a mixture (6).

A general procedure is to use roughly a 100-fold molar excess of CNBr over the methionine residues. Either solid CNBr is added directly to the protein solution or it

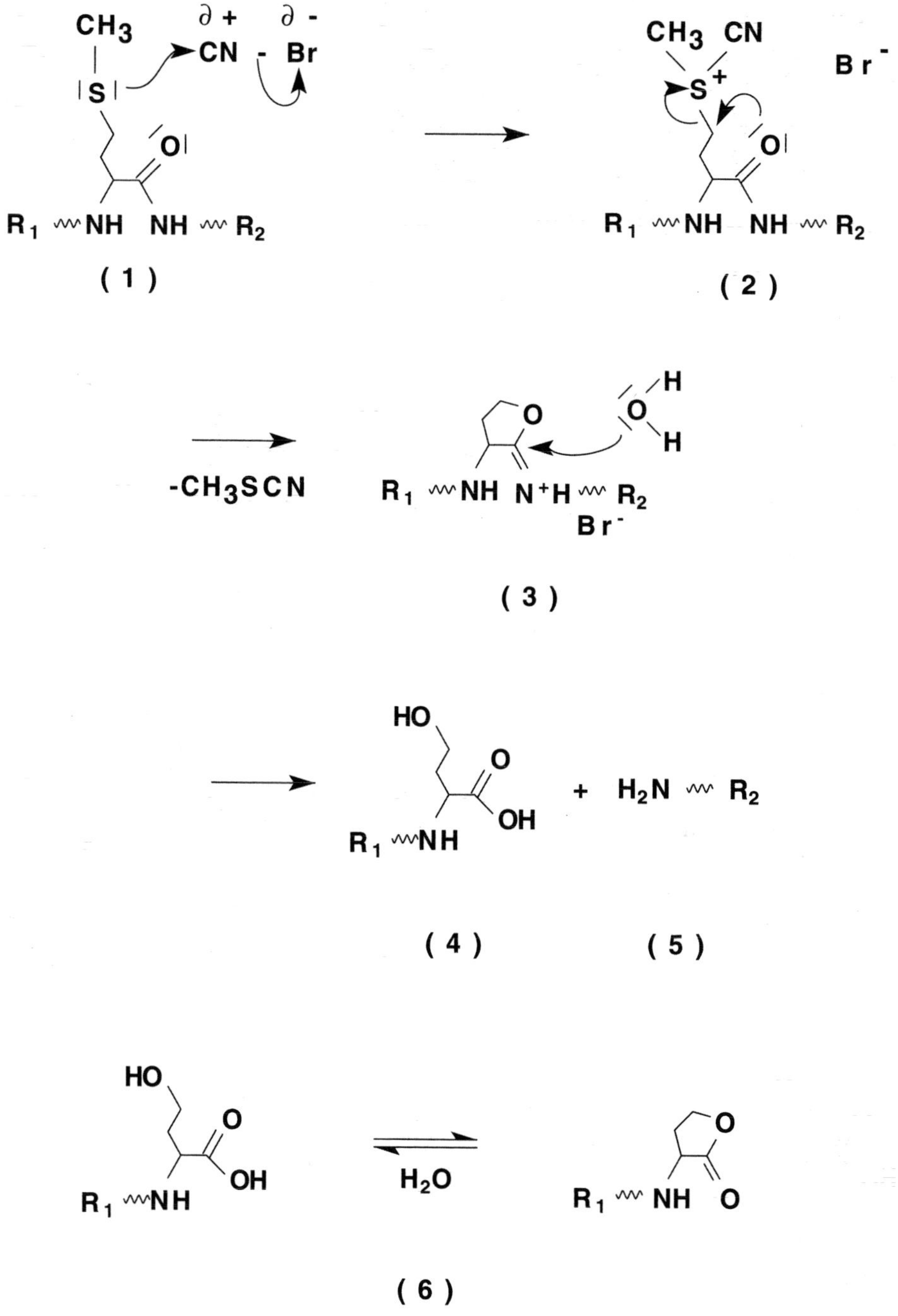

Figure 4. Cyanogen bromide cleavage.

is dissolved in acetonitrile. The reaction is performed under acidic conditions; generally 70 % formic acid is used. The mixture needs to be flushed with nitrogen and is incubated in the dark for 4–24 h. Then water is added (5–10 volumes) and the excess reagent is removed by lyophilization (but remember the toxicity of CNBr and use KOH to neutralize the vapor). Repeat this step twice.

A 1000-fold excess, as described in several applications, should on the whole be avoided because of additional side reactions. A short incubation time is also favorable because additional cleavages at Trp residues occur during a prolonged cleavage. Together with two products arising from the homoserine and the homoserine lactone form of a peptide, a CNBr digest yields quite a complex fragmentation pattern in HPLC.

Problems can arise with Met-Ser and Met-Thr bonds where the hydroxyl group of the side chain may interfere with the ring opening of the iminolactone. The use of 70 % trifluoroacetic acid instead of formic acid has been reported to significantly improve these difficulties [Titani 1972].

A CNBr cleavage can be performed on targets which were found to be N-terminally blocked after automated Edman degradation. For this purpose the sequencer is stopped after a few cycles and the reaction cartridge is disassembled. An aliquot of CNBr (20 μL 70 % formic acid containing an equivalent of a 20-fold excess of CNBr over protein, w/w) is spotted onto the glass filter containing the blocked protein. Then the sample is placed in a vacuum dessiccator and left over a solution of CNBr in formic acid. After 16 h the membrane is air-dried, reassembled in the sequencer and sequence analysis is allowed to continue [Simpson 1984]. This procedure does not separate any resulting fragments. Therefore, this method is of use if only 1–3 methionines are present and/or the amino acid sequence just needs to be confirmed.

PVDF-blotted proteins and polyacrylamide-embedded samples may also be cleaved by CNBr. PVDF membranes are submerged in 70 % formic acid and the reagent is added. After incubation the peptide fragments are extracted and separated [Yuen 1989]. Cleavage in the gel slice and separation of fragments in a second SDS-PAGE [Nikodem 1979] or separation by RP-HPLC [Jahnen 1990] helps to map large fragments.

The fragmentation of blotted proteins by CNBr and a subsequent tryptic digest is reported to be advantageous for the recovery of large fragments. The following strategy has been applied for internal sequence analysis of unknown proteins [Stone 1992]: (1) CNBr cleavage of PVDF blotted proteins; (2) extraction of the fragments; (3) reduction/alkylation of the CNBr fragments; (4) tryptic digest; (5) separation and sequence analysis of CNBr/tryptic peptides. As little as 50 pmol PVDF-blotted protein could be handled and thus proved the applicability of CNBr cleavage on the microscale-level.

4.2 Partial Acid Hydrolysis

Partial acid hydrolysis is useful for particular samples where other fragmentations have failed. A fragmentation of Asp-Pro bonds is what is primarily expected. This technique has not gained widespread application because of its lack of specificity. The rate of hydrolysis of peptide bonds ranges from a specific Asp-Pro cleavage to complete hydrolysis (similar to what is done for amino acid analysis). The extent of fragmentation is hard to predict.

Three parameters can be varied: concentrated or diluted acid solutions, temperature, and time. Some examples [Inglis 1983] are:

- 11 M HCl, 37 °C, 4 d: Xaa-Ser and Xaa-Thr are preferentially hydrolyzed.
- 0.03 M HCl, 110 °C, 6-18 h: Xaa-Asp-Xaa are preferentially hydrolyzed.
- 75 % formic acid, 37 °C, 48 h: Asp-Pro is preferentially hydrolyzed.

Cleavages vary considerably depending on the particular protein. However, side reactions are observed only to a small extent; amide groups can be hydrolyzed and tryptophan is partially destroyed.

The link of SDS-PAGE and partial acid hydrolysis has been described [Vanfleteren 1992]. Gel slices containing Ponceau S or Coomassie stained proteins were subjected to in matrix cleavages using 20 % formic acid at 112 °C for 4 h. Fragments were separated by HPLC and microsequenced with starting amounts of as low as 100 pmol protein.

4.3 Hydroxylamine Cleavage of Asn-Gly Bonds

Asparagine and aspartic acid residues are involved in several spontaneous processes of nonenzymatic protein modification and degradation in vivo. The intermediate formation of a five-membered succinimide ring plays a key role in all these reactions [Stephenson 1989; Geiger 1987].

The side-chain carbonyl group of Asn is attacked by the α-amino group of the C-terminally neighbored residue (1) forming a succinimide intermediate (2) (Figure 5). Under the nucleophilic attack of hydroxylamine the Asn-Gly peptide bond is hydrolyzed and a mixture of α- (3) and β-aspartyl hydroxamate (4) as well as a new N-terminal Gly (5) is released.

The desamidation reaction at Asn yielding a β-carboxylic acid group and the isomerization and racemization at Asp are significant reactions in protein degradation. The succinimide intermediate may undergo ring opening and form either an aspartate (6) or an isoaspartate residue (7) (isomerization). Isoaspartate residues will block an ongoing Edman degradation. Furthermore, the racemization of the α-carbon of asparagine is possible. Small hydrophilic residues C-terminally adjacent to Asn, especially Gly, support this rearrangement due to the absence of steric hindrance.

The Asn-Gly sequence occurs statistically about every 350 amino acids in a protein, and hence the cleavage of this bond will yield usually a few large fragments. Hydroxylamine was shown to be a useful reagent affecting the hydrolyzation of the succinimide intermediate [Blumenfeld 1965; Bornstein 1977; Kwong 1994]. For this purpose, a 1–2 M NH$_2$OH solution is applied in an alkaline buffer (ca. pH 9-11) at 45–67 °C. For example, β-lactoglobulin contains one Asp-Gly bond and a cleavage yield of only 5 % was reported at 45 °C, compared to 30 % at 67 °C [Saris 1983]. Increasing the reaction time did not extend the cleavage. The cleavage of proteins after blotting onto glass fiber sheets by dipping the blot for 48 h at 25 °C in a solution of 1 M NH$_2$OH adjusted to pH 11 with K$_2$CO$_3$ is described [Bauw 1988].

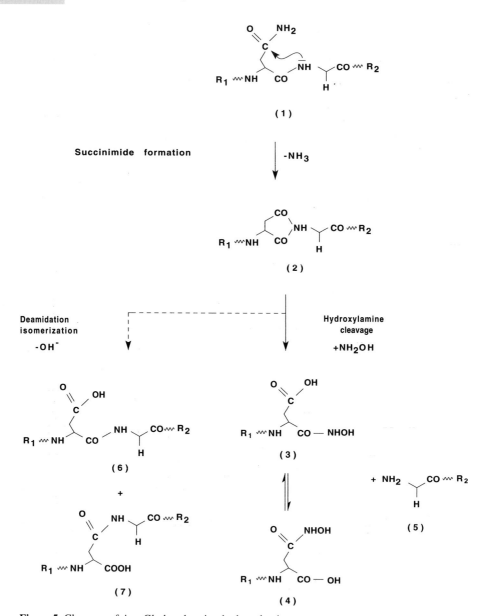

Figure 5. Cleavage of Asn-Gly bonds using hydroxylamine.

4.4 Cleavage at Tryptophan

Tryptophan as a relative rare amino acid is of interest for specific cleavages in order to yield large peptide fragments. Two reagents are recommended to be quite specific and to produce nearly no side reactions. BNPS-skatole (bromine adduct of 2-(2-nitrophenylsulfonyl-3-indolenine) [Omenn 1970] is applied in a 10–50 fold molar excess in 80 % acetic acid and the reaction runs for 24 h at 37 °C. Yields may be up to 70 %. Most experiments have been performed with concentrations of 10 mg/mL protein and applications on the microscale are rarely seen. In one example Trp cleavage is carried out by dipping PVDF blotted proteins in a BNPS-skatole solution (0.25 mg/mL in 80 % acetic acid) for 24 h at 25 °C [Bauw 1988]. The blot is rinsed first with water and then with butylchloride to remove excess reagent.

For the fragmentation by o-iodosobenzoic acid the reagent is dissolved in 80 % acetic acid, 4 M guanidine hydrochloride and first pretreated with p-cresol (20 µL per mL reaction mixture). The additive prevents the formation of o-iodoxybenzoic acid which cleaves unspecifically at tyrosine residues. A 2-3 fold excess (w/w) of reagent then is added to protein and the reaction proceeds at room temperature in the dark for 24 h. The yield of cleavage is reported to be ca. 95 % [Mahoney 1981]. The fragmentations yield C-terminal lactones which can be conveniently coupled to supports for solid-phase sequencing.

Several other procedures to cleave proteins at tryptophan positions with less specificity are described and reviewed by Fontana [Fontana 1986].

4.5 Cleavage at Cysteine

Even though cysteine is an attractive cleavage site because of its rarity, no really useful procedure is available. Cleavage via the thiocyanate derivate using 2-nitro-5-thiocyanobenzoic acid (NTCB) creates N-terminally blocked fragments and requires a Raney nickel treatment in addition [Jacobson 1973; Otieno 1978].

5 References

Aebersold, R.H., Leavitt, J., Saavedra, R.A., Hood, L.E., Kent, S.B.H. (1987) *Proc. Natl. Acad. Sci. USA* 84, 6970–6974. Internal amino acid sequence analysis of proteins separated by one- or two-dimensional gel electrophoresis after in situ protease digestion on nitrocellulose.

Allen, G. (1989) *Sequencing of Proteins and Peptides*. Elsevier, Amsterdam.

Amons, R. (1987) *FEBS Letters 212*, 68–72. Vapor-phase modification of sulfhydryl groups in proteins.

Bauw, G., Van den Bulcke, M., Van Damme, J., Puype, M., Van Montagu, M., Vandekerckhove, J. (1988) *J. Prot. Chem. 7*, 194–196. Protein-electroblotting on polybase-coated glass-fiber and polyvinylidene difluoride membranes: an evaluation.

Bauw, G., Van Damme, J., Puype, M., Vandekerckhove, J., Gesser, B., Ratz, G., Lauridsen, J., Celis, J. (1989) *Proc. Natl. Acad. Sci. USA 86*, 7701–7705. Protein-electroblotting and -microsequencing strategies in generating protein data bases from two-dimensional gels.

Blumenfeld, O.O., Rojkind, M., Gallop, P.M. (1965) *Biochemistry 4*, 1780–1788. Subunits of hydroxylamine-treated tropocollagen.

Bornstein, P., Balian, G. (1977) *Methods Enzymol. 47*, 132–145. Cleavage at Asp-Gly bonds with hydroxylamine.

Brune, D.C. (1992) *Anal. Biochem. 207*, 285–290. Alkylation of cysteine with acrylamide for protein sequence analysis.

Cleland, W.W. (1964) *Biochemistry 3*, 480–482. Dithiothreitol, a new protective reagent for SH groups.

Cleveland, D.W., Fischer, S.G., Kirschner, M.K., Laemmli, U.K. (1977) *J. Biol. Chem. 252*, 1102–1106. Peptide mapping by limited proteolysis in sodium dodecyl sulfate and analysis by gel electrophoresis.

Cole, E.G., Mecham, D.K. (1966) *Anal. Biochem. 14*, 215–222. Cyanate formation and electrophoretic behavior of proteins in gels containing urea.

Crestfield, A.M., Moore, S., Stein, W.H. (1963) *J.Biol.Chem. 238*, 622–627. The preparation and enzymatic hydrolysis of reduced and S-carboxymethylated proteins.

Darbre, A. (1986) *Practical Protein Chemistry.* Wiley, Chichester.

Eckerskorn, C., Grimm, R. (1996) *Electrophoresis 17*, 899–906. Enhanced in situ gel digestion of electophoretically separated proteins with automated peptide elution onto mini-reversed phase columns.

Eckerskorn, C., Mewes, W., Goretzki, H., Lottspeich, F. (1988) *Eur. J. Biochem. 176*, 509–519. A new siliconized-glass fibre as support for protein-chemical analysis of electroblotted proteins.

Eckerskorn, C., Lottspeich, F. (1989) *Chromatographia 28*, 92–94. Internal amino acid sequence analysis of proteins separated by gel electrophoresis after tryptic digestion in polyacrylamide matrix.

Farries, T.C., Harris, A., Auffret, A.D., Aitken, A. (1991) *Eur. J. Biochem. 196*, 679–685. Removal of N-acetyl groups from blocked peptides with acylpeptide hydrolase.

Fernandez, J., Andrews, L., Mische, S. (1994) A one-step enzymatic digestion procedure for PVDF-bound proteins that does not require PVP-40. In: *Techniques in protein chemistry V* (Crabb, J.W.; ed.), pp. 215–222, Academic Press, San Diego.

Ferrara, P., Rosenfeld, J., Guillemot, J., Capdevielle, J. (1993) Internal peptide sequence of proteins in-gel after one- or two dimensional gel electrophoresis. In: *Techniques in protein chemistry IV* (Angeletti, R.; ed.), pp. 379–387, Academic Press, San Diego.

Fiedler, K., Parton, R.G., Kellner, R., Etzold, T., Simons, K. (1994) *EMBO J. 13*, 1729–1740. VIP36, a novel component of glycolipid rafts and exocytic carrier vesicles in epithelial cells.

Findlay, J.B., Geisow, M.J. (1989) *Protein Sequencing.* IRL Press, Oxford.

Fischer, W., Park, M., Lucero, L., Vale, W. (1991) In: *Techniques in protein chemistry II* (Villafranca, J.; ed.), pp. 163–170, Academic Press, San Diego. Evaluation of specific proteases for digestion of memebrane-bound proteins.

Fontana, A. (1972) *Meth.Enzymol. 25*, 419–423. Modification of tryptophan with BNPS-skatole (2-(2-nitrophenylsulfenyl)-3-methyl-3-bromoindolenine).

Fontana, A., Gross, E. (1986) Fragmentation of polypeptides by chemical methods. In: *Practical protein chemistry* (Darbre, A.; ed.), pp. 67–120, Wiley, Chichester.

Fountoulakis, M., Juranville, J.-F., Manneberg, M. (1992) *J. Biochem. Biophys. Methods 24*, 265–274. Comparison of the Coomassie brilliant blue, bicinchoninic acid and Lowry quantitation assays, using non-glycosylated and glycosylated proteins.

Geiger, T., Clarke, S. (1987) *J. Biol. Chem. 262*, 785–794. Deamidation, isomerization, and racemization at asparaginyl and aspartyl residues in peptides.

Gevaert, K., et al. (1997) *Electrophoresis 18*, 2950–60. Peptides adsorbed on RP-chromatographic beads as targets for femtomole sequencing by post-source-decay MALDI-RETOF-MS.

Glazer, A.N., DeLange, R.J., Sigman, D.S. (1975) *Chemical modification of proteins.* North-Holland, Amsterdam.

Gross, E., Witkop, B. (1961) *J. Am. Chem. Soc. 83*, 1510–1511. Selective cleavage of the methionyl peptide bonds in ribonuclease with cyanogen bromide.

Houthaeve, T., Gausepohl, H., Mann, M., Ashman, K. (1995) *FEBS Letters 376*, 91–94. Automation or micro-preparation and enzymatic cleavage of gel electrophoretically separated proteins.

Hsieh, F., Wang, H., Elicone, C., Mark, J., Martin, S., Regnier F. (1996) *Anal. Chem. 68*, 455–4622. An automated analytical system for the examination of protein primary structure.

Inglis, A.S. (1983) *Meth. Enzymol. 91*, 324–332. Cleavage at Aspartic acid.

Inglis, A.S., Edman, P. (1970) *Anal. Biochem. 37*, 73–80. Mechanism of cyanogen bromide reaction with methionine in peptide and proteins.

Jacobson, G.R., Schaffer, M.H., Stark, G.R., Vanaman, T.C. (1973) *J. Biol. Chem. 248*, 6583–6591. Specific chemical cleavage in high yield at the amino peptide bonds of cysteine and cystine residues.

Jahnen, W., Ward, L.D., Reid, G.E., Moritz, R.L., Simpson, R.J. (1990) *Biochem. Biophys. Res. Com. 166*, 139–145. Internal amino acid sequencing of proteins by in situ cyanogen bromide cleavage in polyacrylamide gels.

Jansenss, M.E., Kellner, R., Gäde, G. (1994) *Biochem. J. 302*, 539–543. The damselflies Pseudagrion inconspicuum and Ischnura senegalensis contain a novel adipokinetic octapeptide.

Kasper,C.B. (1975) Fragmentation of proteins for sequence studies and separation of peptide mixtures. In: *Protein sequence determination* (Needleman, S.B.; ed.), pp. 114–161, Springer, Heidelberg.

Kennedy, T.E., Gawinowicz, M., Barzilai, A., Kandel, E., Sweatt, J.D. (1988) *Proc. Natl. Acad. Sci. USA 85*, 7008–7012. Sequencing of proteins from two-dimensional gels by using in situ digestion and transfer of peptides to polyvinylidene difluoride membranes: Application to proteins associated with sensitization in Aplysia.

Kurzchalia, T.V., Dupree, P., Parton, R., Kellner, R., Lehnert, M., Simons, K. (1992) *J. Cell Biol. 118*, 1003–1014. Vip21, a 21 KD membrane protein is an integral component of the Trans-Golgi network-derived transport vesicles.

Kwong, M.Y., Harris, R.J. (1994) *Protein Sci. 3*, 147–149. Identification of succinimide sites in proteins by N-terminal sequence analysis after alkaline hydroxylamine cleavage.

Landon, M. (1977) *Meth. Enzymol. 47*, 145–149. Cleavage at Aspartyl-Prolyl bonds.

Liang, S.P., Lee, T.T., Laursen, R.A. (1991) *Anal. Biochem. 197*, 163–167. Single-step electroelution of proteins from SDS-polyacrylamide gels and immobilization on diisothiocyanate-glass beads in prepacked capillary columns for solid-phase microsequencing.

Mahoney, W.C., Smith, P.K., Hermodson, M.A. (1981) *Biochemistry 20*, 443–448. Fragmentation of proteins with o-iodosobenzoic acid: chemical mechanism and identification of o-iodoxybenzoic acid as a reactive contaminat that modifies tyrosyl residues.

Matsudaira, P.T. (1987) *J. Biol. Chem. 262*, 10035–10038. Sequence from picomole quantities of proteins electroblotted onto polyvinylidene difluoride membranes.

Matsudaira, P.T. (1993) *A practical guide to protein and peptide purification for microsequencing.* Academic Press, San Diego.

Means, G., Feeney, R. (1995) *Analyt. Biochem. 224*, 1–16. Reductive alkylation of proteins.

Miller, C. (1994) In: *Methods: A companion to Methods in Enzymology 6*, 315–333.

Mitta, M., Asada, K., Uchimura, Y., Kimizuka, F., Kato, I., Sakiyama, F., Tsunasawa, S. (1989) *J. Biochem. 106*, 548–551. The primary structure of porcine liver acylamino acid-releasing enzyme deduced from cDNA sequences.

Neugebauer, J.M. (1990) Detergents: An overview. In: *Protein purification* (Deutscher, M.P.; ed.), pp. 239–253, Academic Press, San Diego.

Nikodem, V., Fresco, J.R. (1979) *Anal. Biochem. 97*, 382–386. Protein fingerprinting by SDS-gel electrophoresis after partial fragmentation with CNBr.

Omenn, G.S., Fontana, A., Anfinsen, C.B. (1970) *J. Biol. Chem. 245*, 1895–1902. Modification of the single tryptophan residue of Staphylococcal nuclease by a mild oxidizing agent.

Otieno, S, (1978) *Biochemistry 17*, 5468–5474. Generation of a free a-amino group by Raney-Nickel after 2-nitro-5-thiobenzoic acid cleavage at cysteine residues: application to automated sequencing.

Patterson D., Tarr G., Regnier F., Martin S. (1995) *Anal. Chem. 67*, 3971–8. C-terminal ladder sequencing via matrix-assisted laser desorption mass spectrometry coupled with carboxypeptidase Y time-dependent and concentration-dependent digestions.

Pluskal, M.G., Przekop, M.B., Kavonian, M.R., Vecoli, C., Hicks, D.A. (1986) *Biotechniques 4*, 272–283. Immobilon PVDF transfer membrane: a new membrane substrate for western blotting of proteins.

Raftery, M.A., Cole, R.D. (1966) *J. Biol. Chem. 241*, 3457–3461. On the aminoethylation of proteins.

Rasmussen, H., Van Damme, J., Puype, M., Gesser, B., Celis, J., Vandekerckhove, J. (1991) *Electrophoresis 12*, 873–882. Microsequencing of proteins recorded in human two-dimensional gel protein databases.

Rider, M., Puype, M. Van Damme, J., Gevaert, K., De Boeck, S., D'Alayer, J., Rasmussen, H., Celis, J., Vandekerckhove, J. (1995) *Eur. J. Biochem. 230*, 258–265. An agarose-based gel-concentrating system for microsequence and mass spectrometric characterization of proteins previously purified in polyacrylamide gels starting at low picomole levels.

Riviere, L.R., Fleming, M., Elicone, C., Tempst, P. (1991) Study and applications of the effects of detergents and chaotropes on enzymatic proteolysis. In: *Techniques in protein chemistry II* (Villafranca, J.J.; ed.), pp. 171–179, Academic Press, San Diego.

Rojo, M., Pepperkok, R, Emery, G, Kellner, R, Stang, E, Parton, R, Gruenberg, J (1997) *J. Cell Biol. 139*, 1119–35. Involvement of the transmembrane protein p23 in biosynthetic transport via the intermediate compartment.

Rosenfeld, J., Capdevielle, J., Guillemot, J.C., Ferrara, P. (1992) *Anal. Biochem. 203*, 173–179. In-gel digestion of proteins for internal sequence analysis after one- or two-dimensional gel elektrophoresis.

Rüegg, U.T., Rudinger, J. (1977) *Meth. Enzymol. 47*, 111–116. Reductive cleavage of cystine disulfides with tributylphosphine.

Saris, C.J., Van Eenbergen, J., Jenks, B.G., Bloemers, H.P. (1983) *Anal. Biochem. 132*, 54–67. Hydroxylamine cleavage of proteins in polyacrylamide gels.

Schaer, M., Boernsen, K., Gassmann, E. (1991) *Rapid Commun. Mass Spectrom.* 5, 319–326. Fast protein sequence determination with matrix-assisted laser desorption and ionization mass spectrometry.

Simpson, R.J., Nice, E.C. (1984) *Biochem. Int. 8*, 787–791. In situ cyanogen bromide cleavage of N-terminally blocked proteins in a gas-phase sequencer.

Stephenson, R.C., Clarke, S. (1989) *J. Biol. Chem. 264*, 6164–6170. Succinimide formation from aspartyl and asparaginyl peptides as a model for the spontaneous degradation of proteins.

Stone, K.L., McNulty, D.E., Lopresti, M.L., Crawford, J.M., DeAngelis, R., Williams, K.R (1992) Elution and internal amino acid sequencing of PVDF-blotted proteins. In: *Techniques in Protein chemistry III* (Angeletti, R.H.; ed.), pp. 2 3–34, Academic Press, San Diego.

Stoscheck, C.M. (1990) *Meth. Enzymol. 182*, 50-68. Quantitation of protein.

Tempst, P., Link, A.J., Riviere, L.R., Fleming, M., Elicone C. (1990) *Electrophoresis* 11, 537–553. Internal sequence analysis of proteins separated on polyacrylamide gels at the submicrogram level: improved methods, applications and gene cloning strategies.

Tetaz, T., Bozas, E., Kanellos, J., Walker, I., Smith, I. (1993) In situ tryptic digestion of proteins separated by SDS-PAGE: improved procedures for extraction of peptides prior to microsequencing. In: *Techniques in protein chemistry IV* (Angeletti, R.; ed.) pp. 389–397, Academic Press, San Diego.

Titani, K., Hermodson, M.A., Ericsson, L.H., Walsh, K.A., Neurath, H. (1972) *Biochemistry 11*, 2427–2435. Amino acid sequence of thermolysin. Isolation and characterization of the fragments obtained by cleavage with cyanogen bromide.

Vandekerckhove, J., Bauw, G., Puype, M., Van Damme, J., Van Montagu, M. (1985) *Eur. J. Biochem. 152*, 9–19. Protein-blotting on polybrene-coated glass-fiber sheets.

Vanfleteren, J.R., Raymackers, J.G., Vanbun, S.M., Meheus, L.A. (1992) *Biotechniques 12*, 550–557. Peptide mapping and microsequencing of proteins separated by SDS-PAGE after limited in situ acid hydrolysis.

Ward, L.D., Reid, G.E., Moritz, R.L., Simpson, R.J. (1990) Peptide mapping and internal sequencing of proteins from acrylamide gels. In: *Current research in protein chemistry: techniques, structure, and function* (Villafranca, J.J.; ed.), pp. 179–190, Academic Press, San Diego.

Wilm, M., Shevchenko, A., Houthaeve, T., Breit, S., Schweigerer, L., Fotsis, T., Mann, M. (1996) *Nature 379*, 466–469. Femtomole sequencing of proteins from polyacrylamide gels by Nano electrospray mass spectrometry.

Yuen, S.W., Chui, A.H., Wilson,K.J., Yuan, P.M. (1989) *Biotechniques 7*, 74–83. Microanalysis of SDS-PAGE electroblotted proteins.

Section III:
Bioanalytical Characterization

Amino Acid Analysis

Roland Kellner, Helmut E. Meyer and Friedrich Lottspeich

1 Introduction

Amino acid analysis is used routinely to estimate the amount and to determine the composition of proteins, peptides or free amino acids. It involves two stages: complete hydrolysis of proteins and peptides, followed by quantitation of the released amino acids. The steps are laborious and time-consuming, and there is a continuing need to improve the technique. Whereas in the past nanomoles (10^{-9} M) of sample were available for amino acid analysis, researchers today frequently have only picomole amounts (10^{-12} M). At this low level contaminations present in chromatographic buffers, on glass surfaces and in hydrolysis acids make accurate analysis difficult. Todays demands on precision and sensitivity of amino acid analysis are a challenge to automation and instrument design.

The first part of the story of amino acid analysis was written by W. H. Stein and S. Moore. In the 1940s they worked out amino acid separations on starch columns [Moore 1948]. Starch columns had the disadvantage that fluids of high salt content required desalting prior to chromatography and the capacity was relatively low [Stein 1951]. Therefore they turned to the sulphonated polystyrene resin Dowex-50 for analytical experiments (0.9 x 100 cm column). Stein and Moore reported that an amino acid analysis of a protein hydrolysate could be completed in five days and that the synthetic resin eluted in about half the time required with starch [Moore 1951]. As early as in 1951 they discussed lithium und potassium citrate buffers and the effects of variations in resin and eluent. It was only the amount of amino acid (3–6 mg) they analyzed that differed notably from today. Only three years later a synthetic mixture of 50 components (0.05–0.20 mg) was chromatographed using a lower cross-linked resin (Dowex 50-X4, 4 % cross-linking), a pH and ionic strength gradient, a longer column (0.9 x 150 cm) and a modified photometric ninhydrin detection method [Moore 1954 a,b]. Applications with blood plasma, tissue extracts, and protein hydrolysates indicated successful results [Stein 1954; Tallan 1954; Hirs 1954]. In 1958 Moore, Spackman and Stein reported on chromatographic improvements, achieving a separation within 24 h [Moore 1958]. But even more important was the introduction of the first "instrument for automatically recording the ninhydrin color value of the effluent from ion exchange columns" [Spackman 1958]. The quantitative determination of amino acids down to 100 nmol (ca. 0.1 mg) with a precision of 3 % was shown. This became the classical system for amino acid analysis. In 1972 Stein and Moore were awarded the Nobel Prize for Chemistry for their work [Manning 1993].

Amino acid residues are small and polar compounds which are difficult to handle for most separation techniques except for ion exchange chromatography. Furthermore, these molecules possess almost no UV or fluorescence activity that would allow their detection. The specific derivatization of amino acids causes a substantial change in their chromatographical behavior as well as in their detectability. Throughout the 1970s, developments in chromatography and new derivatization chemistries provided alternative methods of amino acid analysis. High performance liquid chromatography gained wide popularity for the analysis of complex organic, synthetic and biological compounds. In particular, the reversed-phase mode initiated work with new amino acid derivatives, and provided an increase in speed and sensitivity. Analysis times were significantly reduced using reversed-phase columns because the rigidity of packing materials and the high resolution of the columns allowed higher flow rates and efficient separations. In addition, the detection characteristics of several new amino acid derivatives enabled far higher sensitivity to be reached.

Nevertheless, both techniques, ion exchange chromatography with ninhydrin detection and precolumn derivatization with reversed-phase HPLC, are of importance for amino acid analysis today and will be discussed in the following pages.

Capillary electrophoresis (CE) was also shown to be very useful in the analysis of amino acids. Attached to laser-induced fluorescence detection it has been used to handle nanoliters of sample in the most sensitive range yet achieved [Cheng 1988]. Micellar electrokinetic capillary chromatography (MECC) is a mode of CE with significant advantage to separate electrically neutral components and gives the background for miniuarized and rapid separations of amino acids. Microchip-based electrophoretic separations allowed resolution of several amino acids in a few seconds [Smith 1997].

2 Sample Preparation

In contrast to Edman degradation, where unwanted contaminants are lost in the first cycles, amino acid analysis has to contend with their presence in every sample. Hence, in analysing in the low picomole range, sample contamination is of primary concern, and the need arises to coordinate the complete work-up procedure, and to optimize handling and chemicals. Therefore, all glassware must be muffled for at least 12 h at 400 °C to destroy all possible amino acid contaminants. All chemicals for hydrolysis, derivatization and chromatography must be tested for high sensitivity amino acid analysis.

2.1 Peptides and Proteins

The first step in the process to determine the amino acid composition of a protein or peptide is the release of single amino acid residues by cleaving the peptide bonds. Hydrolysis of peptide bonds may be done either by enzymatic or chemical cleavage.

2.1.1 Enzymatic Hydrolysis

For complete enzymatic digestion both endo- and exoproteases are required, e.g. incubation with subtilisin is followed by leucine aminopeptidase, prolidase and carboxypeptidase. The proteases used should be of broad specifity and resistant to autolysis (for details see [Kay 1976]). Complete enzymatic hydrolysis is performed only in special cases, and has gained no widespread application.

2.1.2 Acid Hydrolysis

The standard hydrolysis procedure using 6 N HCl at 110 °C for 24 h [Moore 1963] is performed for most purposes. Obviously the procedure can be varied in regard to a number of parameters: acid, temperature, time, and gas- or liquid-phase mode can be used, and different scavengers are available.

Modified hydrolysis methods are of interest to overcome the problems brought about by the different properties of individual amino acids. In general, losses of 5–40 % of serine, threonine and tyrosine, and 50–100 % of cysteine/cystine, tryptophan, amino sugars and phosphorylated amino acids are expected with standard conditions. Asparagine and glutamine are destroyed during hydrolysis due to conversion of the amide group to a carboxy function, yielding aspartic acid and glutamic acid, respectively.

Hydrolysis is the most crucial step and the most susceptible to contamination or loss of sample. A serious amount of background contamination is introduced when digesting a sample in liquid acid because of surface contamination and the quality of chemicals. Thus, the decision whether to perform gas- or liquid-phase hydrolysis is related to the sensitivity of the analysis because the ratio between sample amount and background contamination is limiting: with high sensitivity less contamination can be tolerated. Gas-phase hydrolysis is the method of choice when analysing minute amounts of sample, and even then the HCl used for hydrolysis must be of the highest purity grade available.

Standard hydrolysis conditions are a compromise in terms of time and temperature because there are variations in the rate of peptide bond cleavage depending on the amino acids involved and differences in amino acid stabilities. Sample hydrolysis times at 110 °C can range from 20 h, giving minimal loss of sensitive residues, to 96 h, giving nearly complete release of amino acids from hydrophobic linkages. Hydrolysis time may be shortened by raising the temperature. High-temperature short-time hydrolysis is performed at 145 °C in 4 h [Gehrke 1985] or at 165 °C in 1.5 h [Dupont 1988]. In combination with acid mixtures even a 15 min hydrolysis (160 °C, propionic acid/HCl 1:1 [Westall 1974]) and a 25 min hydrolysis (166 °C, TFA/HCl 1:2 [Tsugita 1982]) have been described.

A time study is needed using different hydrolysis times to determine the exact composition of a protein. For this purpose gas-phase hydrolysis is performed on triplicate samples at 110 °C for 24, 48 and 72 h. Extrapolation to 0 h hydrolysis time is then used for serine, threonine and other instable amino acids, whereas extrapolation to 100 h hydrolysis time is applied for hydrophobic amino acid residues which are slowly released, like valine, isoleucine or leucine.

By using organic acids for hydrolysis a suitable method for recovering tryptophan was developed. Mercaptoethanesulfonic acid [Penke 1974; Maeda 1984], methane sulfonic acid [Simpson 1976; Inglis 1983], and toluene sulfonic acid [Liu 1971] have been used and yields of up to 93 % for tryptophan were obtained. Once again hydrolysis is performed at about 110 °C and recoveries of all the other amino acids are good. When organic acids are used for hydrolysis, evaporation is neither feasible nor necessary. Hydrolysates may be simply neutralized with sodium hydroxide and then directly applied to the column.

Especially with tryptophan it is necessary to ensure that oxidants in the hydrolysis acid are minimized, in particular dissolved oxygen. Tryptophan is decomposed by a free-radical auto-oxidation mechanism [Hunt 1986]. Oxygen can be removed from the hydrolysis vial by evacuation prior to tube sealing and/or by saturation of the acid as well as the hydrolysis vial with inert gas before closing.

The addition of various anti-oxidants to the hydrolysis acid is also helpful. Scavengers like phenol (1 %) [Moore 1963; Muramoto 1990], thioglycolic acid (0.1–1.0 %, v/v) [Matsubara 1969; Yano 1990], 2-mercaptoethanol (0.1 %) [Ng 1987], tryptamine (3-(2-aminoethyl)indole [Simpson 1983] or sodium sulfite [Swadesh 1984] are added to a 6 N HCl solution.

Due to the incompatibility of cysteine, asparagine and glutamine with acid hydrolysis, determination of these residues needs a special pretreatment. Cysteine and cystine must either be converted with performic acid, yielding cysteic acid [Hirs 1963], or prior to hydrolysis the protein to be analysed is reduced with a thiol and then an alkylation is performed either with iodoacetic acid [Hirs 1963] or 4-vinylpyridine [Raftery 1966].

The amide groups from asparagine and glutamine can be converted with bis-(1,1-trifluoroacetoxy)-iodobenzene to their corresponding diaminopropionic and γ-butyric acid residues [Soby 1981]. The number of residues is determined by the difference in aspartic and glutamic acid with and without pretreatment.

Proteins may be isolated by blotting onto a membrane, and the blotted materials used for further characterization such as N-terminal sequencing and/or amino acid composition. Matrices suitable for that purpose are PVDF (polyvinylidene difluoride) or modified glass fiber filter. Hydrolyzing of membrane bound proteins is performed under quite similiar conditions as described above [Nakagawa 1989; Eckerskorn 1988].

2.1.3 Alkaline Hydrolysis

Alkaline hydrolysis is performed in order to facilitate the recovery of tryptophan, and is applied especially with complex samples like food, when hydrolysis with methanesulfonic acid is not suitable. The reagents usually employed are barium, sodium or lithium hydroxide, applied at a 4 M concentration [Hugli 1972; Knox 1970; Spiess 1981]. Once more, hydrolysis is done at 110 °C for between 18 and 70 h depending on the sample. However, this technique does present technical and manipulative difficulties. Special reaction vessels have to be used together with alkali, since ordinary glass tubes are etched, releasing silicates and promoting side reactions. After hydrolysis the reagent itself must be neutralized and barium ions have to be removed before analysis, either by precipitation as the carbonate or as the sulfate. Losses of amino acids due to incurring adsorption to the precipitated barium salt will occur. Sodium or lithium hydroxide may be simply neutralized with hydrochloric acid; however, this increases the volume of the amino acid solution and the application to the analyzer column may become difficult. The method also causes partial or complete destruction of tryptophan and other amino acids.

Base hydrolysis is also used in the study of phosphoproteins. The phospho-amino acids of serine, threonine and tyrosine show a different stability under alkaline conditions. Especially phosphoserine will mostly be destroyed by β-elimination, whereas phosphothreonine and phosphotyrosine are relatively stable. So alkaline stability can be used as a criteria for identifying phosphoproteins. ^{32}P-labelled proteins fixed either in polyacrylamide gels [Cooper 1981] or bound on PVDF [Kamps 1989] are incubated in 1 M KOH at 55 °C for 2 h, and autoradiographic exposures before and after that treatment are compared. The detection of phosphohistidine was shown after hydrolysis using 3M KOH for 4 h at 105 °C [Noiman 1995].

2.2 Free Amino Acids

Determination of free amino acid content is of interest for complex physiological samples (see Figure 1), for foods, and for bioresearch and bioindustrial use. Samples are mostly a complex mixture of different substances, such as proteins, lipids, etc. These compounds interfere with the analysis, since they may bind to the stationary phase, thereby limiting capacity, or even may block the column. These analyses are among the most demanding, since especially with physiological samples up to 50 components have to be separated and quantitated. In order to improve chromatographic separation, especially for glutamine, asparagine, and glutamic acid, the eluting buffer is changed from a sodium to a lithium salt. Complex matrices like plasma, foodstuff, or fermentation media can contain high salt concentrations or enzymes. Methods such as precipitation, filtration, and centrifugation are utilized to remove them prior to amino acid analysis.

The most common method is the precipitation of protein with 5-sulfosalicylic acid and separation of the sample by subsequent centrifugation [Godel 1984]. Alternatively, high molecular weight compounds may be removed by ultrafiltration, which is a very smooth sample preparation technique.

3 Derivatization

Amino acids require chemical derivatization to improve their chromatographic behavior and/or detectability. There are two approaches: first amino acids can be separated and then derivatized and detected (post-column technique), or derivatization is done prior to chromatography (pre-column technique). An ideal derivatization reagent has the following characteristics:

- It reacts with all primary and secondary amino acids.
- It gives a quantitative and reproducible reaction.

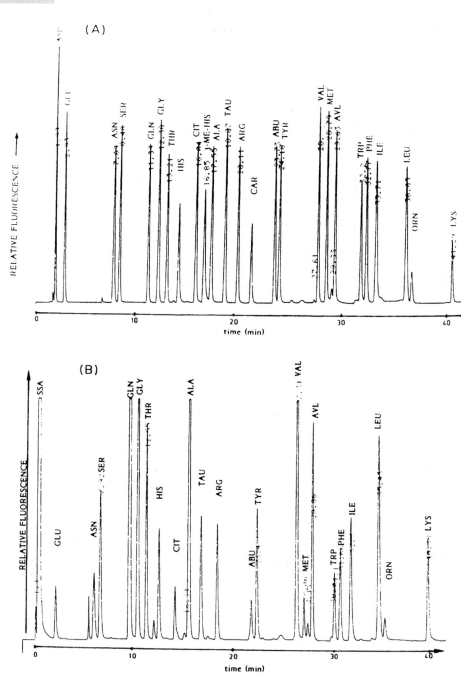

Figure 1. Separation of a physiological mixture of amino acids.

- It yields a single derivative of each amino acid.
- It needs mild reaction conditions.
- It gives stable derivatization products.
- It gives highly UV-absorbing or fluorescent derivatives.
- By-products and excess reagent do not interfere with chromatography.

An important characteristic of a derivatization technique is the amount of sample required for detection. It must be stressed that the detection limit for a derivatization method is an optimized result, and it should be distinguished from the amount that is routinely detected with unknown samples.

The structures of the derivatization reagents used in amino acid analysis are shown in Figure 2.

3.1 Post-Column Derivatization

The post-column derivatization technique separates free amino acids by ion exchange chromatography and then the reagent is introduced into the effluent stream from the column. This mixture passes through a reaction coil and finally appropriate detection is done. Ninhydrin is the classical reagent for post-column derivatization. Alternatively, orthophthaldialdehyde (OPA) and fluorescamine can be used.

3.1.1 Ninhydrin

Since the 1950s when Stein and Moore developed ion exchange separation and post-column derivatization using ninhydrin, remarkable improvements in speed, sensitivity and automation have been achieved, but the fundamental technique has remained the same.

Ninhydrin reacts quantitatively with primary and secondary amino acids as the separated amino acid residues and reagent pass through the post-column reaction coil, which is maintained at a temperature of 100–135 °C. There are no interfering by-products or multiple derivatization products. Amino acid derivatives are formed with UV absorption maxima at 570 nm (primary amino acids) and 440 nm (secondary amino acids). Samples are applied to the ion exchange column (a spherical resin 10 % DVB cross-linked polystyrol, 4 x 125 mm) in citrate buffer at pH 2.2. Separation is achieved with a multi-step pH and ionic strength gradient. The detection limit is about 50 pmol (ca. 6 ng) amino acid derivative.

3.1.2 Orthophthaldialdehyde

When orthophthaldialdehyde (OPA) was introduced for amino acid analysis in 1971 it was used exclusively in the post-column mode [Roth 1971], but today most applications are performed with the pre-column technique. OPA reacts with primary amino acids in the presence of thiol to give highly fluorescent 1-alkylthio-2-alkyl-substituted isoindoles [Simmons1978; Alvarez 1989]. The derivatization is fast (1–3 min)

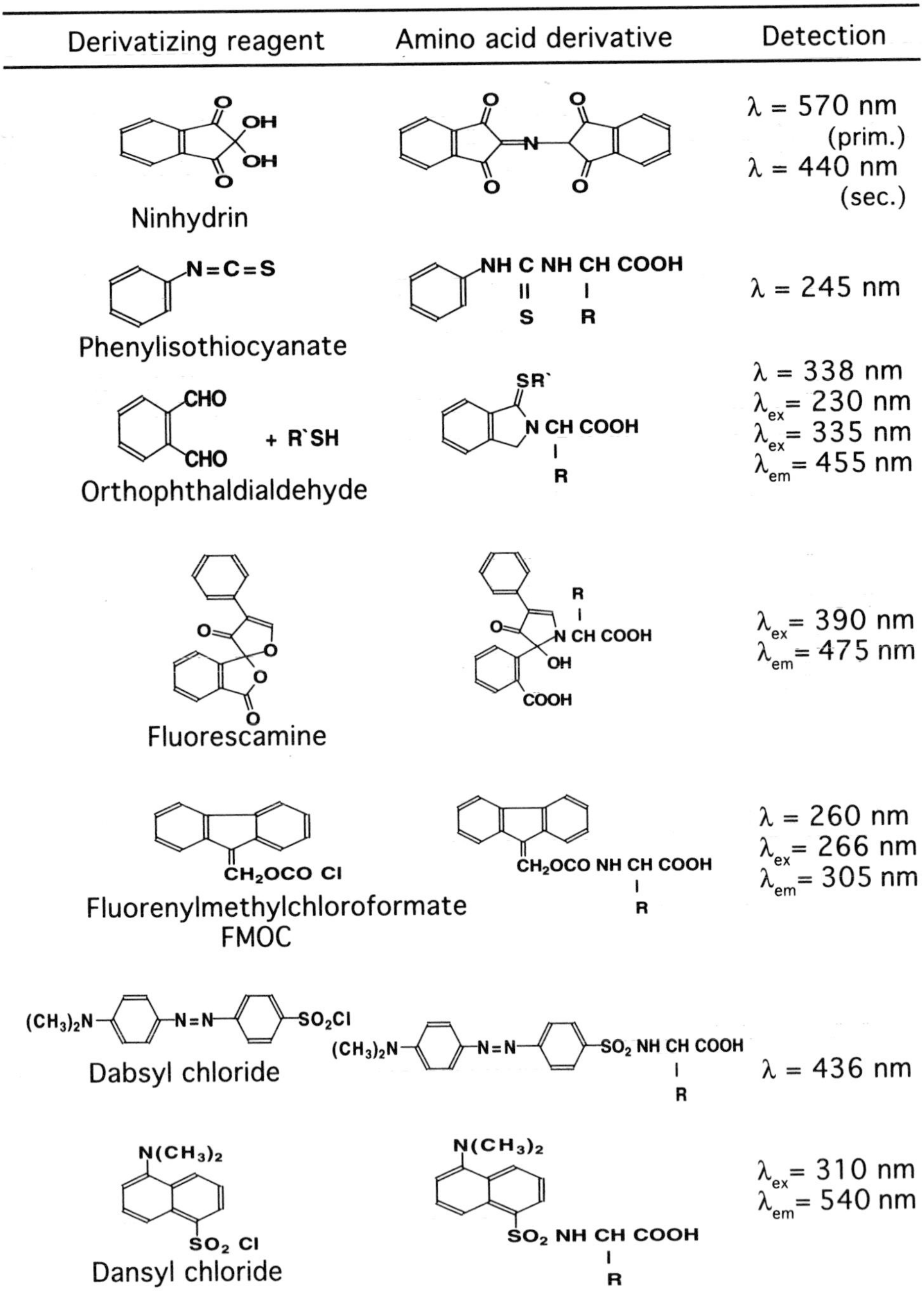

Derivatizing reagent	Amino acid derivative	Detection
Ninhydrin		λ = 570 nm (prim.) λ = 440 nm (sec.)
Phenylisothiocyanate	NH C NH CH COOH | | S R	λ = 245 nm
Orthophthaldialdehyde + R'SH	SR' N CH COOH | R	λ = 338 nm λ_{ex}= 230 nm λ_{ex}= 335 nm λ_{em}= 455 nm
Fluorescamine	R | N CH COOH OH COOH	λ_{ex}= 390 nm λ_{em}= 475 nm
Fluorenylmethylchloroformate FMOC CH₂OCO Cl	CH₂OCO NH CH COOH | R	λ = 260 nm λ_{ex}= 266 nm λ_{em}= 305 nm
Dabsyl chloride	$(CH_3)_2N$—〇—N=N—〇—SO_2 NH CH COOH | R	λ = 436 nm
Dansyl chloride	$N(CH_3)_2$ SO_2 NH CH COOH | R	λ_{ex}= 310 nm λ_{em}= 540 nm

Figure 2. Structures of reagents and derivatives used for amino acid analyses.

and is perfomed at room temperature in alkaline buffer pH 9.5. However, OPA amino acids are not stable.

OPA derivatives can either be detected by UV absorption or fluorescence emission. The reagent itself is not fluorescent. The UV spectra and the fluorescence excitation spectra are symmetrical. There are two maxima, one at 230 nm which is 5 times more intensive than the second one at 335 nm. However, the detection at 230 nm can cause trouble because of UV absorbing contaminants or the intrinsic UV absorption of the buffers. Fluorescence emission occurs at 455 nm and is generally not plagued by background problems.

Secondary amino acids will not react with OPA and must be converted to primary amines, i.e. by NaOCl or chloramine T, prior to derivatization [Bohlen 1979]. These reagents may be added either continuously or when the secondary amino acid is eluting from the column. A continuous addition gives a more stable baseline. The post-column technique gives a detection limit for OPA amino acids of about 10 pmol (ca. 1.2 ng).

3.1.3 Fluorescamine

Fluorescamine was the first reagent tested as an replacement for ninhydrin in order to improve sensitivity. In alkaline media it forms a fluorescent derivative with primary amines, but not with secondary amines [Udenfried 1972; Weigele 1972; Castell 1979]; the reaction is very fast (seconds) and fluorescamine itself is not fluorescent. The fluorescence is recorded at 475 nm, after excitation at 390 nm. There are a number of drawbacks with this reagent; fluorescamine is not stable in aqueous solutions and furthermore the amino acid derivatives have a fluorescence optimum at pH 9, which is difficult to coordinate with ion exchange chromatography. These problems have prevented fluorescamine from gaining significance in amino acid analysis.

3.2 Pre-Column Derivatization

Developments in high performance liquid chromatography (HPLC), namely the reversed-phase method, have enabled the use of a number of derivatizing reagents which alter the amino acids chromatographic behaviour prior to separation. The polar amino acids become more hydrophobic after coupling to an aromatic compound, and reversed-phase HPLC is then ideally suited to separate this derivatized mixture. High efficiency chromatographic packings together with improved LC equipment make it feasible to separate derivatized amino acid mixtures in as little as 10 min.

There are pre-column derivatizing reagents available that can be detected at very low concentrations (femtomoles, 10^{-15} M; ca. 100 pg) either by UV or fluorescence detection. Pre-column fluorescent derivatives provide the highest sensitivity in the range down to 50 fmol, more or less on a routine basis. It must be stressed that contaminating background levels of amino acids are a severe problem at this level of sensitivity; the presence of only a few bacteria in such a sample can make the analysis useless. In other words, it is the sample preparation that is the limiting factor and not the sensitivity of detection.

3.2.1 Phenylisothiocyanate

The reaction of phenylisothiocyanate (PITC) with amino acids is well understood since it is the first step in the protein sequencing reaction by Edman chemistry [Edman 1950]. PITC reacts with primary and secondary amines under alkaline conditions within about 20 min [Heinrickson 1984; Bidlingmeyer 1984]. The resulting phenylthiocarbamyl (PTC) derivatives are moderately stable and there are almost no interfering by-products. There are numerous examples given for separating PTC amino acids on RP-HPLC [Ebert 1986]. A UV absorbance maximum occurs at 245 nm, but no fluorescence activity is obtained. The detection limit using this method is at about 1 pmol (ca. 0.1 ng). PTC amino acid analysis is the most often used pre-column derivatization method.

3.2.2 Orthophthaldialdehyde

Orthophthaldialdehyde may be used either for the pre- or post-column technique and a description of the chemical reaction is already given above. There are different thiols that may be used for derivatization (2-mercaptoethanol, ethanethiol, 3-mercaptopropionic acid). They affect the hydrophobicity and stability of the resulting OPA derivatives, and therefore chromatographic parameters vary (stationary phase, eluting buffer). In general the separation is achieved with an acetonitrile gradient and 12.5 mM sodium phosphate buffer at pH 7.2.

Pre-column derivatization with OPA and subsequent separation by microbore chromatography provides a sensitivity in the low picomole range (UV detection) or in the 100 fmol range (fluorescence detection).

3.2.3 Fluorenylmethyl Chloroformate

After its introduction as a base-sensitive amino acid protecting group [Carpino 1970] and its use for solid-phase peptide synthesis, the 9-fluorenylmethyloxycarbonyl group (FMOC) has been used as a derivatization reagent for amino acid analysis since 1983 [Einarsson 1983; Miller 1990]. Derivatization is fast (<1 min) for both primary and secondary residues, and there is a stability maximum at pH 4.2. The reagent FMOC chloride is rapidly hydrolysed under these conditions and at least a partial removal, e.g. by extraction, is necessary, because the fluorescent by-product interferes with the elution and detection of the FMOC amino acids. The amino acid derivatives show UV absorption at 260 nm and fluorescence activity with excitation at 266 nm and emission at 305 nm, and they can be detected down to about 50 fmol.

3.2.4 Dabsyl Chloride

Dabsyl chloride (4-dimethylaminoazobenzene sulfonyl chloride, DABS-Cl, dabsyl chloride) was first described for use in amino acid analysis in 1975 [Lin 1975; Chang 1983]. Derivatization occurs with primary and secondary residues in about

15 min at 70 °C at pH 9, and the resulting derivatives absorb in the visible range at 436 nm [Vendrell 1986]. The ability of DABS derivatives to be detected in the visible range provides a stable baseline with a large variety of solvents and gradient systems. Dabsyl amino acids are very stable and can be stored for weeks – which can be advantageous if only a few samples have to be analyzed, as the samples can be collected for batchwise processing. The dabsylation step is still waiting to be automated; there is so far no commercial fully automated DABS system available, only reaction kits to perform the derivatization. The main difficulty with dabsylation is the need to use about a four-fold excess of reagent over the total amount of amino acids, and therefore the approximate amount of sample needs to be known. The detection limit is in the sub-picomole range.

3.2.5 Dansyl Chloride

Dansyl chloride (1-dimethylnaphthalin-5-sulfonyl chloride) was originally used for the end group determination of peptides and proteins and its use as a reagent for amino acid analysis was first described in 1981 [Tapuhi 1981]. It reacts with primary and secondary amino acids to give highly fluorescent dansyl residues; however, the reactivity of the reagent is relatively low and derivatization takes about 60 min. Furthermore, there are multiple products and the reaction is not quantitative. Dansyl chloride has not found widespread use as a reagent for amino acid analysis.

3.2.6 Chiral Reagents

The occurrence of enantiomeric amino acids is of particular interest in the quality control of amino acids for peptide synthesis, of peptide drugs or of food proteins. The analysis of enantiomers can be achieved by forming diastereomeric complexes prior to or during chromatography. For amino acid analysis it is advantageous to use a pre-column derivatization with a chiral reagent because all the above described features of this method are then available.

For diastereomer formation an optically pure reagent is required. The fluorescent (+)-1-(9-fluorenyl)ethyl chloroformate (FLEC) forms diastereomers with DL-amino acids. The reaction is analogous to that with FMOC. For example, the sample reacts in a borate buffer at pH 6.8 within 4 min at room temperature. Excess reagent is extracted with pentane. The HPLC separation uses C8 or C18 columns and a fluorescence detection with excitation at 260 nm and emission at 315 nm, or UV at 254 nm [Einarsson 1987].

Another means of enantiomeric analysis is derivatization by the OPA method using chiral thiols to form diastereomeric isoindolyl derivatives. The use of different N-substituted L-cysteines as a chiral modifier has also been described [Brückner 1989].

HPLC separation of DL-amino acids has also been obtained after derivatization with N_2-(5-fluoro-2,4-dinitrophenyl)-L-alanine amide (FDNP-Ala-NH$_2$, "Marfey's reagent") [Marfey 1984]. This reagent corresponds to a chiral variant of Sanger's reagent. Novel FDNP-amino acid amides using side chain branched L-Val, L-Phe or L-Pro showed even larger differences in elution times for the formed diastereomers

[Brückner 1990]. Derivatization of DL-amino acids is performed at slightly basic pH in mixtures of acetone and ammoniumbicarbonate for 1–2 h at 40 °C. For separation the diastereomers are applied to a C18 column and eluted with a triethylammonium phosphate buffer pH 3.0 and an acetonitrile gradient. The derivatized amino acids are detected at λ = 340 nm. Some mono- and disubstituted amino acids are formed and because of the large retention time differences, mixtures of all residues can become difficult to interpret. A combination of Marfey's method with LC/MS used 1-fluoro-2,4-dinitrophenyl-5-L-leucinamide (L-FDLA) for derivatization and ESI-MS and frit-FAB-MS for identification [Fujii 1997] (Figure 3).

A further approach to determining amino acid compositions and their enantiomeric forms is separation of the trifluoroacetylated amino acid N-propyl esters on glass

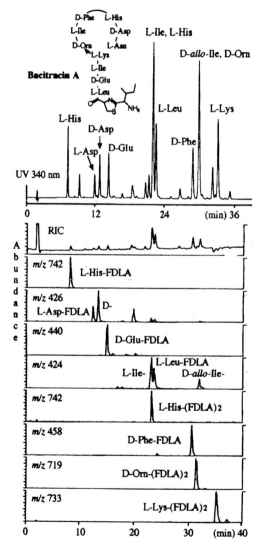

Figure 3. Determination of the absolute configuration of amino acids using Marfey's method. HPLC chromatogram and reconstructed ion and mass chromatograms (under negative ion mode) of L-FDLA derivatives of the hydrolyzed bacitracin A [from: Fujii 1997].

capillaries coated with a chiral phase using a gas chromatograph [Frank 1978]. Amino acids are derivatized in a two-step procedure:
(1) Esterification; n-propanol/4 N HCl reacts for 30 min at 110 °C.
(2) Acylation; trifluoroacetic anhydride reacts for 11 min at 150 °C.

Histidine needs to be modified by a treatment for 10 min at 110 °C with chloroformic acid isobutylester to give the corresponding ethoxycarbonyl derivative. Samples are injected after drying in a stream of nitrogen and diluting in organic solvent. Capillary columns typically used are 20 m x 0.28 mm coated with a 0.25 mm film of Chirasil-Val. The use of a chiral stationary phase allows separation of all amino acids together with their enantiomers, and thus determination of enantiomeric purity. Calibration is achieved by enantiomer labeling, whereby the amino acids are quantitated by adding their D-forms as internal standard [Frank 1978]. Enantiomer labelling has the advantage that target compound and internal standard with identical chemical and physical properties undergo derivatization and analysis. Detection is performed by a flame ionization detector (FID) or alternatively a nitrogen-selective detector (NSD). GC-MS coupling gives an additional feature, and mass-specific detection is especially useful in identifying unknown or unusual residues.

The major drawback here is the large amount of sample required for automatic handling. Detection limits are 100 pmol (FID) or 5 pmol (NSD), but due to the ratio of injection to total sample volume only small aliquots of the sample are used for analysis and starting amounts in the nanomole range are necessary.

4 Data Evaluation

Amino acid analysis is used to obtain the following information:

- Characterization and quality control of peptides and proteins: e.g. synthetic peptides.
- Quantitation. The most important application of protein structural analysis is the determination or at least estimation of the amount of material present. The interpretation of sequencing results relies on the quantitative information whether the amount was insufficient or a N-terminal blockage is given. The exact determination of sample concentrations is needed together with spectroscopic studies.
- Identification. It has been shown that amino acid analysis is capable of identifying known proteins in a protein database even with an error of 10–20 % per residue [Eckerskorn 1988; Sibbald 1991; Shaw 1993]. Internet resources are available, e.g. http://expasy.hcuge.ch/ch2d/aacompi.html and http://www.embl-heidelberg.de/aaa.html. (See also chapter IV.1).

For amino acid analysis the term "sensitivity" is often used with two different meanings: either for the smallest amount of sample (protein or peptide) necessary to perform all steps involved in the analysis, or the smallest amount of sample (derivatized amino acids) that can be detected. Calculations should be based on the amount of protein analysed, paying attention to the yields from hydrolysis, sample handling, and derivatization.

Table 1. Unavoidable troublesome effects connected with amino acid analysis.

Amino Acid	Effect
All methods:	
Glu, Ser, Gly	Tend to be high due to contamination
Thr, Ser, GlcN, GalN	Partly destroyed during hydrolysis
Trp, Cys, Asn, Gln	Almost completely destroyed during hydrolysis
Tyr, Met	Easily oxidized during hydrolysis
Val, Ile, Leu	May be low due to incomplete hydrolysis of hydrophobic peptide bonds
Post-column method:	
Pro	Low colour yield, quantitation difficult
Pre-column method:	
Asp, Glu	Very strong peaks, difficult to quantify
	Often secondary products are seen
	Matrix effects (salts, detergents, dyes) have severe impact on results

For each method there are certain problematic amino acids as far as derivatization behaviour or detectability is concerned (multiple peaks, incomplete derivatization; Table 1).

Some of these problems can be minimized by introducing correction factors; e.g. a time course study for the destruction of amino acids during hydrolysis or completeness of hydrolysis, blank analysis to estimate levels of contamination, or an internal standard (norleucine, sarcosine, β-alanine). A multilevel calibration is necessary for peak calculation, i.e. the range for calibration and analysis must correspond.

However, if there is only a small amount of material available for analysis one has to accept that there are only a few amino acids one can really trust in their overall yield, such as alanine, leucine or phenylalanine.

In general, with amounts greater 100 pmol per amino acid in the hydrolysate it should be possible to keep the error for most amino acids within + 10 %; with smaller amounts usually the error is between 10–20 % per residue.

5 Instrumentation

Amino acid analysis systems are configured to work either in the pre- or in the post-column mode. According to collaborative amino acid analysis studies [Crabb 1990], the number of instruments used in core facilities are about the same for either technique. Ninhydrin (27 out of 30 post-column systems) and PITC (25 out 32 pre-column systems) are most frequently used for derivatization. No fully automated system comprising hydrolysis, derivatization and analysis is commercially available. Semi-automated gas-phase hydrolysis can be performed with the help of dedicated workstations. This equipment allows one to flush hydrolysis vials with inert gas, evacuate the head space and set the hydrolysation temperature.

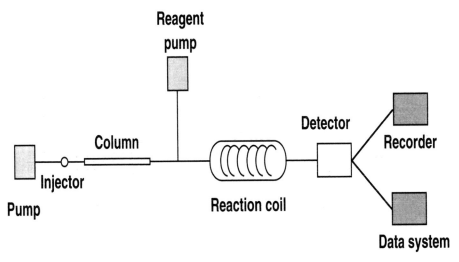

Figure 4. Post-column derivatization.

Post-column derivatization is achieved in general by dedicated systems and usually ninhydrin chemistry is used. The essential components of a post-column derivatization amino acid analysis system are shown in Figure 4. Routinely these instruments will enable detection down to 100 pmol with ninhydrin derivatization, and down to 50 pmol per amino acid in the hydrolysate with OPA derivatization.

Pre-column instrumentation is based on conventional HPLC equipment with varying extents of automation (Figure 5). In general every standard HPLC system is capable of being used for amino acid analysis if the derivatization is performed manually. However, the chromatography has to be optimized regarding the eluent, column and gradient. It is not possible to discuss here either HPLC equipment or the numerous separations for amino acids which are given in the literature (see references for the derivatization method of interest).

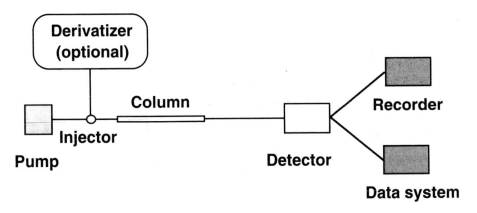

Figure 5. Pre-column derivatization.

6 Discussion

The commonly used derivatization reagents are ninhydrin, PITC, OPA, FMOC, and DABS-Cl. Ninhydrin gives visible-light-absorbing amino acid derivatives detectable in the picomole range; UV-absorbing PITC derivatives can be detected down to the low picomole range; fluorescent OPA and FMOC residues are easily detected in the femtomole range; and DABS-Cl residues absorb visible light also with a detection limit in the femtomole range.

Ninhydrin reacts with all amino acids but is restricted to post-column derivatization. PITC gives derivatives with primary and secondary amino acids. Its drawback is the limitation in sensitivity imposed by the UV detection and the instability of the derivatives. OPA only reacts with primary amino acids. In order to detect secondary amino acids such as proline there is the need to use hypochlorite as an oxidizing reagent. OPA derivatives are not stable and so the experiment needs precise time control for reagent delivery and on-line injection. FMOC reacts with both primary and secondary residues yielding stable derivatives. The main problem here is the presence of reagent by-products in the product mixture which interfere with detection and quantitation of some amino acids. An additional extraction step is therefore necessary, which can lead to inaccuracies in the analysis due to losses of hydrophobic residues with the extraction solvent. DABS derivatives are stable for weeks. But there is no automated system available, and for quantitative analysis the stochiometry of the derivatization reaction must be controlled.

The post-column derivatization procedure has been in use since the 1950s, and a great deal of experience and applications have accumulated. The instrumentation is sophisticated and close attention is being paid to the requirements for good mixing of the effluent stream of the analytical column with the reagent stream and dual wavelength detection. The post-column method has the drawback of long analysis times (about 1 h) and of limited sensitivity. These are caused by the ion exchange mode with its inherent separation and equilibration times. An additional problem is peak broadening during mixing in the reactor prior to detection, which causes a loss in sensitivity. Furthermore, high ionic strength buffers are necessary and they represent a source of contaminations.

Pre-column derivatization has been closely connected to the development of HPLC and the reversed-phase mode. Modern HPLC equipment has a number of advantages over the classical ion exchange technique, i.e. highly efficient columns, and low dead volume pumping and detection systems. These analytical techniques require smaller sample amounts and permit higher sensitivity. Elution buffers still limit detection levels in this equipment when post-column derivatization is utilized. The effect of buffers is less important when pre-column amino acid derivatization is chosen. Using chiral derivatizing reagents and analysing the enantiomeric composition is an important application in this mode.

Nevertheless pre-column derivatization also has disadvantages. Instrumentation and methodology are complex because of the need to optimize derivatization and chromatography. Matrix effects influence derivatization and quantitation. For example, ammonium salts have no effect on the analysis; but sodium buffers, in particular sodium phosphate, reduce yield and reproducibility when determining amino acid composition. Detergents may interfere, giving artifact peaks co-eluting with some residues [Dupont 1989].

Optimized chromatography with fast analysis times of 15–20 min also means a separation of standard amino acids that is crowded, and unusual residues may easily be missed. Co-elution with a standard amino acid may prevent identification or poor separation may hinder quantitation.

A further problem is the performance of reversed-phase HPLC columns, which often vary from batch to batch. They also lose resolution and develop high back pressure faster than ion exchange media.

Obviously, optimizing pre-column techniques is a process still under development.

When looking for a convenient system for amino acid analysis, the user has to consider the instrumentation as well as the derivatization method, and since there is no one perfect system available one has to compare benefits and limitations with specific demands. The most important factors influencing the decision which system to choose for amino acid analysis are likely to be the sensitivity range, the sample throughput, and the cost. If high sensitivity and accuracy are required for unknown samples as in the field of protein and peptide sequencing, it is essential to have the most sensitive system. On the other hand, a large sample throughput and a system that is not labor-intensive might be of interest when sample amounts are not limiting, as in quality control or analysis of synthetic peptides.

The Association of Biomolecular Resource Facilities (ABRF) [Williams 1988; Crabb 1990; ABRF 1998] worked on a collaborative amino acid analysis study with the aim of appraising the accuracy and precision of methods, hydrolysis conditions, and the quantitation of cysteine [Crabb 1990; Ericsson 1990]. Interestingly, half of the responding core facilities utilized post-column derivatization and the other half pre-column derivatization methods. As far as the derivatization chemistry used was concerned, 47 % out of 43 laboratories employed ninhydrin-based amino acid systems and 44 % PITC-based instrumentation. In cases where two analysers were placed in a core facility generally both pre- and post-column derivatization techniques are in operation. The latest results on sample studies and interesting notes on biological techniques can be found on ABRFs'.

Only minimizing contamination will enable routine analysis of femtomole amounts of derivatized amino acids. To use extremely small sample amounts, researchers must get used to working in closed environments, similar to those used in the manufacture of integrated circuits in the electronic industry. Only then can the ubiquitous amino acid background be reduced. Amino acid analysis still remains a challenge to automation and instrumentation, even though it was the first biochemical technique to be automated over 30 years ago.

7 References

ABRF (1998) http://www.abrf.org.

Alvarez-Coque, M.C.G., Hernandez, M.J.M., Camanas, R.M.V., Fernandez, C.M. (1989) *Anal. Biochem.178*, 1–7. Formation and instability of o-phthalaldehyde derivatives of amino acids.

Bidlingmeyer, B.A., Cohen, S.A., Tarvin, T.L. (1984) *J. Chromatogr. 336*, 93–104. Rapid analysis of amino acids using pre-column derivatization.

Boehlen, P., Mellet, M. (1979) *Anal. Biochem. 94*, 313–321. Automated fluorometric amino-acid analysis – Determination of proline and hydroxyproline.

Brückner, H., Wittner, R., Godel, H. (1989) *J. Chromatogr. 476*, 73–82. Automated enantioseparation of amino acids by derivatizatio with o-phthaldialdehyde and N-acylated cysteines.

Brückner, H., Keller-Hoehl, C. (1990) *Chromatographia 30*, 621. HPLC separation of DL-amino acids derivatized with N2-(5-fluoro-2,4-dinitrophenyl)-L-amino acid amides.

Carpino, L.A., HA, G.Y. (1970) *J. Am. Chem. Soc. 92*, 5748–5749. The 9-fluorenyl-methoxycarbonyl function, a new base-sensitive amino-protecting group.

Castell, J.V., Cervera, M., Marco, R. (1979) *Anal. Biochem. 99*, 379–391. A convenient micromethod for the assay of primary amines and proteins with fluorescamine. A reexamination of the conditions of reaction.

Chang, J.Y., Knecht, R., Braun, D.G. (1983) *Meth. Enzymol. 91*, 41–48. Amino acid analysis in the picomole range by precolumn derivatization and high-performance liquid chromatography.

Cheng, Y.F., Dovichi, N.J. (1988) *Science 242*, 562–564. Subattomole amino acid analysis by capillary zone electrophoresis and laser-induced fluorescence.

Cooper, J.A., Hunter, T. (1981) *Mol. Cell. Biol. 1*, 165–178. Changes in protein phosphorylation in Rous Sarcome Virus-transformed chicken embryo cells.

Crabb, J.W., Ericsson, L., Atherton, D., Smith, A.J., Kutny, R. (1990). A collaborative amino acid analysis study from the association of biomolecularresource facilities. In: *Current research in protein chemistry* (Villafranca, J.J.; ed.) pp. 49–61, Academic Press, New York.

Dupont, D., Keim, P., Chui, A., Bozzini, M., Wilson, K. (1988). Gas-phase hydrolysis for PTC-amino acid analysis. *Applied Biosystems Users Bulletin No. 2*, Foster City.

Dupont, D.R., Keim, P.S., Chui, A., Bello, R., Bozzini, M., Wilson, K.J. (1989). A comprehensive approach to amino acid analysis. In: *Techniques in protein chemistry* (Hugli, T.E.; ed.) pp. 284–294, Academic Press, New York.

Eckerskorn, C., Jungblut, P., Mewes, W., Klose, J., Lottspeich, F. (1988) *Electrophoresis 9*, 830–838. Identification of mouse brain proteins after two-dimensional electrophoresis and electroblotting by microsequence analysis and amino acid composition analysis.

Ebert, R.F. (1986) *Anal. Biochem. 154*, 431–435. Amino acid analysis by HPLC: Optimized conditions for chromatography of Phenylthiocarbamyl derivatives.

Edman, P. (1950) *Acta Chem. Scand. 4*, 283–293. Method for determination of the amino acid sequence in peptides.

Einarsson, S., Josefsson, B., Moeller, P., Sanchez, D. (1987) *Anal. Chem. 59*, 1191–1195. Separation of amino acid enantiomers and chiral amines using precolumn derivatization with (+)-1-(9-fluorenyl)-ethyl chloroformate and reversed-phase liquid chromatography.

Einarsson, S., Josefsson, B., Lagerkvist, S. (1983) *J. Chromatogr. 282*, 609–618. Determination of amino acids with 9- fluorenylmethyl chloroformate and reversed-phase high-performance liquid chromatography.

Ericsson, L.H., Atherton, D., Kutny, R., Smith, A.J., Crabb, J.W. (1991). Realistic expectations for amino acid analysis. In: *Methods in protein sequence analysis* 1990 (Jörnvall, H., Höög, J.O.; eds.) pp.143–150, Birkhäuser, Basel.

Frank, H., Nicholson, G.J., Bayer, E. (1978) *J. Chromatogr. 146*, 197–206. Gas chromatographic-mass spectrometric analysis of optical active metabolites and drugs on a novel chiral stationary phase.

Fujii, K., Ikai, Y., Oka, H., Suzuki, M., Harada, K. (1997) *Anal. Chem. 69*, 5146–51. A nonempirical method using LC/MS for determination of lthe absolute configuration of constituent amino acids in a peptide.

Gehrke, C.W., Wall, L.L., Absheer, J.S., Kaiser, F.E., Zumwalt, R.W. (1985) *J. Assoc. Off. Anal. Chem. 68*, 811–821. Sample preparation for chromatograpy of amino acids: Acid hydrolysis of proteins.

Godel, H., Graser, T., Földi, P., Pfander, P., Furst, P. (1984) *J. Chromatogr. 297*, 49–61. Measurement of free amino acids in human biological fluids by high-performance liquid chromatography.

Heinrikson, R.L., Meridith, S.C. (1984) *Anal. Biochem. 136*, 65–74. Amino acid analysis by reverse-phase high-performance liquid chromatography: Precolumn derivatization with phenylisothio-cyanate.

Hirs, C.H.W. (1967) *Meth. Enzymol. 11*, 197–199. Performic acid oxidation.

Hirs, C.H.W., Stein, W.H., Moore, S. (1954) *J. Biol. Chem. 211*, 941–950. The amino acid composition of ribonuclease.

Hugli, T.E., Moore, S. (1972) *J. Biol. Chem. 247*, 2828–2834. Determination of the tryptophan content of proteins by ion exchange chromatography of alkaline hydrolysates.

Hunt, S. (1986). Degradation of amino acids accompanying in vitro protein hydrolysis. In: (Darbre, A.; ed.) pp. 376–398, Wiley, Chichester.

Inglis, A.S. (1983) *Meth. Enzymol. 91*, 26–36. Single hydrolysis method for all amino acids, including cysteine and tryptophan.

Kamps, M.P., Sefton, B.M. (1989) *Anal. Biochem. 176*, 22–27. Acid and base hydrolysis of phospho-proteins bound to Immobilon facilitates analysis of phosphoamino acids in gel-fractionated proteins.

Kay, J. (1976) *Internat. J. Peptide Protein Res. 8*, 379–383. Complete enzymic digestion of acidic proteins.

Knox, R., Kohler, G.O., Palter, R., Walker, H.G. (1970) *Anal. Biochem. 36*, 136-143. Determination of tryptophan in feeds.

Lin, J.K., Chang, J.Y. (1975) *Anal. Chem. 47*, 1634–1638. Chromophoric labeling of amino acids with 4-dimethylamino-azobenzene-4-sulphonyl chloride.

Liu, T.Y., Chang, J.Y. (1971) *J. Biol. Chem. 246*, 2842–2848. Hydrolysis of proteins with p-tuloene-sulfonic acid.

Manning, J.M. (1993) *Protein Science 2*, 188–1191. The contributions of Stein and Moore to protein science.

Matsubara, H., Sasaki, R.M. (1969) *Biochem. Biophys. Res. Commun. 35*, 175-181. High recovery of tryptophan from acid hydrolysates of proteins.

Miller, E.J., Narkates, A.J., Niemann, M.A. (1990) *Anal. Biochem. 190*, 92–97. Amino acid analysis of collagen hydrolysates by reversed-phase high-performance liquid chromatography of 9-fluore-nylmethyl chloroformate derivatives.

Moore, S., Spackman, D.H., Stein, W.H. (1958) *Anal. Chem. 30*, 1185–1190. Chromatography of amino acids on sulfonated polystyrene resins.

Moore, S., Stein, W.H. (1948) *J. Biol. Chem. 176*, 367–388. Photometric ninhydrin method for use in the chromatography of amino acids.

Moore, S., Stein, W.H. (1951) *J. Biol. Chem. 192*, 663–681. Chromatography of amino acids on sulfonated polystyrene resins.

Moore, S., Stein, W.H. (1954) *J. Biol. Chem. 211*, 907–913. A modified ninhydrin reagent for the photometric determination of amino acids and related compounds.

Moore, S., Stein, W.H. (1954) *J. Biol. Chem. 211*, 893–906. Procedures for the chromatographic determination of amino acids on four per cent cross-linked sulfonated polystyrene resins.

Moore, S., Stein, W.H. (1963) *Methods Enzymol. 6*, 819–831. Chromatographic determination of amino acids by the use of automatic recording equipment.

Muramoto, K., Kamiya, H. (1990) *Anal. Biochem. 189*, 223–230. Recovery of tryptophan in peptides and proteins by high-temperature and short-term acid hydrolysis in the presence of phenol.

Nakagawa, S., Fukuda, T. (1989) *Anal. Biochem. 181*, 75–78. Direct amino acid analysis of proteins electroblotted ontopolyvinylidene difluoride membrane from sodium dodecyl sulfate-polyacrylamide gels.

Ng, L.T., Pascaud, A., Pascaud, M. (1987) *Anal. Biochem. 167*, 47–52. Hydrochloric acid hydrolysis of proteins and determination of tryptophan by reversed-phase high-performance liquid chromatography.

Noiman, S., Shaul, Y. (1995) *FEBS Lett. 364*, 63–66. Detection of histidine-phospho-proteins in animal tissues.

Penke, B., Ferenczi, R., Kovacs, K. (1974) *Anal. Biochem. 60*, 45-50. A new acid hydrolysis method for determining tryptophan in peptides and proteins.

Raftery, M.A., Cole, R.D. (1966) *J. Biol. Chem. 241*, 3457. On the aminoethylation of proteins.

Roth, M. (1971) *Anal Chem. 43*, 880–882. Fluorescence reaction for amino acids.

Shaw, G. (1993) *Proc. Natl. Acad. Sci. 90*, 5138–5142. Rapid identification of proteins.

Sibbald, P.R., Sommerfeld, H., Argos,P. (1991) *Anal. Biochem. 198*, 330–333. Identification of proteins in sequence databases from amino acid composition data.

Simpson, R.C., Spriggle, J.E., Veening, H. (1983) *J. Chromatogr. 261*, 407-414. Off-line liquid chromatographic-mass spectrometric studies of ortho-phthalaldehyde-primary amine derivatives.

Simpson, R.J., Neuberger, M.R., Liu, T.Y. (1976) *J. Biol. Chem. 251*, 1936–1940. Complete amino acid analysis of proteins from a single hydrolysate.

Simons, S.S., Johnson, D.F. (1978) *J. Org. Chem. 43*, 2886–2891. Reaction of o-phthalaldehyde and thiols with primary amines: formation of 1-alkyl(and aryl)thio-2-alkylisoindoles.

Smith, J.T. (1997) *Electrophoresis 18*, 2377–92. Developments in amino acid analysis using capillary electrophoresis.

Soby, L.M., Johnson, P. (1981) *Anal. Biochem. 113*, 149–153. Determination of asparagine and glutamine in polypeptides using bis(I,I-trifluoroacetoxy)iodobenzene.

Spackman, D.H., Stein, W.H., Moore, S. (1958) *Anal. Chem. 30*, 1190–1206. Automatic recording apparatus for use in the chromatography of amino acids.

Spiess, J., Villareal, J., Vale, W. (1981) *Biochemistry 20*, 1982–1988. Isolation and sequence analysis of a somatostatin-like polypeptide from ovine hypothalamus.

Stein, W.H., Moore, S. (1951) *J. Biol. Chem. 190*, 103–106. Electrolytic desalting of amino acids. Conversion of arginine to ornithine.

Stein, W.H., Moore, S. (1954) *J. Biol. Chem. 211*, 915–926. The free amino acids of human blood plasma.

Swadesh, J.K., Thannhauser, T.W., Scheraga, H.A. (1984) *Anal. Biochem. 141*, 397–401. Sodium sulfite as an antioxidant in the acid hydrolysis of bovine pancreatic ribonuclease A.

Tallan, H.H., Moore, S., Stein, W.H. (1954) *J. Biol. Chem 211*, 927–940. Studies on the free amino acids and related compounds in the tissues of the cat.

Tapuhi, Y., Schmidt, D.E., Lindner, W., Karger, B.L. (1981) *Anal. Biochem. 115*, 123–129. Dansylation of amino acids for high-performance liquid chromatography analysis.

Tsugita, A., Scheffler, J.J. (1982) *Eur. J. Biochem. 124*, 585–588. A rapid method for acid hydrolysis of protein with a mixture of trifluoracetic acid and hydrochloric acid.

Udenfried, S., Stein, S., Boehlen, P., Dairman, W., Leimgruber, W., Weigele, M. (1972) *Science 178*, 871–872. Fluorescamine: a reagent for assay of amino acids, peptides, proteins, and primary amines in the picomole range.

Vendrell, J., Aviles, F.X. (1986) *J. Chromatogr. 358*, 401–413. Complete amino acid analysis of proteins by dabsyl derivatization and reversed-phase liquid chromatography.

Weigele, M., Blount, J.F., Tengi, J.P., Czaijkowski, R.C., Leimgruber, W. (1972) *J. Amer. Chem. Soc. 94*, 4052–4054. Fluorogenic ninhydrin reaction. Structure of the fluorescent principle.

West, K.A., and Crabb, J.W. (1989). Automatic hydrolysis and PTC amino acid analysis: A progress report. In: *Techniques in protein chemistry* (Hugli T.E.; ed.) pp. 295–304, Academic Press, San Diego.

Westall, F., Hesser, H. (1974) *Anal. Biochem. 61*, 610–613. Fifteen-minute acid hydrolysis of peptides.

Wheeler, C.H., et al. (1996) *Electrophoresis 17*, 580–7. Characterisation of proteins from two-dimensional electrophoresis gels by matrix-assisted laser desorption mass spectrometry and amino acid compositional analysis.

Williams, K.R., Niece, R.L., Atherton, D., Fowler, A.V., Kutny, R., Smith, A.J. (1988) *Faseb J. 2,* 3124–3130. The size, operation, and technical capabilities of protein and nucleic acid core facilities.

Yan, J.X., et al. (1996) *J. Chromatrogr. 736*, 291–302. Large-scale amino-acid analysis for proteome studies.

Yano, H., Aso, K., Tsugita, A. (1990) *J. Biochem. 108,* 579–582. Further study on gas phase acid hydrolysis of protein: Improvement of recoveries for tryptophan, tyrosine, and methionine.

Chemical Methods for Protein Sequence Analysis

Friedrich Lottspeich, Tony Houthaeve and Roland Kellner

1 The Edman Degradation

The sequential degradation of proteins and peptides was studied by the Swedish scientist Pehr Edman from 1949 on and became known as "Edman degradation" [Edman 1949]. Edman described not only a highly efficient reagent and the mechanism of its chemical reaction with proteins, but also gave detailed instructions for experimental use [Edman 1950, 1956]. Additionally, he recognized the capability to automate this method and introduced the first automated protein sequencer in 1967 [Edman 1967].

Edman degradation cleaves the N-terminal amino acid from a peptide or protein backbone and prepares the derivatized residue for its identification. In this way the amino acid sequence of a protein is determined by repetitive chemical reaction. Edman degradation comprises three individual steps: the coupling, the cleavage and the conversion. Finally, a derivatized amino acid, the PTH amino acid, is produced and identified.

Latest sequencer models are equipped with up to four reaction chambers and permit multiple sample loadings. Another new feather is the miniaturization of the PTH-analysis which allows to work routinely in the femtomole range (Procise, PE-Applied Biosystems).

1.1 Coupling, Cleavage and Conversion

The first step couples the Edman reagent phenylisothiocyanate (PITC) to the free N-terminal amino group of a peptide chain. The reaction takes place within 15–30 min at a basic pH and a temperature of 40–55 °C in a very high yield. The resulting phenylthiocarbamyl peptide is washed by some hydrophobic solvent (e.g. ethylacetate) to get rid of excessive reagent and reaction by-products.

An unprotonated amine group is required for the coupling reaction and alkaline conditions of at least pH 9 are suitable. A further increase in pH would support reactivity but also increase side reactions, e.g. the alkali-catalysed hydrolysis of PITC, giving aniline. Additionally, the free amine group of aniline reacts with PITC, yielding diphenylthiourea (DPTU), the most intensive by-product of Edman degradation.

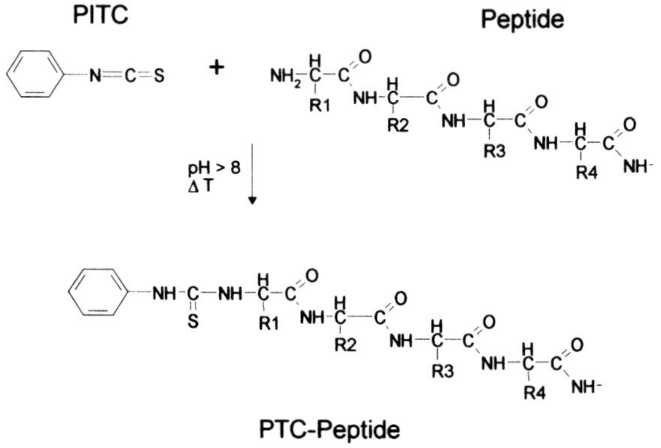

Figure 1. The coupling reaction.

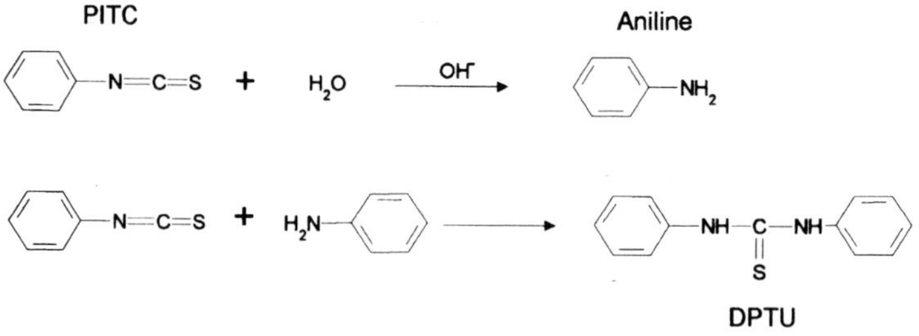

Figure 2. Formation of the by-product DPTU during Edman degradation.

The dried PTC peptide is treated with dehydrated acid, e.g. trifluoroacetic acid. The first amino acid is cleaved as a heterocyclic derivative, an anilinothiazolinone (ATZ) amino acid, after the nucleophilic attack of the sulphur atom to the carbonyl group of the first peptide bond. This emphasizes the importance of the sulphur atom in PITC; its nucleophilicity is needed for the cyclization. The corresponding oxygen-containing reagent, the phenylisocyanate, also has the ability to couple to the amino group, but cyclization and cleavage is not possible. In conclusion, any sulphur/oxygen exchange must be avoided and this is achieved by an inert gas atmosphere during the sequencing reaction.

The small, hydrophobic ATZ amino acid has a significantly different solubility from that of the hydrophilic peptide. It is extracted by a non-polar solvent, e.g. chlorobutane or ethylacetate. In a following step, called the conversion, the instable ATZ amino acid is rearranged, yielding the stable PTH derivative (see below). The shortened peptide comprises a new N-terminus and after drying it can be subjected to another reaction cycle.

The unstable ring structure of the ATZ amino acids is opened by the influence of aqueous acid and an increased temperature (ca. 60 °C). Rearrangement takes place, yielding the more stable PTH amino acid.

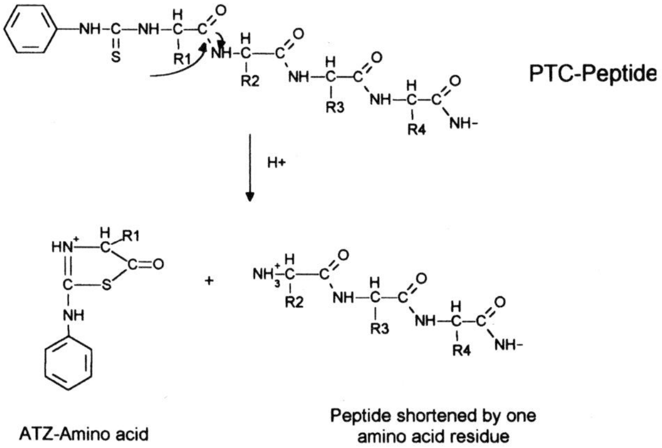

PTC-Peptide

Figure 3. The cleavage reaction.

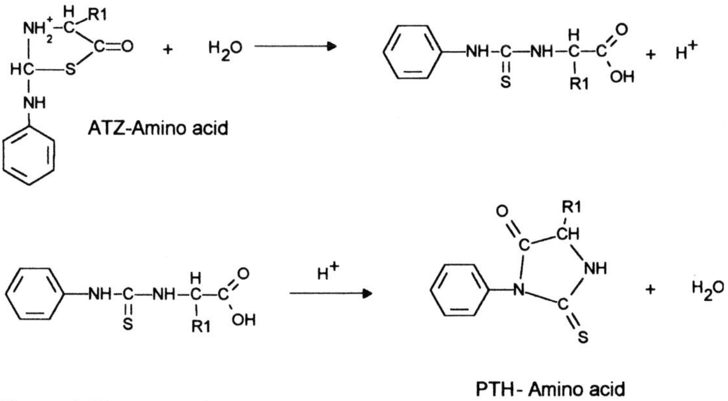

ATZ-Amino acid

Peptide shortened by one
amino acid residue

ATZ-Amino acid

Figure 4. The conversion step.

PTH- Amino acid

1.2 Identification of the PTH Amino Acids

PTH amino acids are analysed with a limit in the upper femtomole range after separation by RP-HPLC (Figure 5). Chromatographic identification and quantitation of the UV signals is done by reference to the retention times and the absorbance of a PTH standard run. The PTH amino acids display characteristic UV spectra with an absorbance maximum at 269 nm and a specific molar absorption coefficient of ca. 33 000 AU/Mol.

Since its introduction in 1950 numerous attempts have been made to improve the Edman reaction and to increase the sensitivity for PTH detection. Various fluorescent isothiocyanates with detection limits in the lower femtomole range have been considered as coupling reagents. However, none of these gained acceptance because of insufficient yield either for coupling or for cleavage. Furthermore, it is difficult to separate these large and bulky fluorophores, because they have very similar chromatographic properties.

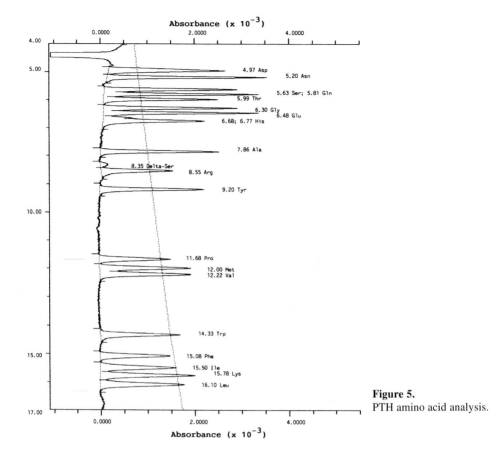

Figure 5.
PTH amino acid analysis.

Another approach is using mass spectrometric identification of amino acid derivatives. Coupling reagents like 3-[4′(ethylene-N,N,N-trimethylamino)phenyl]-2-isothiocyanate result in products which are extremely sensitive for mass analysis [Aebersold 1992]. However, once more the quantitative yield for coupling and cleavage is troublesome.

An interesting idea to take advantage of fluorescent detection came from Tsugita [Tsugita 1989], which however never did succeed. The conventional coupling and cleavage reaction of the Edman degradation are carried out, but the conversion step is modified. The extracted ATZ amino acid is converted by aminolysis instead of hydrolysis. The application of a fluorescent-labelled amine forms an amino acid derivative with detectability in the attomole range. However, until very recently experiments failed to find reaction parameters yielding quantitatively unique products. The main reason for these difficulties was that after the cleavage reaction the rather unstable ATZ amino acid is to a certain extent converted to the PTC and PTH amino acid. This happens especially in the presence of reducing agents that are commonly added to the solvents as antioxidants and are necessary for high repetitive yields on microscale sequencing. Recently, Farnsworth and Steinberg developed conditions where all the post-cleavage derivatives can be converted to a single derivative, the ATZ amino acid, by using an aqueous base [Farnsworth 1993a]. The PTC amino acid can now be completely converted back to the ATZ amino acid with the help of a Lewis acid (borontrifluoride etherate). This homogenous preparation of ATZ amino acid can react with fluorescent amines, resulting in greatly enhanced detectability compared to PTH amino acids [Farnsworth 1993b]. At present this chemistry is implemented in automated sequencers.

2 Instrumentation

The chemical reaction proposed by Edman for the sequential degradation of peptides and proteins has remained nearly unchanged since 1950. However, there has been an increase in sensitivity of about a thousandfold (compare the ca. 10 nmol starting material in the Edman paper 1950 to the ca. 10 pmol nowadays), and this was achieved by two major improvements: PTH amino acid analysis has switched from thin layer chromatography to high pressure liquid chromatography, and a reaction in the gas phase has been implemented for a new "gas phase sequencer".

It was Pehr Edman himself who was the first to publish an automated version of the sequential degradation [Edman 1967]. Automatization improved the quality of the sequence analysis drastically by reducing sample loss which is inheret in manual operation. The quality of sequence analysis is measured using the term "repetitive yield", that is the overall yield of one step in the Edman degradation. The higher this yield, the longer the sequences that can be determined before side products and "out of frame" sequences prevent the correct assignment of the cleaved PTH amino acids. Repetitive yields of >90 % could only be achieved after automatization. Nowadays "standard" yields are in the range of ca. 95 %, giving sequencing results with a length of 30–40 residues (Figure 6).

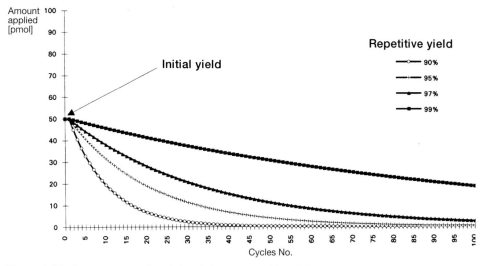

Figure 6. Maximum sequence length in relation to "repetitive yield".

145

The amount of PTH amino acid detected after the first cycle in relation to the applied amount of protein is called the "initial yield". The total amount of protein is not available for sequencing: in general it is only some 50 %, and reasons for this are poorly understood. However, the initial yield depend on the investigated protein, the sequencer and the reaction parameters.

The first Edman sequencer consisted of a solvent delivery system which delivered solvents and reagents by nitrogen pressure to the reaction cartridge via an electronically operated valve block. After coupling and cleavage the released ATZ amino acids were sampled in a cooled fraction collector. Conversion and identification were done off-line.

The same instrumental set up shown by Edman is still to be seen in modern sequencers (Figure 7). The conversion flask was added for on-line transfer of the extracted ATZ amino acids and for the conversion to the PTH derivatives. Additionally, the PTH amino acid analysis was automated and also performed on-line.

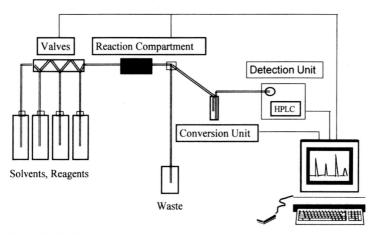

Figure 7. Protein sequencer.

2.1 The Liquid Phase Sequencer

A spinning cup was used as the reaction vessel of the first sequencer in 1967. The protein was held against the inner wall by centrifugal forces [Edman 1967]. The volume of the acids and bases delivered to the spinning cup was precisely measured so as to dampen the protein part only. Excess reagents and solvents were removed by evaporation using a vacuum system. Extraction of by-products was done efficiently by continuous delivery of solvents via a centred tube. A pick-up line was installed in an upper inner ring of the cup to remove the waste. This technique was adapted for a commercial liquid phase sequencer introduced in 1969 by Beckman Inc. and was in use until 1980.

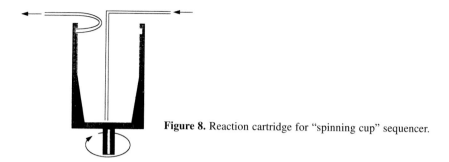

Figure 8. Reaction cartridge for "spinning cup" sequencer.

2.2 The Solid Phase Sequencer

For solid phase sequencing the protein is covalently bound to a solid support like polystyrene or glass beads, and packed into a small column [Laursen 1971]. First of all, this concept avoids sample loss because of wash-out. It enables the use of harsh conditions for sequential degradation and washing steps. Optimized and reproducible reaction parameters for Edman degradation can be used and an excellent performance (repetitive yield >96 %) for solid phase sequencing is achieved. Frequently, very long sequencer runs were performed (e.g. up to 80 residues), and C-terminally coupled peptides could be sequenced completely. However, two major drawbacks prevented its more general use. Firstly, amino acid residues coupling the protein to the matrix can neither be extracted nor be determined as PTH derivatives. Secondly, coupling of the protein to the matrix is necessary prior to the Edman reaction, but the coupling yields are unpredictable and may vary enormously. A fresh attempt to use the solid phase reaction was started recently when new coupling techniques using membranes as solid supports were studied, but due to instrumental problems did not gain widespread acceptance.

Figure 9. Reaction cartridge for solid phase sequencers.

2.3 The Gas Phase Sequencer

The most important achievement in protein sequence analysis was the development of the gas phase sequencer [Hewick 1981]. An increase in sensitivity afforded 1000-fold less protein. This method eliminated most other techniques because of its efficiency and simplicity. For gas phase sequencing the protein is applied onto a chemically inert glass fibre filter, if necessary with the help of a carrier like polybrene. Two chemicals, the coupling base and the cleavage acid, which are capable of dissolving the protein, are delivered in the gas phase. An argon or nitrogen stream transports trimethylamine or trifluoroacetic acid, respectively, to the reaction cartridge. In that way the required pH conditions are achieved without the risk of sample wash-out. The extraction of reaction by-products and of the ATZ amino acids is done by organic solvents again in the liquid phase; however, the protein is not soluble. The gas phase technique in conjunction with the development of zero-dead volume valve blocks created a new generation of protein sequencers which are optimized for sequence analysis of low sample amounts.

Already the first sequencers of this kind enabled sequencing of less than 100 pmol samples. Additional instrumental improvements, e.g. the on-line HPLC separation and optimized identification of PTH amino acids, enable gas phase sequencers to operate in the range of 10 pmol and even below.

Figure 10. Reaction cartridge for gas phase sequencers.

2.4 The Pulsed Liquid Phase Sequencer

The pulsed liquid phase sequencer is very similar to the gas phase sequencer, only the cleavage acid is delivered as a liquid pulse. A small volume is precisely measured and blown onto the target, which is just sufficient to wet the sample. The liquid phase is preferred for the cleavage acid because it ensures a faster cleavage reaction. While reacting the volatile acid evaporates, and after cleavage a stream of nitrogen or argon removes the vapour; sample wash-out is thereby omitted. Cycle times of 30 min and less are achieved together with optimized times for washing and together with variable temperature control of the Edman degradation [Totty 1993].

2.5 The Biphasic Column Sequencer

A rather different approach to immobilizing the protein for Edman chemistry was introduced by Hewlett Packard. An adsorptive biphasic column is used which consists of a hydrophopic and a hydrophilic portion. The protein, even in very large volumes, is applied and retained on top of the hydrophobic portion while inorganic salts and buffers can be washed away. Then the hydrophilic and the hydrophobic column halves are joined together and the resulting biphasic column is mounted in the sequencer. The instrument allows the flow direction of the solvents to be reversed, which helps to prevent sample washout – all aqueous solvent flows are directed against the hydrophobic support holding the protein via hydrophobic interactions. Alternatively, with organic solvents the hydrophilic support retains the protein on the column while hydrophobic contaminants or hydrophobic reaction products (e.g. DPTU or the ATZ amino acid) are eluted. Of course, now the flow direction has to be against the hydrophilic part of the biphasic column.

The sensitivity and repetitive and initial yields are comparable to those of the gas phase sequencers. However, the chromatographic approach for sample immobilization offers some possibilities for automated sample preparation, where the reaction cartridge itself may function as a kind of "sample bus".

Figure 11. Reaction cartridge for a biphasic column sequencer.

3 Difficulties of Amino Acid Sequence Analysis

The chemistry of the Edman reaction is well understood and works with high yields. The instruments of the leading suppliers ensure excellent performance. Nevertheless, amino acid sequence analysis still has to face a number of serious problems which make sequencing difficult if not impossible, especially in the lower picomole range. These problems may belong to two categories – either relating to the sample and its matrix, or to chemical or technical difficulties of the method itself.

3.1 The Sample and Sample Matrices

The investigated protein needs to be an *N-terminally homogeneous sample*. The interpretation of PTH chromatograms derived from pure samples is usually straightforward. In each chromatogram for every cycle only one PTH amino acid is detected. However, non-homogeneous samples will display several PTH signals, and correlation to the originated protein chain is not possible. If, for example, there are two proteins mixed together, an interpretation for a major and a minor sequence is only possible when the individual amounts of the components differ significantly. A sample ratio of even 2:1 makes interpretation uncertain. The major difficulty in interpreting results for mixtures comes from the variable yields for every PTH amino acid. The characteristics of the amino acid side chains influence the results, as do minor changes of reaction parameters and instrumental settings.

Contaminations like salt, detergents and free amino acids interfere even in small amounts with the Edman chemistry or disturb the efficiency of the extraction steps. Because of the mostly polar characteristics of the organic solvents, contaminations in the reaction cartridge are poorly dissolved and disturb sequencing for many cycles. Slightly unpolar contaminations are extracted together with the PTH amino acids and will be detected as artefact peaks in the HPLC chromatogram. This makes identification and quantitation of the PTH amino acids difficult or even impossible. After a couple of cycles unpolar contaminations are washed out and an ongoing successful analysis may still be possible. This is one reason why sequencing results in publications often do not include the first residues.

Solid phase sequencing has clear advantages in handling contaminations because various polar solvents can be used to wash covalently attached proteins and so achieve a reasonably pure sample for sequencing. For gas phase sequencing sample preparation is of major importance. The last step in a protein purification protocol must ensure the sample is free of salts and other impurities. The method of choice for peptides is reversed-phase HPLC using volatile solvents like 0.1 % TFA/acetonitrile. Proteins may not easily be chromatographed and afterwards desalted. Here, a valuable tool is non-covalent immobilization on hydrophobic membranes. This membrane attachment can also be done after polyacrylamide gel electrophoresis by electroblotting the proteins onto an inert membrane such as PVDF, modified glass fibre filter or nitrocellulose [Matsudaira 1987; Eckerskorn 1988]. Here, proteins are bound tight enough that contaminations can be washed away without sample loss.

Applying the sample onto the sequence matrix is a trivial but very important step. However, one is often not aware that small amounts of protein and peptides (µg or ng) are easily lost because of *unspecific adsorption onto surfaces* (e.g. Eppendorf cups, glass vials or tips). Special precautions have to be considered for quantitative analysis like dissolving in >95 % formic acid or coatings for the surface in sample vials.

A fundamental problem for amino acid sequence analysis is the *N-terminal blockage* of the peptide or protein. Edman degradation needs a free amine group at the N-terminus, otherwise PITC coupling is impossible. A huge number of the naturally occurring proteins (ca. 50 %) are N-terminally modified, e.g. by an acetyl, a formyl, or a pyroglutamic acid group. Rarely, chemical or enzymatic deblocking can be achieved. Blocked proteins need to be chemically or enzymatically cleaved and after chromatographic or electrophoretic separation internal peptide fragments are sequenced. N-terminal blockage is not only due to the native properties of a protein; sometimes blockage is introduced during sample purification and preparation. Reasons for such artificial blockage are impure chemicals and detergents reacting with the N-terminal amino function and/ or an increased pH (>9–10) in a purification step.

A most important but difficult aspect is the *quantitation* of a protein to be sequenced. A reasonable idea of the amount of protein present is needed to judge whether N-terminal blockage or inhomogeneity of the sample has occurred. One band in a gel or a symmetrical peak in chromatography cannot guarantee the presence of one single protein component. Therefore, it is important to correlate the amount of sample applied for sequencing and the quantity of PTH amino acids detected. It must be remembered that just about 50 % of a sample is accessible due to the initial yield. If an unexpectedly low amount of PTH amino acid is detected either less protein has been given than claimed, or the sample contains a mixture of the detected protein plus an additional, blocked protein. Because no information about the amount of a blocked component can be derived, it might be the case that the obtained sequence information relates to a minor, non-relevant protein. This finding is very important for the studies being performed. However, overestimation of a protein amount is much more frequent. Methods for the quantitation of proteins are mostly unusable for minute amounts or there is a systematic error of up to a factor of 10. The only method for quantitative studies on small protein amounts is amino acid analysis (see Chapter III.1). However, this is a complicated method and is very susceptible to contamination. Furthermore, a large amount of the sample (0.1–0.5 µg) needs to be dedicated for this determination. For these reasons, it is not routinely used. In general, protein amounts are a guess according to the intensity after electrophoresis and staining. But the intensity of a band is due to the thickness of a gel, the staining technique, protein characteristics and the protein amount itself (non-linearly). Hence, dealing with these parameters requires some experience and great care.

3.2 Difficulties with Edman Chemistry

Difficult residues. Chemical and physical properties of all the amino acid residues are very different. The intensity of the PTH signals varies for identical amounts of different amino acids. Some residues are partly destroyed because of the harsh conditions of the Edman chemistry and eventually give rise to artefact peaks. For example, dehydration occurs with serine (up to 80 %) and threonine (up to 50 %); non-derivatized cysteine is indetectable; tryptophan is destroyed by up to 100 % depending on the neighbouring residues; PTH-histidine and PTH-arginine are poorly extracted because of their polarity. Methionine is easily oxidized and an insufficient quality of the solvents will give very poor yields.

Modified residues may be detected close to or overlapping with known amino acids and therefore cause misinterpretation. Usually, PTH standards of modified residues are not available. Only a few positions of modified residues within the PTH chromatogram have been defined.

The interpretation of PTH chromatograms becomes more and more difficult because of increasing *background* while a sequencer run continues. The background comes up because of labile peptide bonds, especially involving aspartic acid, a small percentage of which are hydrolysed during each cycle. Typically, the background first significantly increases for about 20 cycles. At the maximum the amino acid composition of the protein is nearly reflected in the PTH chromatogram. At the end of the analysis background decreases, because some of the fragments are degraded completely.

There are several *variable parameters* which make every sequencer a unique instrument, such as reaction time, volume and delivery rate of solvents and reagents, drying times and temperature. These determine the performance of a sequencer while influencing the initial yield as well as the repetitive yield. Of course, the sample itself and the sequencing support additionally influence the performance. Standard programs delivered by the manufacturers take this into account. Optimized programs for samples spotted onto polybrene coated glass fibre filters or PVDF blotted proteins are available. There are special programs for serine- or proline-rich proteins, or for synthetic, resin-bound peptides. However, usually a standard program is used which is adjusted for the particular instrument to ensure high repetitive yields and short cycle times.

Chemicals for sequence analysis must be of very high purity; any contamination will cause a loss in the repetitive yield or give artefact peaks in the PTH chromatogram. Because of the many reasons for insufficient yields (see above), in practice troubleshooting is extremely difficult and time consuming. A special purification procedure is applied to continuously guarantee a high quality of reagents and solvents (sequencing grade), and strict quality control must be performed.

The *HPLC equipment for PTH separations* must routinely operate at the highest technical level. Commercially available sequencers are able to handle proteins in the low picomole range and detection limits for PTH amino acids with modern detectors and microbore columns are in the range of about 1 pmol. Twenty PTH amino acids plus some Edman by-products must be separated within 15–25 min gradients and the best reversed-phase columns must be operated at top performance. Even minor changes in gradient or temperature conditions will affect retention time shifts and errors in the PTH identification. A high precision of solvent delivery and extremely precise reproducibility in the gradient formation are essential.

4 C-Terminal Sequence Analysis

The carboxy-terminal (C-terminal) sequence information is important to characterize native and expressed proteins, to understand post-translational processing, to study N-terminally blocked proteins and to assist the design of oligonucleotide primers for gene amplification via the polymerase chain reaction. The information is complementary to the Edman degradation and together the N- and C-terminal sequence analysis are required to confirm a proteins identity.

The sequence analysis of proteins and peptides from their C-terminus face attempts already dating from early this century [Schlack and Kumpf 1926]. The automation of the chemical C-terminal sequence analysis was achieved only a few years ago due to severe difficulties in the handling of one or more of the 20 naturally occurring amino acids. Applicability for the routine is still limited due to an insufficient performance. Sample amounts are required of at least 200–500 pmol of protein and allow the amino acid assignment typically of one to five residues [for reviews see Inglis 1991; Boyd 1992; Bailey 1995].

There are two different methods to obtain C-terminal information:

- chemical degradation methods
- enzymatic degradation methods

and both may be combined either with chromatographic or mass spectrometric identification. In general, chromatography is used to analyze the released C-terminal amino acid residues and mass spectrometry helps to determine the shortened peptide fragments; the calculation of mass differences will give the amino acid residues, respectively the sequence information.

4.1 Chemical Degradation

The search for a chemical method to stepwise remove amino acids from the C-terminal end of proteins began early this century. Schlack and Kumpf [Schlack 1926] applied the method of Johnson and Nicolet [Johnson 1911] for a small peptide and converted acyl amino acids with thiocyanate to acylthiohydantoins and released the resulting thiohydantoin amino acid derivatives (TH-aa) using 1M NaOH. The TH-aa have UV absorption spectra and extinction coefficients similar to those of the PTH amino acids formed by the Edman conversion; RP-HPLC is used for identification.

Since then alternative chemistries have been proposed to modify the Schlack and Kumpf reaction using other thiocyanate-based reagents. The severity of the reaction conditions has been blamed for the lack of success with the Schlack and Kumpf reaction. Optimized chemical procedures where described [Inglis 1991] and were used to build C-terminal sequencers. But none of these sequencers could satisfy the practical needs as compared to N-terminal sequencing.

The reason is given by the poor reactivity of the carboxy group. An additional activation step is required to initiate the C-terminal reaction with a thiocyanate reagent. Harsh chemical conditions will attack amino acid side chains, e. g. Asp and Glu, increasing the probality of undesired side reactions as well as labile peptide bonds will give internal cleavages. The different reaction rates for the particular amino acids is perhaps the most serious shortcoming for all C-terminal chemistries. Therefore larger sample amounts are needed to start a C-terminal degradation and the result will provide only a very limited number of identified residues.

The chemical degradation of proteins from the C-terminus comprises the following steps:

1. Activation of the carboxy group
2. Coupling of a thiocyanate reagent, formation of a thiohydantoin amino acid derivate
3. Cleavage of the C-terminal amino acid
4. Determination of the released amino acid.

There are two recent advancements for the automatization and commercial availability of C-terminal sequencing methodology.

The approach by PE-Applied Biosystems adapted an Edman Sequenator for this purpose [Boyd 1992]. The protein is immobilized onto a membrane support and first the modification with diphenylchlorophosphate/DIEA forms an acyl phosphate at the C-terminus which is then cyclized to a thiohydantoin moiety after exposure to ammonium thiocyanate/TFA (a) (Fig. 12). Subsequently, alkylation with bromomethyl naphthalene under basic conditions (b) makes the alkylated TH-aa a better leaving group and cleavage reaction occurs much faster (c). The TH-aa are determined by reverse-phase HPLC (d). This cleavage is accompanied by the simultaneous formation of a new peptidyl-TH at the C-terminus (e). The reaction cyclus can thus be repeated (Figure 12).

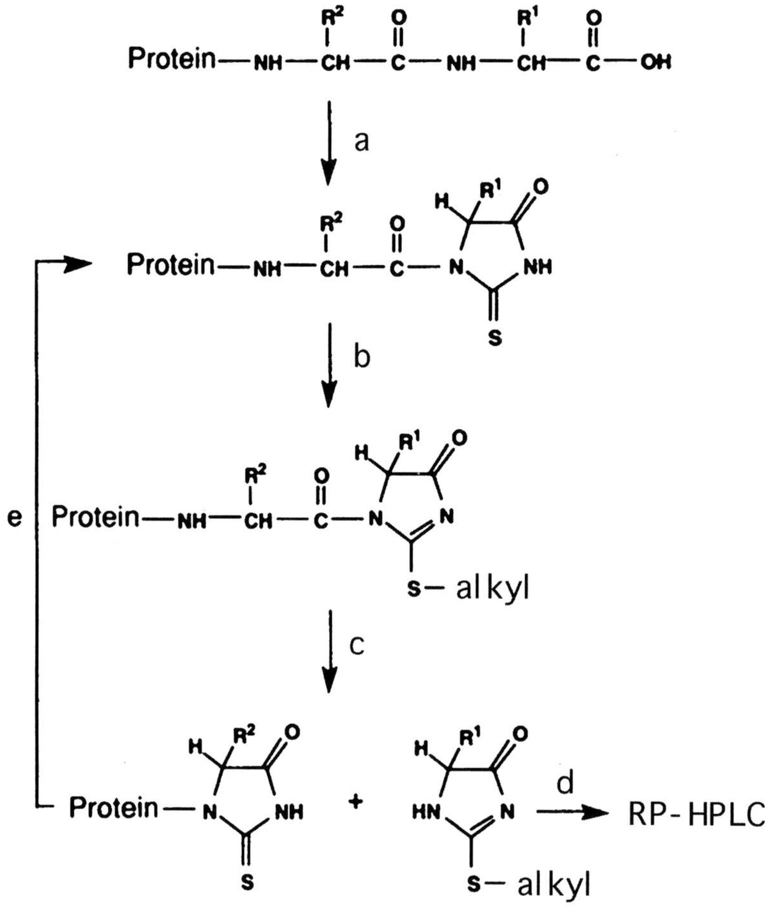

Figure 12. Alkylation chemistry according to [Boyd 1992].

An important advantage in this scheme is that the C-terminus is activated only in the initial step. Any newly created C-terminal fragments created by side reactions will not enter the reaction cycle. For this reason there is no build up of amino acid background peaks as sequencing progresses – as it is seen with N-terminal Edman sequencing.

The approach by Hewlett-Packard immobilizes the protein onto a Zitex membrane [Bailey 1995]. Sequencing chemistry applies diphenyl phosphoroisothiocyanatidate (DPP-ITC) in ethyl acetate/heptane as coupling reagent. The intermediate peptidyl-phosphoryl anhydride cyclizes in the presence of pyridine to the C-terminal peptidylthiohydantoin. Cleavage with potassium trimethylsilanolate (KOTMS) in alcoholic solvent generates a shortened peptide and a TH-aa. After acidification with water/TFA during hydrolysis also C-terminal proline could be analyzed. The TH-aa derivatives are identified by reverse phase HPLC and UV detection at 269 nm

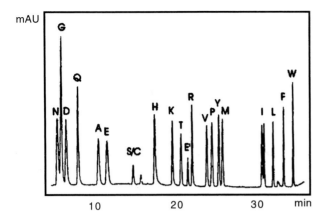

Figure 13. Thiohydantoin-amino acid standard chromatogram after separationon reverse-phase HPLC according to [Bailey 1995]

(Figure 13). The shortened peptide is ready for continued rounds of sequencing. The cycle time for one step is about 1 h.

4.2 Enzymatic Methods

The first use of carboxypeptidase for the determination of C-terminal residues was described by [Grassmann 1930]. A crude preparation of enzyme was used to remove the C-terminal glycine residue from glutathione. Since then carboxypeptidases have been applied in almost every protein sequence determination project. Several types of this exopeptidase (designated e.g. carboxypeptidase A) are commercially available; they comprise different pH-optima and various enzyme specificities (Table 1).

Basically, C-terminal end group determination or sequence analysis with carboxy-peptidases involves digestion of the protein with the appropriate enzyme or mixture of enzymes and withdrawing aliquots of the digested sample at various (logarithmic) time intervals. The released amino acids present in these aliquots are identified and quantified by amino acid analysis; their relative amounts are used to obtain the C-terminal sequence information.

Table 1. Carboxypeptidases used for C-terminal sequence analysis.

Enzyme	pH-optima	Cleavage specificity
Carboxypeptidase A	8.5	fast after hydrophobic aa not after Pro, Arg
Carboxypeptidase B	8.5	fast after Lys, Arg not after Pro
Carboxypeptidase P	5.5	all except Ser, Gly
Carboxypeptidase Y	5.5	after hydrophobic aa and Pro

Different release rates for the various amino acid residues represent a major difficulty inherent to this method (Figure 14). Misinterpretations for the C-terminal sequence may occur when a fastly cleaved residue is determined in the same quantity than the slowly released amino acid from penultimate position. Another difficulty are proteins with ragged C-terminal ends when more than one amino acid residue is present [Horn 1995]. For these reasons it is difficult to deduce more than 4–5 residues of a C-terminal sequence using carboxypeptidases. However, the enzymatic approach is more sensitive compared to chemical methods and sample amounts of about 100 pmol protein can be analyzed. Furthermore, the enzymatic method gives flexibility to manipulate the digest parameters individually; e.g. if the release rate is slow, the enzyme/substrate ratio or the temperature can be increased or vice versa [Lu 1988]. Attempts for automated enzymatic C-terminal determinations have been published [Thoma 1991].

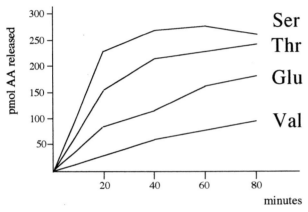

Figure 14. Release of amino acid residues by carboxypeptidase digestion (exemplified for Ser, Thr, Glu and Val).

4.3 Methods Combined with Mass Spectrometry

With the advancement of mass spectrometric techniques, alternative approaches towards C-terminal sequence determination have been developed. Mass spectrometry makes it possible to determine the shortened peptide fragments instead of the released amino acid residues. Thereby difficulties related to the release or the quantification of single amino acids can be circumvented. The mass of the peptide fragments and the calculated mass differences unequivocally derive a C-terminal sequence. However, mass accuracy becomes a critical feature for larger peptides or proteins because single amino acid residues must be discriminated which is difficult in relation to the large protein. Applications could demonstrate the use of MALDI MS [Brown 1995; Patterson 1995; Bonetto 1997] as well as ESI MS [Smith 1993; Takamoto 1995] for proteins of up to 20 kDa.

5 References

Aebersold, R., Bures, E.J., Namchuk, M., Goghari, M.H., Shushan, B., Covey, T.C. (1992) *Protein Science 1*, 494–503. Design, synthesis, and characterization of a protein sequencing reagent yielding amino acid derivatives with enhanced detectability by mass spectrometry.

Bailey, J., Tu, O., Issai, G., Ha, A., Shively, J. (1995) *Anal. Biochem. 224*, 588–596. Automated carboxy-terminalsequence analysis of polypeptides containing C-terminal Proline.

Bonetto, V., Bergmann, A., Jörnvall, H., Sillard, R. (1997) *Anal. Chem. 69*, 1315–1319. C-terminal sequence analysis of peptides and proteins using carboxypeptidases and mass spectrometry after derivatization of Lys and Cys residues.

Boyd, V., Bozzini, M., Zon, G., Noble, R., Mattaliano, R. (1992) *Anal. Biochem. 206*, 344–352. Sequencing of peptides and proteins from the carboxy terminus.

Brown, R., Lennon, J. (1995) *Anal. Chem. 67*, 3990–99. Sequence-specific fragmentation of matrix-assisted laser-desorbed protein/peptide ions.

Crimmings, D.L., Grant, G.A., Mende-Mueller, L.M., Niece, R.L., Slaughter, C., Speicher, D.W., Yüksel, K.Ü. (1992). Evaluation of protein sequencing core facilities: Design, characterization, and results from a test sample (ABRF-91SEQ). In: *Techniques in Protein Chemistry III* (Angeletti, R.H.; ed.), pp. 35–43, Academic Press, New York.

Eckerskorn, C, Mewes, W., Goretzki, H., Lottspeich, F. (1988) *Eur. J. Biochem. 176*, 509–519. A new siliconized glass fiber as support for protein chemical analysis of electroblotted proteins.

Edman, P. (1949) *Arch. Biochem. 22*, 475. A method for the determination of the amino acid sequences in peptides.

Edman, P. (1950) *Acta Chem. Scand. 4*, 283–290. Method for determination of the amino acid sequence in peptides.

Edman, P. (1956) *Acta Chem. Scand. 10*, 761. On the mechanism of the phenyl isothiocyanate degradations of peptides.

Edman P., Begg, G. (1967) *Eur. J. Biochem. 1*, 80–91. A Protein Sequencer.

Farnsworth, V., Steinberg, K. (1993a) *Anal. Biochem. 215*, 200–210. The generation of phenylthiocarbamyl or anilinothiazolinone amino acids from the postcleavage products of the Edman degadation.

Farnsworth, V., Steinberg, K. (1993b) *Anal. Biochem. 215*, 190–199. Automated subpicomole protein sequencing using an alternative postcleavage conversion chemistry.

Grassmann, W., Dyckerhoff, H., Eibeler, H. (1930) *Hoppe Seyler's Z. Physiol. Chem. 189*, 112–120.

Hewick, R.M., Hunkapiller, M.W., Hood, L.E., Dreyer, J. (1981) *J. Biol. Chem. 256*, 7990–7997. A gas-liquid-solid-phase peptide and protein sequencer.

Horn, M.J., Mayhew, J.W., ÓDea, K.C. (1995) *Protein Sci. 4* (1), 155. A method for the charaterization of the C-terminal of peptides and proteins.

Inglis, A. (1991) *Anal. Biochem. 195*, 183–196. Review: Chemical procedures for C-terminal sequencing of peptides and proteins.

Johnson, T., Nicolet, B. (1911) *J. Am. Chem. Soc. 33*, 1973–78.

Laursen, R.A. (1971) *Eur. J. Biochem. 20*, 89–102. Solid phase Edman degradation; An automated peptide sequencer.

Lu, H.S., Klein, M.L., Lai, P.H. (1988) *J. Chromatogr. 447*, 351–364. Narrow-bore high performance liquid chromatography of phenylthiocarbamyl amino acids and carboxypeptidase P digestion for protein C-terminal sequence analysis.

Matsudaira, P. (1987) *J. Biol. Chem. 262*, 10035–10038. Sequence from picomole quantities of proteins electroblotted onto polyvinylidene difluoride membranes.

Patterson, D., Tarr, G., Regnier, F., Martin, S. (1995) *Anal. Chem. 67*, 3971–78. C-terminal ladder sequencing via MALDI MS coupled with carboxypeptidase Y time-dependent and concentration-dependent digestions.

Schlack, P., Kumpf, W. (1926) *Hoppe Seyler's Z. Physiol. Chem. 154*, 125–170.

Smith, C., Duffin, K. (1993) Carboxy-terminal protein sequence analysis using carboxypeptidase P and electrospray mass spectrometry. I: Techniques in protein chemistry IV (Angeletti, R.; ed.) pp. 419-426, Academic Press, New York.

Takamoto, K., Kamo, M., Kubota, K., Satake, K., Tsugita, A. (1995) *Eur. J. Biochem. 228*, 362–372. Carboxy-terminal degradation of peptides using perfluoroacyl anhydrides: A C-terminal sequencing method.

Thoma, R., Crimmins, D. (1991) *J. Chromatogr. 537*, 153–1655. Automated phenylthiocarbamyl amino acid analysis of carboxypeptidase/aminopeptidase digests and acid hydrolysates.

Totty, N.F., Waterfield, M.D., Hsuan, J.J. (1993) *Protein Science 1*, 1215–1224. Accelerated high-sensitivity microsequencing of proteins and peptides using a miniature reaction cartridge.

Tsugita, A., Kamo, A., Jone C.S., Shikama, N. (1989) *J. Biochem. 106*, 68–75. Sensitization od Edman amino acid derivatives using the fluorescent reagent, 4-aminofluorescein.

Wang, R., Chait, B., Kent, S., (1993) Protein ladder sequencing: a conceptually novel approach to protein sequencing using cycling chemical degradation and one step readout by matrix-assisted laser desorption mass spectrometry. In: *Techniques in protein chemistry IV* (Angletti, R.; ed.) pp. 409–418, Academic Press, San Diego.

Analyzing Post-Translational Protein Modifications

Helmut E. Meyer

1 Introduction

The majority of all proteins contain post-translationally modified amino acids. These modifications are either permanent to allow a stable three dimensional structure of the protein, to target a protein to a distinct compartment in the cell or are transient to confer different states of activity or function.

One main modification is protein phosphorylation which is mainly involved in signal transduction from the outer cell membrane to distinct targets inside the cell. Many other modifications are known like methylation, acetylation, myristylation, sulfatation, hydroxylation, isoprenylation, glycosylation, phosphatidylinositol glycan membrane anchoring and many others [Krishna 1992].

This article will show some examples, demonstrating how to analyze post-translational modifications by Edman degradation and/or mass spectrometry and will give some hints for the right strategy to reach the desired goal.

In analyzing and localizing post-translational modifications in a protein or a peptide by the classical Edman degradation method [Edman 1949] one has to consider the chemical stability of such modifications during purification and during the following analyses. Each modification exhibits special characteristics and needs proper techniques for its successful analysis.

To get a better overview a simple classification of post-translational modifications is introduced with respect to the before mentioned points. Afterwards example(s) are discussed for each of these classes and how to analyze them. Additionally, some examples will be given of how mass spectrometry can be helpful in analyzing post-translational modifications.

2 Classification of Post-Translational Modifications According to their Behavior during Purification and Edman Degradation

2.1 Modifications: Stable during Purification and Edman Degradation

- Detectable by its PTH derivative using the gas-phase sequencing protocol, i.e. methyl-histidine (Figure 1), methyl-lysine, hydroxy-proline etc.
- Not detectable by its PTH derivative using conventional gas-phase sequencing protocol and on-line PTH analysis, i.e. phospho-tyrosine, glyco-asparagine, glyco-threonine, glyco-serine, farnesyl-cysteine, gluco-arginine, lanthionine, methyl-lanthionine, stable N-terminal blocking groups like N-acetyl or N-acyl groups etc.
  - but detectable by its PTH derivative using a solid-phase sequencing protocol after covalent attachment of the peptide and on-line PTH analysis, i.e. glyco-asparagine (Figure 2), glyco-threonine (Figure 2), glyco-serine or lanthionine
  - or detectable by its PTH derivative using a solid-phase sequencing protocol like before 1) but using a special detection method, i.e. phospho-tyrosine (Figure 3)
  - or detectable after chemical removal of special N-acyl groups, i.e. N-terminal pyruvate or α-oxo-butyric acid (Figure 4)
  - not detectable by Edman degradation, i.e. with N-acetyl- or N-acyl groups and other N-terminal blocking groups Edman degradation does not start. There are some special techniques described in the literature on how to remove some of those blocking groups successfully [Tsunasawa 1993]
  - not detectable by Edman degradation but can be determined by mass spectrometry, i.e. gluco-arginine (Figure 5) or farnesyl-cysteine (Figure 6).

2.2 Modifications: Stable during Purification but Unstable during Edman Degradation

- Detectable
  - by a specific PTH derivative of the destructed parent amino acid, i.e. phospho-serine (Figure 7), phospho-threonine
  - or after chemical modification to a stable amino acid derivative before starting sequence analysis, i.e. phospho-serine, phospho-threonine (Figure 8), lanthionine (Figure 4).
- Due to the splitting of the functional group only the PTH derivative of the unmodified amino acid is detectable, i.e. sulfo-tyrosine, phospho-histidine, etc.
- Not detectable by a specific PTH derivative, i.e. dehydroalanine, dehydro-α-amino-butyric acid

- Edman degradation stops at this position, no further degradation is possible. However, in this case the block is removable, see Figure 4.
- or the unsaturated amino acid is chemically modified to a stable derivative (Figure 4).

2.3 Modifications: Unstable during Purification and Edman Degradation

Aspartyl-phosphate, phospho-histidine, phospho-cysteine, and some other very unstable post-translational modifications belong to this group. Special techniques have to be applied for their analysis. For phospho-cysteine such a procedure was described by [Weigt et al. 1995] and for phospho-histidine by [Medzihradszky et al. 1997].

3 Examples

3.1 1-Methyl-Histidine

In looking for the autophosphorylation site of skeletal muscle myosin light chain kinase we isolated a phosphorylated peptide which contained a histidine residue in position 14 of this peptide [Meyer 1987]. During sequence analysis this histidine was found to be accompanied reproducibly by an unknown byproduct eluting just before the expected elution time for PTH-proline (Figure 1) but after PTH-histidine. To prove the identity of this byproduct we evaluated the elution position of PTH-3-methyl-histidine and PTH-1-methyl-histidine and found that the peptide contains a 1-methyl-histidine. This kind of histidine methylation was shown for the first time in this protein. All other known histidine methylations in proteins hitherto known take place in the 3-position. As stated before methyl-histidine is a stable amino acid derivative and can be determined without any change in the usual protocol for gas-phase Edman degradation.

3.2 Glyco-Asparagine, Glyco-Threonine

Glyco-amino acids belong to the group of post-translationally modified amino acids which are very hydrophilic and are not soluble in the organic solvents like butylchloride or ethylacetate, usually used as transfer solvents in gas-phase Edman degradation. Therefore, without modification of the degradation protocol, a gap in the sequence analysis occurs at those positions where these glyco-amino acids are expected. With glyco-serine sometimes a small amount of the DTT-adduct of dehydroalanine is detectable which is formed by partial β-elimination of the glyco-moiety [Stöcker 1990]. To gain direct access to their PTH derivatives during sequence analysis, the carbohydrate-containing peptides will be covalently attached to a modified PVDF-membrane which contains arylamine groups allowing the coupling via the side

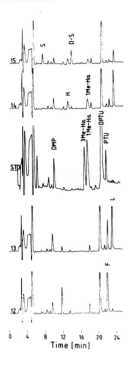

Figure 1. Sequence analysis of 610 pmole of the N$^\pi$-methyl-histidine containing peptide of myosin light chain kinase around the modified histidine. The chromatograms corresponding to degradation cycles 12–15 are shown. The middle chromatogram shows a standard mixture containing PTH-N$^\pi$-methyl-histidine and PTH-Nt-methyl-histidine, only. D-S is the PTH-dithio-threitol-adduct of dehydroalanine, a byproduct of PTH-serine. DMPTU, dimethylphenylthiourea; DPTU, diphenyl-thiourea; PTU, phenylthiourea. Taken from [Meyer 1987] with permission.

chain(s) and C-terminal carboxy group. This technique makes the application of liquid trifluoroacetic acid as a transfer solvent applicable, a very polar solvent which transfers the glycosylated amino acid derivatives quantitatively. Andrew Gooley [Gooley 1991] and co-worker show this kind of analysis with different N- and O-glycosylated peptides. In the examples given there is a glyco-asparagine in the upper picture of Figure 2 and a glyco-threonine in the lower picture. Due to the heterogeneity of the carbohydrate moiety not only a single signal is detected but in both cases several distinct PTH-derivatives eluting in the same region of the chromatogram are visible.

3.3 Phospho-Tyrosine

Like the glycosylated amino acids, phospho-tyrosine is a very hydrophilic molecule and needs solid phase conditions after covalent attachment to provide extraction from the conversion flask by liquid trifluoroacetic acid [Meyer 1991a].

An additional difficulty is that the on-line PTH-analysis gives no clear signal for PTH-phospho-tyrosine. This is due to the bad adsorption and concentration of this compound at the top of the reversed phase column, and PTH-phosho-tyrosine gives a smear-signal which is only visible when performing the analysis with a higher amount of peptide. As a solution for this problem we developed a capillary electrophoresis (CE) method which allows the unambiguous identification of PTH-phospho-tyrosine in the low picomol range.

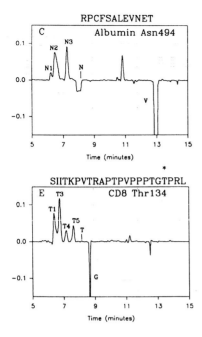

Figure 2. PTH-amino acid analysis chromatograms of glycopeptides from mutant human albumin Casebrook and rat CD8α. 2 nmol of the albumin tryptic glycopeptide and 1 nmol of the CD8α chain were dissolved in 20 % acetonitrile and covalently linked via the carboxyterminal amino acid to aryl-amine activated PVDF membranes (Sequelon-AA) by water soluble carbodiimide activation (Sequelon-AA are available from MilliGen/Biosearch, Division of Millipore). The peptides were subjected to automated Edman degradation using a MilliGen ProSequencer 6600 using the standard cycle program recommended by the manufacturer. The PTH-derivatives of amino acids and glyco-amino acids were identified by on-line reversed-phase HPLC. Chromatogram *C* shows the processed chromatogram for albumin Casebrook glycopeptide, cycle 10, PTH-Asn494(Sac). In this Asn(Sac) peptide, three PTH-derivatives for Asn(Sac) are identified, all elute before genuine PTH-Asn and the major broad peak, N2, co-elutes with PTH-Asp. Chromatogram *E* shows cycle 4 of the CD8α glycopeptide which shows PTH-Thr at 8.5 min and the dehydro-threonyl group at 11.5 min as well as four unique PTH-derivatives marked with T1, T2, T3, T4 and T5. Taken from Gooley 1991, with permission.

An example of such an analysis is shown in Figure 3. On the left there are the chromatograms of the on-line PTH-amino acid analysis, on the right there are the chromatograms of the off-line CE-chromatography.

Today, a combination of mass spetrometry and Edman degradation is the method of choice. An example for this strategy is given in [Lehr et al. 1998].

3.4 N-Pyruvyl or N-α-Oxo-Butyric Acid

This kind of N-terminal blocking group is known to be present in some peptide anti-biotics called lantibiotics. The same groups are formed by hydrolytic cleavage of pep-tides containing dehydroalanine or dehydro-α-aminobutyric acid.

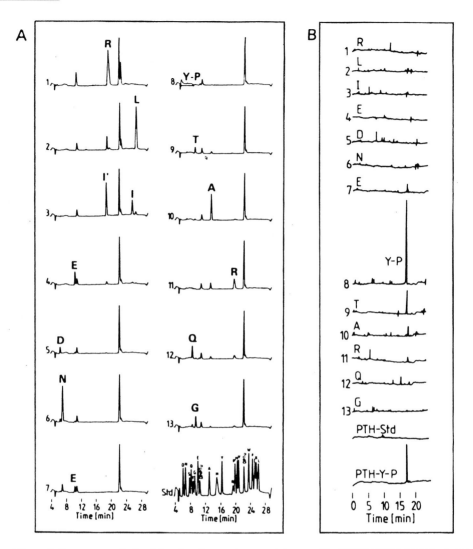

Figure 3. Sequence analysis of the phosphotyrosine containing peptide RLIEDNEYpTARQG. A) We used 1 nmol of the phosphopeptide for covalent attachment to the Sequelon-AA membrane. 65 % of the peptide is covalently bound and degraded by solid-phase sequence analysis. 50 % of each degradation sample is directly analyzed by the on-line PTH-amino acid analyzer. The chromatograms of the PTH-amino acid analyses are shown. Y-P denotes the PTH-derivative of phosphotyrosine. B) The residual 50 % of each degradation sample is dried under vacuum and dissolved in 10 µL of 50 mM sodium chloride containing 20 % acetonitrile and 0.35 % trifluoroacetic acid. 60 nL is analyzed by capillary electrophoresis. The chromatograms of these analyses are shown. Conditions for capillary electrophoresis: electrophoresis buffer: 20 mM sodium citrate, pH 2.5; capillary: fused silica, 72 cm (50 cm to detector), 50 µm inner diameter; injection time (vacuum): 20 sec = 60 nL injection volume; voltage: -25 kV; temperature: 30 °C; detector rise time: 2 sec; absorption units full scale 0.008 A; polarity: negative (minus mode); detection: 261 nm. Please notice that under these conditions only negatively charged PTH-amino acids will be detected. Therefore, no signals are obtained from the standard PTH-amino acids. Taken from [Meyer 1993b] with permission.

In 1962 McGregor & Carpenter described a procedure to cleave the α-ketoacyl group by treatment with hydrogen peroxide, a reaction first discovered by Bunton in 1949. We modified this method for our purpose [Meyer 1994] to liberate the N-terminus of Pep5, a lantibiotic of known structure, containing an N-α-oxo-butyric acid as an N-terminal blocking group [Kellner 1991]. After covalent attachment of the peptide via its side chain NH_2O-groups of the lysine residues, we first modified the dehydroamino acids by addition of ethanethiol to prevent unwanted cleavage of the peptide at these sites. The modified peptide was then treated with trifluoroperacetic acid in the gas phase for 1 h at room temperature. Traces of the oxidizing reagent have to be removed very carefully to allow Edman degradation later. As a final step, treatment with an alkaline ethanethiol mixture was carried out. After intensive washings with ethanol and water the specimen was placed in the protein sequencer.

As can be seen in Figure 4 the sequence analysis started with the expected alanine in the first position and can be followed to the C-terminal end (only cycles 1–21 are shown here). Thus the N-terminal blocking group was removed successfully.

3.5 Gluco-Arginine

Gluco-arginine is like phospho-tyrosine a very hydrophilic compound and cannot be extracted from the reaction cartridge with ethylacetate usually applied for this purpose.

When we isolated the tryptic ^{14}C-labelled glucopeptide from a selfglucosylating protein from corn (this was a collaborative effort with David Singh and B. Whelan, University of Miami, USA and J.W. Metzger, University of Tübingen, Germany) we found the following sequence by Edman degradation: E G A N F V X G Y P F S L R. No radioactivity was detected in cycle 7 where the attachment site for the glucose residue was suspected. By performing amino acid analysis with a 20 pmol aliquot we found the following composition: D(1.0), E(1.0), S(1.2), G(2.8), R(2), A(1.2), P (n.d.), Y(1.1), V(1.0), L(1.5). Thus we suspected that the unidentified residue X might be a modified arginine.

Electrospray mass spectrometry performed with 150 pmol of the purified peptide revealed a molecular mass of 889.0 Da for the double protonated molecular ion, confirming the presence of the suspected gluco-arginine residue. MS/MS analysis by collision-induced dissociation (Figure 5) shows the stripping of the glucose residue giving an additional signal at m/z 807.5. The mass difference is exactly the expected value for the glucose residue. In this way the gluco-arginine residue was discovered as a new post-translationally modified amino acid in the self-glucosylating protein from corn.

This is an example of the combination of different techniques – Edman degradation, amino acid analysis and mass spectrometry – in elucidating the structure of an unknown peptide.

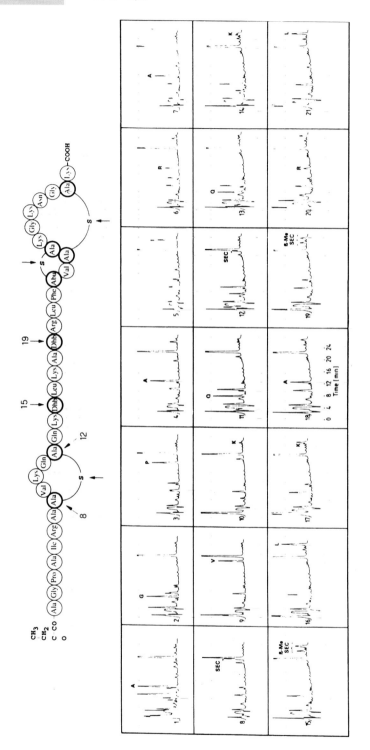

Figure 4. Solid-phase sequence analysis of modified Pep5. The peptide (4 nmol) was covalently attached to an activated DITC membrane. Pep5 was derivatized with alkaline ethanethiol reagent and then deblocked with gaseous trifluoroperacetic acid for 1 h. A second treatment with alkaline ethanethiol was used to replace the sulphons formed during the oxidation step by the thioethers. The known structure of Pep5 is shown at the top of the figure [Kellner 1991]. The chromatograms 1–21 of the on-line PTH-amino acid analyses are shown. Amino acids are designated with the one-letter code. DSER denotes the PTH-dithiothreitol adduct of dehydroalanine. PTH-β-methyl-S-ethyl-cysteine (β-Me-SEC) is clearly visible as an signal doublet in cycles 15 and 19. PTH-S-ethyl-cysteine (SEC) can be seen in cycles 8 and 12, all marked by arrows.

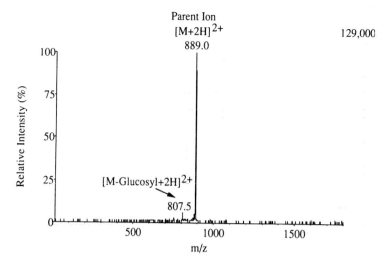

Figure 5. Daughter ion spectrum of the gluco-arginine-containing peptide of corn starch. The mass spectrum was established by collisional activation of the $(M+2H)^{2+}$ ion (m/z 889). About 50 pmol of the peptide was consumed to establish the tandem mass spectrometric experiment. The mass spectrum was recorded on a Sciex (Toronto) API III triple-quadrupole mass spectrometer with 2400 Da mass range equipped with an ion spray source. The mass spectrometer was operated under unit-mass resolution conditions. Ion spray voltage was 5 kV. All other details are as given in [Heilmeyer1992].

3.6 Farnesyl-Cysteine

This kind of post-translational modification was first discovered in fungal and yeast mating factors [Anderegg 1988]. A C-terminal CAAX-motif (C is cysteine, A are aliphatic amino acids and X is any uncharged amino acid) is the recognition signal for a protein polyisoprenyltransferase.

After we had completed the determination of the primary structure of the α and β subunit of the phosphorylase kinase from rabbit skeletal muscle, we identified the two C-terminal Lys-C peptides of both subunits as post-translationally modified peptides [Heilmeyer 1992]. Since both carry the motif for farnesylation we suspected that they might be farnesylated at the single cysteine in the fourth last position. Sequence analysis of these peptides revealed a gap in both cases and is most probably due to the very hydrophobic structure of the farnesyl-cysteine residue, preventing the emergence of the PTH-derivative during the usual time of chromatography.

Therefore, we applied mass spectroscopic techniques in this case, too. Figure 6 shows at the top the ion spray mass spectrum of the C-terminal Lys-C peptide of the α subunit, performed with 100 pmol of the purified peptide. The double protonated molecule yields the most intense signal and its m/z value of 1092.5 demonstrates the presence of a peptide with a molecular mass of 2184 Da. This is 204 Da higher than the calculated mass for the unmodified peptide. However, the residual mass of a farnesyl residue is 204 Da. To prove the location of this farnesyl group at the cysteine residue we performed tandem MS/MS spectrometry with the double charged

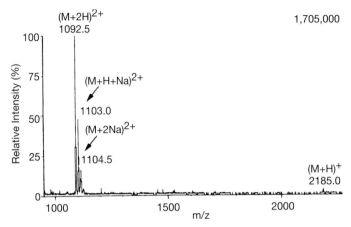

Figure 6. (a) Ion spray mass spectrum of the C-terminal peptide from the α subunit. An aliquot of the purified C-terminal Lys-C peptide of the a subunit of phosphorylase kinase was flow injected (5 μL/min) into the ion spray mass spectrometer. About.100 pmol of the peptide was employed to establish the spectrum. The total number of ions obtained in the spectrum is given in the upper right corner.

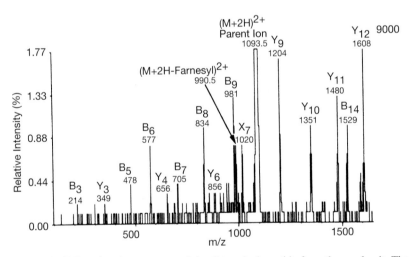

Figure 6. (b) Daughter ion spectrum of the C-terminal peptide from the α subunit. The spectrum was obtained by collisional activation. Approximately 50 pmol of the peptide was employed to carry out the tandem MS/MS experiment of the $(M+2H)^{2+}$ = 1092.5 ion as described in [Heilmeyer 1992]. N- or C-terminal fragments, formed during collision-induced dissociation, are designated A, B, C and X, Y, Z, respectively: two dashes represent addition of two hydrogen atoms. The main fragment species formed belong to the B and Y" series reflecting splitting of the peptide bond. Taken from [Heilmeyer 1992] with permission.

molecular ion. Figure 6b shows the mass spectrum of this experiment performed with 50 pmol of the peptide. There are two fragments Y3 and Y4 showing the presence of a peptide fragment comprising the last three or last four amino acids. The mass difference (656.0–349.0) of 307 Da demonstrates that the cysteine carries the farnesyl group.

3.7 Phospho-Serine

Phospho-serine is one of those amino acid derivatives which are stable during peptide purification, but which will be destroyed during Edman degradation [Meyer 1991b]. However, as shown in Figure 7, phospho-serine yields a characteristic PTH derivative detectable in the reversed-phase separation of the PTH amino acids. The detailed mechanism by which the product, PTH derivative of the dithiothreitol adduct of dehydroalanine, is formed during Edman degradation is shown in [Meyer 1988].

On the left of Figure 7 there is the phospho-peptide analyzed by Edman degradation and on the right the dephosphoform of the same peptide. Very characteristically, in the case of phospho-serine the PTH-derivative of the dithiothreitol-adduct of dehydroalanine is formed exclusively, whereas an unmodified serine yields a mixture of this derivative and PTH serine.

3.8 Phospho-Threonine

As an alternative, phospho-serine and phospho-threonine can be chemically modified to give stable derivatives which can be detected during sequence analysis by their specific derivatives [Meyer 1993a]. Figure 8 shows, as an example, the sequence analysis of a phospho-threonine-containing peptide after transformation of the phospho-threonine to β-methyl-S-ethyl-cysteine. This reaction proceeds quantitatively treating the phospho-peptide with an alkaline ethanethiol reagent at 50 °C for 1 h. The details of this procedure are given in [Meyer 1993].

As can be seen in Figure 8, the phospho-threonine in position 5 of the peptide is transformed to β-methyl-S-ethyl-cysteine which elutes from the reversed phase column as its PTH derivative. The signal doublet is to be expected, since β-methyl-S-ethyl-cysteine contains two asymmetric carbon atoms and we get a racemic mixture of the D/L forms in both positions. Thus, what we see is the separation of the diastereomeric forms of two pairwise enantiomers.

3.9 Screening for Phospho-Serine/Threonine Containing Peptides by HPLC/MS

A powerful new technique in searching for phosphopeptides is HPLC coupled with an electrospray mass spectrometer which permits the direct identification of phosphopeptides containing phospho-serine or phospho-threonine [Meyer 1993b].

Figure 9 gives an example of this strategy. Purified insertin, an actin-inserting domain of the protein tensin from chicken gizzard muscle [Weigt 1992], is fragmented by tryptic digestion. We separated these fragments by reversed phase HPLC connected on-line to an electrospray triple quadrupole mass spectrometer. The mass spectrometer works like a highly sophisticated detector, sampling a complete mass spec-

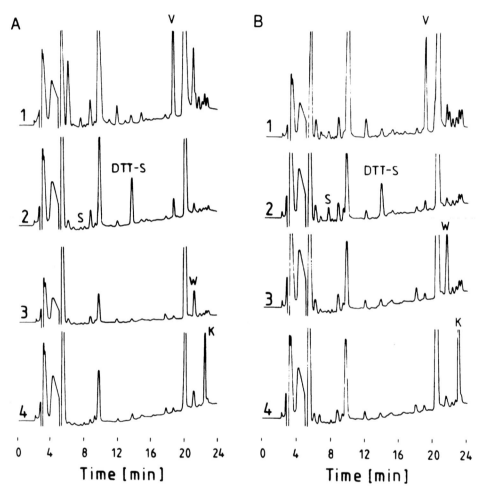

Figure 7. Sequence analyses of the phospho- and dephosphopeptide Eβ2 from the β-subunit of phosphorylase kinase without S-ethyl-cysteine modification. In *A*, about 100 pmol of the phosphopeptide Eβ2 (E refers to glutamic acid residue number 2 of the whole β subunit sequence) is applied onto the gas-phase sequencer and sequenced without S-ethyl-cysteine modification. The PTH chromatograms of degradation steps 1 to 4 are shown. The chromatogram of cycle 2 demonstrates that the phosphoserine present in this position is quantitatively transformed to the dithiothreitol adduct of dehydroalanine (DTT-S). In *B*, 60 pmol of the same peptide in the nonphosphorylated form is analyzed. A clear difference between the phosphoserine in *A* and the serine in *B* can be seen. This demonstrates that phosphoserine can be located unambiguously in such a case without S-ethyl-cysteine modification in the gas-phase sequencer. Taken from [Meyer 1991b] with permission.

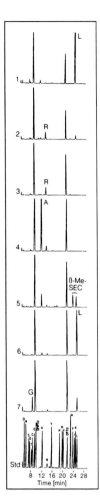

Figure 8. Sequence analysis of the phosphopeptide LRRATpLG after modification with ethanethiol. 2 nmol of the peptide was used for the modification and was directly applied onto the polybrene treated glass fiber disk of the gas-phase sequencer. Instead of butylchloride, used in the older model 470 gas-phase sequencer, ethylacetate was used as transfer solvent. The chromatograms of the on-line PTH amino acid analyses are shown. PTH-β-methyl-S-ethylcysteine (β-Me-SEC) is clearly visible in cycle 5 as a signal doublet. All other details are as given in [Meyer 1993a].

trum every few seconds. By summing up all the detected ions, a total ion chromatogram can be reconstructed (Figure 9b). All expected and identified tryptic peptides are labelled in the total ion chromatogram (Figure 9b). Two phosphopeptides, marked P1 and P2, exhibit the characteristic behavior of a phosphoserine/threonine-containing peptide. P2 is identical with the phospho-form of the tryptic peptide T27. The calculated molecular mass of P2 is 2592 Da, 80 Da higher than the molecular mass of T27 due to the additional phosphate group. We extracted the doubly protonated molecular ions from the total ion chromatogram (Figure 9b) with the expected m/z values of 1296 and 1257 (Figure 9c) to demonstrate the presence of the phospho- (P2) and dephospho-form of the tryptic peptide T27. In the lower trace of figure 9c (m/z 1257) two signals separated by 0.2 min are visible. The earlier eluting component represents the double protonated molecular ion of the phosphopeptide P2 (m/z 1296 upper trace) which has lost its phosphate group by collision with nitrogen before entering the mass spectrometer. The later eluting component demonstrates the

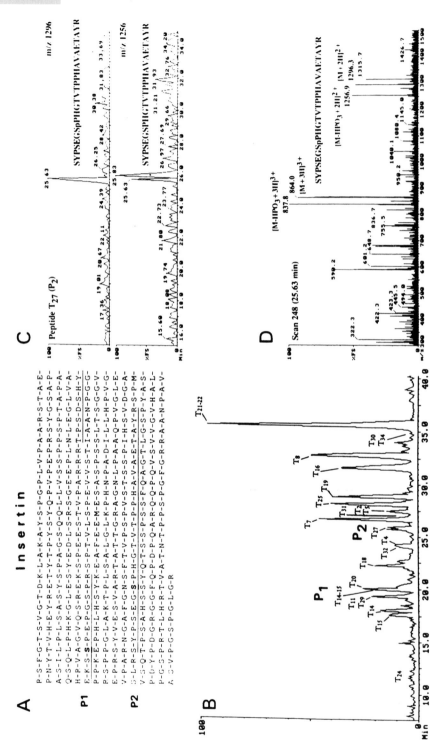

presence of the unphosphorylated peptide T27 which is more hydrophobic and elutes later from the reversed-phase column after its phosphorylated counterpart. The mass spectrum of the phosphorylated peptide P2 (Figure 9d) reveals that the double protonated molecular ion $[M+2H]^{2+}$ has an m/z value of 1296.3 and the triple protonated molecular ion $[M+3H]^{3+}$ has an m/z value of 864.0. The molecular mass calculated from these m/z values is 2592 Da, which corresponds to the tryptic peptide T27 carrying one additional phosphate group (+80 Da). This phosphopeptide looses some of the phosphate moiety during electrospray mass spectrometry due to collision with the drying gas before entering the mass spectrometer, generating additional molecular ions with m/z values of 1257.0 and 837.8, respectively, which are also present. This indicates the neutral loss of the phosphate group which is very characteristic for phosphoserine- and phosphothreonine-containing peptides. Using this technique the identification of phosphopeptides is already completed during peptide separation and in some cases the direct sequencing by tandem mass spectrometry allows the localization of the phosphoamino acid.

However, in most cases electrospray mass spectrometry or tandem mass spectrometry demand assistant protein chemical techniques to localize the phosphoamino acid in the primary structure of a phosphopeptide as shown before.

3.10 Lanthionine, 3-Methyl-Lanthionine, Dehydroalanine, Dehydro-α-aminobutyric Acid

As shown in Figure 4, the lantibiotic peptides contain in addition to N-α-ketoacyl groups some unusual amino acids like lanthionine, 3-methyl-lanthionine, dehydroalanine, and dehydro-α-aminobutyric acid.

When applying Edman degradation directly to those peptides, lanthionine and 3-methyl-lanthionine do not allow a normal course of sequencing, presumably due to the tension and difficult accessibility of the ring structures formed by these amino acids. A heavy drop and high chemical lag occur during sequence analysis. Furthermore, sequencing finally stops at dehydroalanine or dehydro-α-aminobutyric acid [Kellner 1988].

Figure 9. On-line HPLC electrospray analysis of a tryptic digest of the phosphoprotein insertin. *A* Amino acid sequence of insertin. Phosphopeptides P1 and P2 are underlined. Identified phosphoserine residues are printed in boldface. *B* Total ion current chromatogram of separated peptides from the insertin molecule. The peptides identified are labelled with their tryptic fragment number. Identified phosphopeptides are marked P1 and P2, respectively. *C* Single ion chromatogram of the doubly protonated ion of peptide T27(P2) in phosphorylated form (*upper trace*) and dephospho-form (*lower trace*) extracted from the total ion current chromatogram *B*. Notice the signal doublet in the lower trace. The m/z 1256 ion detected at 25.63 min is due to the neutral loss of the phosphate group from the phosphorylated peptide. *D* Single scan spectrum of the phosphopeptide P2. The spectrum is taken at 25.63 min the point of the highest signal intensity for this ion. Notice both the presence of the phosphorylated form (m/z 1296.3 and 864.0, respectively) and of the dephosphorylated form (m/z 1256.9 and 837.8, respectively) of the double and triple protonated molecular ions which is due to the neutral loss of the phosphate group from the phosphorylated peptide. Taken from [Meyer 1993b] with permission.

As shown in Figure 4, all these difficulties are overcome by alkaline ethanethiol treatment, trifluoroperacetic acid oxidation followed by a second treatment with alkaline ethanethiol. Lanthionine and 3-methyl-lanthionine are thereby transformed to a mixture of cysteine/S-ethylcysteine or cysteine/β-methyl-S-ethylcysteine, respectively. This way, Edman degradation takes its normal course and most of the primary structure of these lantibiotics can be elucidated in a single sequence analysis [Meyer 1994].

4 References

Anderegg, R.J., Betz R., Carr, S.A., Crabb, J.W., Duntze W. (1988) *J. Biol. Chem. 263*, 18236–18240. Structure of *Saccharomyces cerevisiae* mating hormone α-factor.

Bunton, C. A. (1949) *Nature 163*, 444. Oxidation of α-Diketones and α-Keto-Acids by Hydrogen Peroxide.

Edman, P. (1949) *Arch. Biochem. 22*, 475–476. A Method for the Determination of the Amino Acid Sequence in Peptides.

Gooley, A.A., Classon, B.J.,Marschalek, R., Williams, K.L. (1991) *Biochem. Biophys. Res.Comm. 178*, 1194–1201. Glycosylation sites identified by detection of glycosylated amino acids released from Edman degradation: The identification of Xaa-Pro-Xaa-Xaa as a motif for Thr-O-glycosylation.

Heilmeyer, Jr., L.M.G., Serwe, M., Weber, C., Metzger, J., Hoffmann-Posorske, E., Meyer, H.E. (1992) *Proc. Natl. Acad. Sci. USA 89*, 9554–9558. Farnesylcysteine, a constituent of the α and β subunits of rabbit skeletal muscle phosphorylase kinase: Localization by conversion to S-ethylcysteine and by tandem mass spectrometry.

Kellner, R., Jung, G., Hörner, T., Zähner, H., Schnell, N., Entian, K.D., Götz, F. (1988) *Eur. J. Biochem. 177*, 53–59. Gallidermin, a new lanthionine containing polypeptide antibiotic.

Kellner, R., Jung, G., Sahl, H.G. (1991) Structure elucidation of the tricyclic lantibiotic Pep5 containing eight positively charged amino acids. In: *Nisin and novel lantibiotics* (Jung, G., Sahl, H.G.; ed.), pp. 141–158, Escom, Leiden.

Krishna, R. G., Wold, F. (1992) Post-translational modifications: unique amino acids in proteins. In: *Frontiers and New Horizons in Amino Acids Research* (Takai, K.; ed.), Elsevier, Amsterdam.

Lehr, S., Kotzka, J., Klein, E., Siethoff, Ch., Knebel, B., Tennagels, N., Affüpper, M., Klein, H.W., Brüning, J.C. Meyer, H.E., Krone, W., and Müller-Wieland, D. (1998) *Biochemistry*, in press. Human Gab-1 Protein: Identification of Tyrosine Phosphorylation Sites by EGF Receptor Kinase *in vitro*.

McGregor, W.H., Carpenter, F.H. (1962) *Biochemistry 1*, 53–60. Alkaline Bromine Oxidation of Amino Acids and Peptides: Formation of α-Ketoacyl Peptides and their Cleavage by Hydrogen Peroxide.

Medzihradszky, K.F., Phillipps, N.J., Senderowicz, L., Wang, P., Turck, C.W. (1997) *Protein Sci. 6*, 1405–1411. Synthesis and characterization of histidine-phosphorylated peptides.

Meyer, H.E., Mayr, G.W. (1987) *Biol. Chem. Hoppe-Seyler 368*, 1607–1611. N_π-Methylhistidine in Myosin-Light-Chain-Kinase.

Meyer, H.E., Hoffmann-Posorske E. Kuhn C.C., Heilmeyer, Jr., L.M.G. (1988) Microcharacterization of Phosphoserine Containing Proteins. Localization of the Autophosphorylation Sites of Skeletal Muscle Phosphorylase Kinase. In: *Modern Methods in Protein Chemistry*, Vol. 3 (Tschesche, H.; ed.), pp. 185–212, de Gruyter, Berlin.

Meyer, H.E., Hoffmann-Posorske, E., Donella-Deana, A., Korte, H. (1991a) Sequence Analysis of Phosphotyrosine-Containing Peptides. In: *Methods in Enzymoloy, Vol. 201 Protein Phosphorylation* (T. Hunter, B. M. Sefton; eds.), pp. 206–224, Academic Press, San Diego.

Meyer, H.E., Hoffmann-Posorske E., Heilmeyer, Jr., L.M.G. (1991b) Determination and Location of Phosphoserine in Proteins and Peptides by Conversion to S-Ethyl-Cysteine. In: *Methods in Enzymology, Vol. 201 Protein Phosphorylation* (T. Hunter and B. M. Sefton eds.), pp. 169–185, Acadaemic Press, San Diego.

Meyer, H.E., Eisermann, B., Donella-Deana, A., Perich, J.W., Hoffmann-Posorske, E., Korte, H. (1993a) *Protein Sequences & Data Analysis 5,* 197–200. Sequence analysis of phosphothreonine-containing peptides by modification to β-methyl-S-ethylcysteine.

Meyer, H.E., Eisermann, B., Heber, M., Hoffmann-Posorske, E., Korte, H., Weigt, C., Wegner, A., Donella-Deana, A., Perich, J.W., (1993b) *FASEB J. 7,* 776–782. Strategies for non-radioactive methods in the localization of phosphorylated amino acids in proteins.

Meyer, H.E., Heber, M., Eisermann, B., Korte, H., Metzger, J.W., Jung, G. (1994) *Anal. Biochem.,* submitted. Sequence analysis of lantibiotics: novel derivatization procedures allow a fast access to complete Edman degradation.

Stöcker, G., Meyer, H.E., Wagener, C., Greiling H. (1990) *Biochemical J. 274,* 415–420. Purification and N-terminal amino acid sequence of a chondroitin sulphate/dermatan sulphate proteoglycan isolated from intima/media preparations of human aorta.

Tsunasawa S., Hirano H. (1993) Deblocking and subsequent microsequence analysis of N-terminally blocked proteins immobilized on PVDF membrane. In: *Methods in Protein Sequence Analysis* (Imahori, K., Sakiyama, F.; eds.), pp. 45–53, Plenum Press, New York.

Weigt, C., Gaertner, A., Wegner, A., Korte H., Meyer, H.E. (1992) *J. Mol. Biol. 227,* 593–595. Occurrence of an actin-inserting domain in tensin.

Weigt, C., Korte, H., von Strandmann, R.P., Hengstenberg, W., Meyer, H.E. (1995) *J. Chromatogr. A, 712,* 141–147. Identification of phosphocysteine by electrospray mass spectrometry combined with Edman degradation.

Analysis of Biopolymers by Matrix-Assisted Laser Desorption/ Ionization (MALDI) Mass Spectrometry*

Ute Bahr, Michael Karas and Franz Hillenkamp

1 Introduction

During the two past decades, important achievements in bioorganic mass spectrometry have been made by the development of new ionization techniques for the analysis of biopolymers such as proteins and carbohydrates. Traditional mass spectrometric methods, which proved so useful for analyzing compounds with low molecular masses, were of little use for measuring underivatized compounds with high molecular masses. The general problem to be solved was to convert the polar, nonvolatile biopolymer macromolecules into intact, isolated ionized molecules in the gas phase.

The so-called desorption/ionization techniques use different physical approaches for the conversion; field desorption [Beckey 1977] applies a high electric field to the sample; in fast atom bombardment [Barber 1981] and ^{252}Cf plasma desorption [Macfarlane 1976] the sample is bombarded by highly energetic ions or atoms; thermospray ionization [Blakely 1980] and electrospray ionization [Fenn 1990] form ions directly from small, charged liquid droplets. Laser desorption [Hillenkamp 1992; Cotter 1987] and the newly developed version of this method, matrix-assisted laser desorption/ionization (MALDI) [Hillenkamp 1991] make use of short, intense pulses of laser light to induce the formation of intact gaseous ions.

Electrospray ionization, as well as matrix-assisted laser desorption/ionization have already demonstrated their capabilities for mass spectrometric analysis of biopolymers in the molecular mass range between thousand and a few hundred thousand Daltons. In this article, the authors will provide an overview on MALDI mass spectrometry, particularly its principles, instrumentation and application to biopolymer analysis.

* Reprint from Fresenius, J. Anal. Chem. (1994) 348,783-791. With the kind permission of Springer-Verlag, Berlin, Heidelberg, New York.

2 Development of MALDI

First attempts to use laser light as mass spectrometric ionization method for organic molecules date back to the 1970s [Vastola 1970; Posthumus 1978; Stoll 1979; Cotter 1981; Hardin 1981]. Among the variety of lasers tested for laser desorption two types of lasers proved to be successful: CO_2-lasers with a wavelength of 10,6 μm in the infrared (IR) and lasers emitting in the far ultraviolet (UV). With both laser types an efficient and controllable energy transfer to the sample by resonance absorption of the sample at the irradiation wavelength is possible, in the IR by excitation of rovibrational states, in the UV by electronic excitation. To avoid thermal decomposition of labile organic molecules lasers with pulse widths on the nanosecond timescale, such as Q-swiched Nd-YAG, excimer or TEA-CO_2-laser are used to transfer the energy within a very short time. The pulsed desorption of ions favours the combination of the laser desorption ion source with a time-of-flight (TOF) mass analyzer [Hillenkamp 1975; Van Breemen 1983] or a Fourier transform ion cyclotron resonance (FT-ICR) mass analyzer [Weller 1990; Koster 1992], both make it possible to record complete mass spectra for each laser shot. All early experiments on laser desorption of organic ions were restricted to the analysis of molecular masses below 2000 Da. Results obtained in UV-laser desorption revealed a "soft" desorption of molecular ions only from highly absorbing molecules. The limitation in mass range was believed to result from energy transfer also into photodissociation channels. Samples which cannot be resonantly excited at the laser wavelength used need very high irradiances (laser power/area, W/cm^2) for ion production which inevitably destroy large organic molecules. These limitations, which prohibited a general use of laser desorption for organic mass spectrometry, promoted the search for a matrix enabling the analysis of high molecular mass biopolymers [Karas 1987]. The high mass molecules are finely dispersed in a matrix, consisting of small highly absorbing species. In this way they can be desorbed and ionized irrespective of their individual absorption characteristics [Tanaka 1988; Karas 1988, 1989a].

2.1 Mechanisms of Matrix-Assisted Laser Desorption/Ionization

The matrix is believed to serve three major functions:

1. Absorption of energy from the laser light. The matrix molecules absorb the energy from the laser light and transfer it into excitation energy of the solid system. Thereby an instantaneous phase transition of a small volume (some molecular layers) of the sample to gaseous species is induced. In this way the analyte molecules are desorbed together with matrix molecules, with limited internal excitation. Different models for desorption are discussed in the literature [Vertes 1990; Johnson 1991].

2. Isolation of the biomolecules from each other. The biomolecules are incorporated in a large excess of matrix molecules, strong intermolecular forces are thereby reduced (matrix isolation). Incorporation of analyte into matrix crystals taking place upon evaporation of the solvent forms an essential prerequisite for successful MALDI analysis, it moreover provides an in-situ cleaning of the sample and is the reason for a high tolerance against contaminants.

3. Ionization of the biomolecules. An active role of the matrix in the ionization of the analyte molecules by photoexcitation or photoionization of matrix molecules, followed by proton transfer to the anlyte molecules is likely, though not proven unequivocally to date. A lot of work was already done and is still going on to find substances useful as matrices for MALDI analysis. Depending on the used laser wavelength, the solubility of the analyte and the class of compounds, different matrix compounds or mixtures of matrix compounds are used. A list of the most commonly used substances is given in Table 1.

Table 1. Most commonly used matrices for MALDI MS.

Matrix	Wavelength [nm]	Comments
2,5-Dihydroxy benzoic acid (DHB)	337, 355	
DHB + 10 % 5-methoxy salicylic acid	337, 355	used for masses > 20 000 Da
Sinapinic acid	337, 355	
α-Cyano-4-hydroxycinnamic acid	337, 355	used for "surface preparation"
Nicotinic acid	266	
3-Hydroxy picolinic acid	337, 355	used for oligonucleotides
Succinic acid	2.94, 10.6 μm	
Glycerol	2.94, 10.6 μm	liquid matrix

3 Instrumentation

3.1 Time-of-Flight (TOF) Mass Spectrometers

MALDI of large molecules is usually coupled to TOF mass analysis, although several applications have been performed on FT-ICR [Buchanan 1993; Castoro 1993], magnetic sector analyzers [Hill 1991; Annan 1992], and with quadrupole ion trap mass spectrometer [Schwartz 1993; Chambers 1993; Jonscher 1993]. In TOF analyzers the mass to charge ratio of an ion is determined by measuring their flight time. After acceleration of the ions in the ion source to a fixed kinetic energy, they pass a field free drift tube with a velocity proportional to $(m_i/z_i)^{-1/2}$ (m_i/z_i is the mass-to-charge ratio of a particular ion species). Due to their mass-dependent velocities, ions are separated during their flight. A detector at the end of the flight tube produces a signal for each ion species. Typical flight times are between a few microseconds and several 100 μs. Figure 1a shows a diagram of a linear TOF mass spectrometer. Acceleration voltages are typically 1–30 kV and the flight path lengths range from 0.5–3 m.

a.

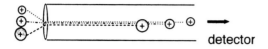

detector

b.

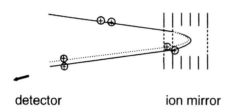

detector   ion mirror

Figure 1. Schematic diagram of time-of-flight mass analyzers (TOF). a. Linear TOF: ions are separated according to their mass dependant velocities. b. Reflectron TOF: the initial velocity distribution of ions of same mass are largely corrected.

One problem in TOF mass analysis results from the energy distribution of the ions due to the desorption/ionization process. This initial energy spread leads to a peak broadening at the detector. Thereby the mass resolution m/Δm, which is a measure of an instruments capability to produce separate signals from ions of similar mass and thus a measure of the performance of the instrument is limited. The peak broadening can be reduced by using a reflectron (ion mirror) TOF mass analyzer as shown in Figure 1b. The ions are decelerated in the reflectron and turn around at different locations in the reflecting electric potential gradient, thus ions of higher kinetic energy spend a longer time in the ion mirror. If the geometry and the voltages of the reflectron are arranged properly, the arrival time spread can be largely corrected for at the plane of the detector and increased mass resolution is obtained. For reflectron TOF mass analyzers in combination with MALDI ion sources resolutions of up to 6000 (hwfm; half width full maximum) have been reported, whereas linear TOF analyzer are limited to about 500 mass resolution. A drastic enhancement of mass resolution is achieved by "delayed ion extraction" [Colby 1994; Brown 1995; Vestal 1995; Lennon 1995]. The ions are produced in a field-free region or a weak electrical field and after a mass-dependent delay time extracted by application of a high voltage pulse. This technique enhances mass resolution in linear as well as in reflector instruments. Depending on the length of the analyzer mass resolutions between 4000 for angiotensin (m/z 1293) in a 1.2 m linear TOF and 10000–15000 for 1–6 kDa peptides in a reflector TOF (6.4 m) can be reached routinely [Vestal 1995; Edmondson 1996]. Figure 2 shows the mass spectrum of a peptide mixture with a mass resolution of more than 13000 for all peptides. Using a long delay time and a weak extraction field strength also for high mass ions the resolution is enhanced to about 900 for proteins of about 50000 Da [Bahr 1997].

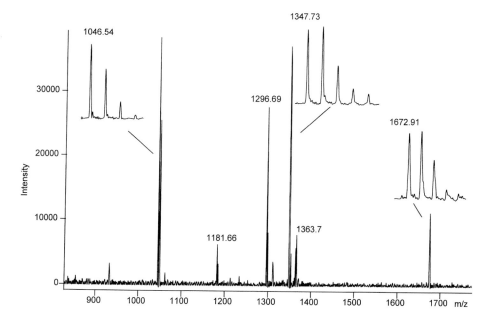

Figure 2. High resolution MALDI spectrum of a peptide mixture obtained with a reflectron TOF mass spectrometer with delayed extraction (Voyager STR, with kind permission of Perseptive Biosystems, Wiesbaden, Germany).

3.2 Laser Desorption Ion Source

Laser ion sources in use for MALDI may differ in some technical details, but all comprise some common features [Feigl 1983; Beavis 1989a; Salehpour 1989; Spengler 1990]. They all use pulsed lasers with pulse durations of 1–200 ns. Most commonly used are nitrogen N_2-lasers emitting at 337 nm or Nd-YAG lasers, whose emission wavelength of 1064 nm has been transferred to 355 nm or 266 nm by frequency tripling or quadrupling using nonlinear optical crystals. Moreover, excimer lasers (193, 248, 308 and 351 nm), frequency doubled, excimer pumped dye lasers (220–300 nm) have been used for MALDI, as well as dye lasers (wavelength in the visible) and infrared lasers (TEA CO_2: 10,6 μm and Er-YAG: 2,94 μm). For the required irradiances in the range of 10^6 to 10^7 W/cm^2 the laser beams are focussed to values between 30 and 500 μm by suitable optical lenses. The angle of incidence of the laser beam on the sample surface varies between 15–70°. The irradiance at the sample surface is a critical parameter, the minimum (threshold) irradiance to produce ions is well defined and best results are obtained for a laser irradiance no more than ca 20% above the threshold. Thus the intensity of the laser beam on the sample has to be carefully adjusted. This can be done by neutral density filters, angle-dependent reflection attenuaters or polarizers. The position of the laser focus on the sample surface can be changed either by moving the sample towards a fixed laser spot or by steering the laser beam.

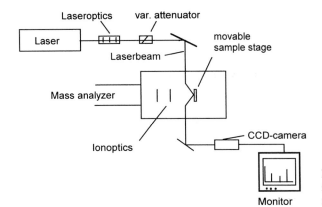

Figure 3. Schematic drawing of the ion source of a laser desorption mass spectrometer.

It should be emphasized, that optical control of the sample is a valuable means to yield optimal MALDI results (see: sample preparation). In Figure 3 the diagram of a laser ion source with movable sample stage and a video-microscope for observation is shown.

3.3 Ion Detection and Data Collection

To detect ions from matrix-assisted laser desorption/ionization secondary electron multiplier are used. The high mass ions produce either electrons or low-mass ions at the conversion dynode of the multiplier. These particles are than used to start the multiplication cascade in an electron multiplier. The surfaces used for the conversion are either copper-beryllium or the lead glass inner surface of a microchannel plate. The yield of secondary electrons and ions from the conversion dynode is a function of the velocity of the ions to be detected. MALDI instruments using low ion acceleration energy need postacceleration of high mass ions in order to compensate for the lower detection efficiency at lower ion velocities. This is achieved by a separated dynode held at a potential of typically 20 kV placed in front of the multiplier. The detector signal is either amplified with a fast linear amplifier or directly digitized by a digital oscilloscope (transient recorder) with a sampling rate of 100×10^6 sample/s. The data are then transferred to a PC for spectrum averaging, mass calibration and storage. Typically 10–50 spectra, each from a single laser shot, are summed to improve the signal-to-noise ratio and to allow for an accurate mass determination.

4 Applications

4.1 Sample Preparation

The preparation of samples for MALDI analysis is quite simple and fast. A 5–10 g/L solution of the matrix material is prepared in either pure water or a mixture of water and organic solvent (acetonitrile, ethanol); a mixture of water acidified by trifluoro-acetic acid (0.1 %) and acetonitrile (2:1) is a well-suited solvent. 10^{-5}–10^{-7} M solutions of the analyte are prepared in the same solvent as the matrix. Small amounts of both solutions (between 0,5–10 µL) are then mixed together on the metal sample support (usually stainless steel) to give a final analyte concentration of 0.005–0.05 µg/µL in the mixture. The solvent is evaporated and the sample tranferred to the vacuum chamber of the mass spectrometer.

Depending on the matrix used finely dispersed small crystallites or extended crystalline areas at the rim of the droplet can be observed under microscopic inspection. The most intense ion signals are usually associated with these well-developed crystalline regions of the sample [Strupat 1991; Beavis 1993], proving that incorporation of the analyte into the crystalline matrix is essential.

Another approach for sample preparation is described by Vorm [Vorm 1994a]: matrix solution is applied to the target in a solvent that evaporates very fast and leads to the formation of a thin layer of microcrystals of matrix. Most frequently α-cyano-4-hydroxycinnamic acid solved in acetone is used for this "surface preparation" procedure. A small volume of analyte solved in a polar solvent is than placed on top of the matrix surface and the solvent is allowed to evaporate slowly. Inpurities, like salts, can be washed away from the surface and a better sensitivity, reproducibility and mass accuracy can be achieved.

The time needed for one analysis is only some minutes, including sample preparation and transfer to the mass spectrometer. Additionally, some minutes are needed for mass calibration.

4.2 Ion Fragmentation

Metastable fragmentation of ions can occur and strongly depend on the kind and mass of analyte, the matrix and instrumental parameters as field strength in the acceleration region [Spengler 1995; Karas 1996] or composition and pressure of the residual gas [Spengler 1992]. In linear TOF instruments fragments from metastable decay appear at the detector at the same time than the intact ions, in reflector instruments most of them are either sorted out by the reflector or they appear at lower flight times in the spectrum leading to a broadening of the peaks [Karas 1995; Karas 1996]. Particularly oligonucleotides and nucleic acids as well as several carbohydrate moieties, e.g. sialic acids and fucoses in glycoproteins show a high tendency towards fragmentation [Nordhoff 1992; Karas 1995]. These substances should be either analyzed in a linear TOF or very "soft" matrices should be applied like 3-hydroxy picolinic acid.

Metastable fragmentation can, however, be used for structure analysis by a method called "post source decay" (PSD) analysis which is described in detail in Chapter III.5 in this book.

4.3 Molecular Weight Determination of Proteins and Glycoproteins

In MALDI mass spectra generally the most intense signal is the singly charged molecular ion. Additionally, doubly and triply charged molecular ions as well as singly and multiply charged cluster ions appear. Figure 4 shows a typical MALDI spectrum from carbonic anhydrase (bovine) with a molecular weight of 29024 Da.

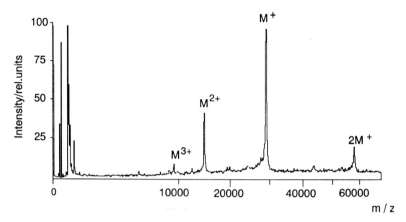

Figure 4. MALDI spectrum of carbonic anhydrase with a molecular weight of 29 024 Da. DHB was used as matrix, the laser wavelength was 337 nm.

In the low mass range ($<$ 500 Da) ion signals from the matrix are to be seen. The base peak is the singly charged molecular ion, with lower intensity the doubly and triply charged molecular ions as well as the singly charged dimer is registered. The relative abundances of these ions in the spectrum depend on matrix and the concentration of analyte used.

Proteins are detected as the protonated species for positive ions and deprotonated ones for negative ions. Both positive and negative ion spectra can usually be obtained at comparable intensities. Fragment ions due to the loss of small neutral molecules such as H_2O, NH_3 or HCOOH from protonated molecular ions are of low intensity. Fragmentation which can be assigned to the cleavage of covalent bonds in the protein backbone is not prominent under standard conditions.

Hundreds of different proteins have been measured in the mean time by MALDI mass spectrometry. No limitations of the application caused by primary, secondary, or tertiary structure has yet been discovered. Proteins with different solution-phase properties, including proteins that are insoluble in ordinary aqueous solutions and glycoproteins that contain large proportions of carbohydrate, can be analyzed. Figure 5a shows a MALDI spectrum of a glycoprotein, native Penicillium lipase. The spectrum shows the singly and doubly protonated molecule. Comparison of the peak width of molecular ion peaks from this substance and that from the peptide cytochrome C (CC, added to the sample as calibrant) shows that the glycoprotein peak is extremely broad because of the heterogeneity in the carbohydrate part of the molecule. In this mass range the mass resolution of TOF instruments in not sufficient to resolve individual glycoprotein components. However, the width of the peak gives a measure for the heterogeneity of the compound. Two approaches are practicable to determine the carbohydrate content of a glycoprotein. If the amino acid sequence is known, the difference between the calculated mass of the peptide backbone and the average mass of the molecule determined from the centroid of the peak gives the average mass of the carbohydrate attached to the protein. The second possibility is the treatment of the compound with glycosidases, enzymes which chemically cleave the carbohydrates,

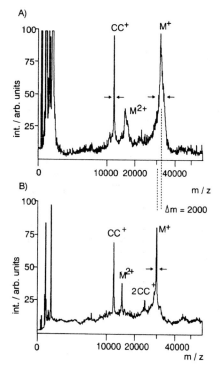

Figure 5. MALDI spectra of Penicillium Lipase. a. Spectrum of the native lipase. Cytochrome C (CC) is added as mass calibrant. b. Spectrum of the declycosylated (Endoglycosidase H) lipase. From the mass shift of 2000 Da the carbohydrate content of the native substance can be calculated.

and subsequent measurement of the molecular mass. In Figure 5b the mass spectrum of the deglycosylated (with endoglycosidase H) lipase is shown. The molecular ion peak has considerably shifted to smaller mass. From the mass difference of 2000 Da a carbohydrate content of 6.3 % can be calculated. A detailed description is given in [Hedrich 1993].

High-mass molecules such as monoclonal antibodies (MoAb) with molecular masses up to 150 000 Da can easily be analyzed by MALDI [Siegel 1991]. Figure 6 shows a schematic drawing of an IgG monoclonal antibody. It has a Y-shaped form consisting of two identical heavy chains (HC) with masses of about 50 000 Da and two identical light chains (LC) with masses of about 24 000 Da. The heavy chains contain a small amount of carbohydrate. MoAb are currently investigated for the targeting of anticancer drugs or imaging reagents to tumor sides. The imaging reagents (e.g. radioactive metal ions) bind to chelating agents which are chemically bound either to lysine or the carbohydrate of the MoAb. They have typical masses between 300 -1500 Da and MALDI has been used to determine the molecular mass of the pure and conjugated MoAb as well as the average mass of the carbohydrate moiety and the average number of the chelator molecules. Figure 7 shows the MALDI spectra of pure Chimeric B MoAb (a), the deglycosylated molecule (b) and the conjugated with DTPA (diethylene triamine-penta-acetic acid, mw 375.3 Da) [Siegel 1993].

From the mass differences determined an average carbohydrate content of 3 574 Da was found and a loading value of 5.9 DTPA.

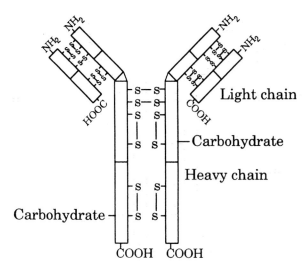

Figure 6. Schematic structure of a monoclonal antibody of IgG class.

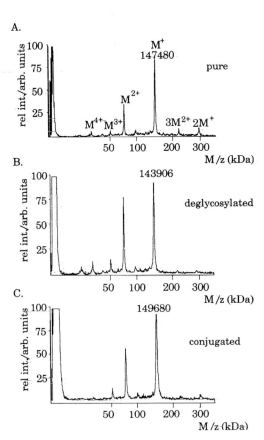

Figure 7. MALDI spectra of the monoclonal antibody chimeric B 72.3 ([Hedrich 1993] with permission of Finnigan MAT, Bremen). a. Native substance; b. after cleavage of the carbohydrate; c. conjugated with DTPA.

One important feature of the MALDI technique is the high tolerance against inorganic and organic contaminants. Some buffers and salts normally used in biochemical procedures do not have to be removed before analysis. Using sinapinic acid as matrix inorganic salts and denaturation agents like urea and guanidine hydrochloride up to 2 mM in the protein solution and buffering agents like citrate, glycine, hepes, tris, ammonium bicarbonate and ammonium acetate up to 200 mM do not strongly affect the analysis [Beavis 1989, 1990a]. With DHB as matrix up to 10 % sodium dodecyl sulfate (SDS) in the protein solution is tolerable [Strupat 1991].

4.3.1 Accuracy of Mass Determination

For mass determination the mass scale has to be calibrated. This is usually done with well-defined reference compounds either internally by adding them to the analyte-matrix mixture or externally by measuring their masses in a separate analysis, preferentially from a second spot on a support enabling the loading of multiple samples. A mass accuracy of 0.01 % (one part in ten thousand) for proteins up to 30 000 Da can be obtained [Beavis 1989b, 1990b]. Applying "delayed extraction" a mass accuracy of peptides with molecular weights of 1–4 kDa is 10–15 ppm by using external calibration and better than 5 ppm by using internal calibration [Edmondson 1996; Russel 1997]. Decrease of mass accuracy in the higher mass range is often due to non-resolved adducts between analyte and matrix. Heterogeneity of the protein under investigation, if not resolvable into individual signals, is another reason. Here the use of delayed extraction is helpful to improve mass resolution and thus mass accuracy. For carbonic anhydrase (m/z 29025 Da) a mass accuracy of 0.5 Da was reached with internal calibration [Bahr 1997].

4.3.2 Sensitivity and Mass Range

The outstanding sensitivity achieved under standard preparation conditions is a strength of the MALDI technique. Typically, about 1 pmol of analyte is loaded for analysis. 1–10 fmol of protein have been shown to suffice for analysis [Strupat 1991; Karas 1989b]. Using the "surface preparation" procedure described above spectra from sub-femtomole amounts can be obtained [Vorm 1994a]. The amount of material consumed for the analysis is much smaller than the total amount loaded onto the sample support, nearly the entire sample can be recovered afterwards. Another approach to reduce the sample amount for mass analysis is the use of picoliter vials [Jespersen 1994]. These are etched depressions on chip surfaces which are filled with 250 pL of protein and matrix solution. Peptide spectra from sample amounts in the low attomole range are detected. The amount of material consumed for the analysis is much less than the total amount loaded onto the sample support, nearly the entire sample can be recovered after analysis. The mass range accessible for MALDI using a time-of-flight instrument extends to more than 300 000 Da. The largest functional protein entity detected so far is the trimer of urease at 272 000 Da [Hillenkamp 1989].

4.4 Analysis of Oligonucleotides

MALDI has been used to ionize a variety of oligoribonucleotides and oligo(deoxy)-ribonucleotides [Hillenkamp 1990; Parr 1992; Nordhoff 1992; Keough 1993; Tang 1993; Wu 1993; Pieles 1993]. The largest nucleic acids measured are a 500mer DNA [Tang 1994] and a 461mer RNA [Kirpekar 1994] with masses around 150 000 Da. A comprehensive review is given by Nordhoff [Nordhoff 1996].

Oligo(deoxy)ribonucleotides yield stronger signals and better mass resolution for negative ions than for positive ions. Only for higher molecular mass tRNAs and rRNAs the signal intensity is higher for the positive ions. New matrices were required to yield high-quality DNA- and RNA-MALDI results, escpecially 3-hydroxy picolinic acid proved to be very useful [Wu 1993]. Figure 8 shows the UV-MALDI spectrum of a 18mer DNA using this matrix showing the protonated and sodium-attached molecular ions. Good results have also been obtained using an infrared wavelength (2.94 μm) and succinic acid as matrix [Nordhoff 1993]. In MALDI mass spectrometry nucleic acids are prone to form metal attached molecular ions, mainly from alkali salts present in the solution or from metal contaminants at the sample holder. Multiple salt formation leads to a broad distribution of pseudomolecular ions. Figure 9a shows the negative IR-MALDI spectrum of d5′-CGC GAT ATC GCG-3′ containing 20 mM KBr in the analyte solution.

Deprotonated, potassium attached molecular ions $[M-(n+1)H+nK]^-$ are formed with n ranging from 0–10. Additionally small signals from sodium- instead of potassium-attachment appear. In this mass range the signals are well resolved but with increasing mass of the oligonucleotide, this effect results in a broad unresolved peak and the molecular weight can not be determined correctly. This can be avoided by

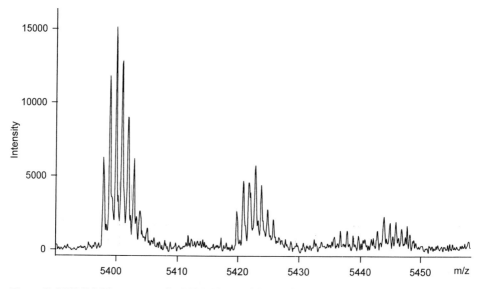

Figure 8. UV-MALDI spectrum of a DNA 18mer with 3-hydroxy picolinic acid as matrix (Voyager STR, with kind permission of Perspective Biosystems, Wiesbaden, Germany).

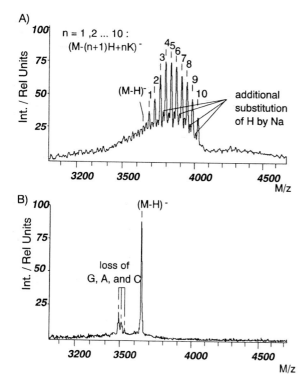

Figure 9. Negative IR-MALDI spectra of d5'-CGC GAT ATC GCG-3' with succinic acid as matrix. a. The analyte solution contains 20 mM KBr. The spectrum shows multiple potassium-attached molecular ion peaks. b. Spectrum after exchangment of alkali cations against NH_4^+. Only the deprotonated molecular ion peak remains.

exchanging the metal cations against ammonium ions [Nordhoff 1992]. A very efficient and simple way to do this is to add some suitable cation exchange polymer beads to the sample solution or directly to the sample droplet on the support. Figure 9b shows the spectrum of the same sample after this procedure. Only the deprotonated molecular ions (beside some small fragments) are to be seen. No negative influence on the desorption process could be observed.

4.5 Analysis of Glycans and Glycoconjugates

Successful ionization of underivatized oligosaccharides by MALDI was demonstrated using different matrices [Mock 1991; Stahl 1991; Harvey 1993]. Spectra of oligomeric distributions of dextrins and dextrans with molecular weights up to 15 000 Da have been obtained. Native and permethylated glycosphingolipids [Egge 1991] and gangliosides [Juhasz 1992] have been analyzed with detection limits of about 10 fmol. Figure 10 shows MALDI spectra of a native glycosphingolipid (GSL_{nat}, rabbit erythrocytes) (a) and from a mixture of permethylated glycosphingolipids (GLS_{perm}) (b). The general structure of these compounds is shown in Figure 10c. All glycans show mainly sodium or potassium-attached molecular ions. The sample amounts required and the accuracy of mass determination are in the same order as for peptides and proteins. Larger oligosaccharides appear to be more difficult to ionize

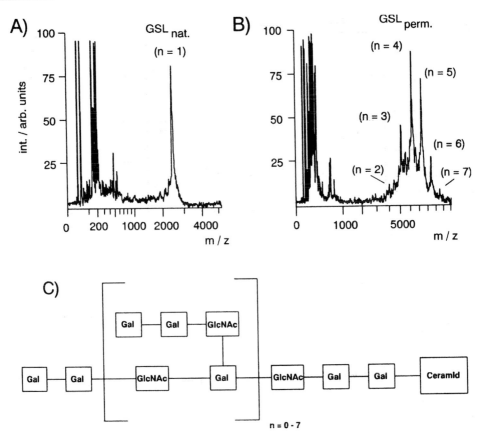

Figure 10. a. MALDI spectrum of a native glycophigolipid (rabbit erythrocytes). b. MALDI spectrum of a mixture of permethylated glycoshingolipids (rabbit erythrocytes). c. General structure of GSL of rabbit erythrocytes.

than those with lower molecular weight, the mass range has, as yet been extended to only about 15 000 Da. Future work is necessary to exploit the potential for MALDI of these compounds.

5 Combination of MALDI with Biochemical Methods

5.1 Peptide Mapping of Digested Proteins by MALDI

MALDI mass analysis of peptide mixtures produced by enzymatic or chemical diges-tion of proteins is a strategy for elucidation of protein structure. The absence of frag-ment ions and the dominance of singly protonated molecular ions as well as the high sensitivity, the accuracy of mass determination and the tolerance against impurities and contaminants make MALDI very well-suited for fast analysis of peptide mixtures and the interpretation of the spectra straightforward. MALDI analysis of chemical or enzymatic digests or of hydrolysis products have been reported [Schär 1991; Caprioli 1991; Billeci 1993; Thiede 1995; Vorm 1994c]. Using different matrices of matrix mixtures all peptides in a mixture can be detected. A new strategy of protein identification has potential to become the method of choice: protein fragments are analyzed by mass spectrometry and the protein is identified by searching the masses against a protein sequence data base [Mann 1993; Griffin 1995; Fenyö 1996]. Often a few peptide masses are sufficient to uniquely identify the protein.

5.2 Combination of MALDI and Gel Electrophoresis

Sodium dodecyl sulfate-polyacrylamide gel electrophoresis (SDS-PAGE) and high-resolution two-dimensional polyacrylamide electrophoresis (2-DE) are widely used methods for separation of small quantities of proteins. For further protein chemical analysis like sequencing or amino acid composition analysis the most rapid and efficient method is the electrotransfer of these proteins onto a suitable membrane (electroblot). For accurate mass determination of these proteins a MALDI analysis directly from the blot membrane has been performed [Eckerskorn 1992]. Gel electrophoretically separated proteins were electroblotted onto polyvinylidene difluoride (PVDF) or polyamide membranes. Pieces of membrane containing the protein of interest were soaked in matrix solution and transferred to the mass spectrometer to be analyzed by MALDI. Figure 11 shows a MALDI spectrum of trypsin inhibitor with a molecular mass of 19 978 Da blotted onto a PVDF membrane.

Best results were obtained using the infrared wavelength from an Er-Yag-Laser. The spectrum has nearly the same quality than MALDI spectra from a normal preparation. Staining of proteins on the gel can sometimes lead to peak broading because of multiple attachment of dye molecules to the analyte, but staining by colloidal pigments was found to be compatible to a subsequent MALDI analysis [Strupat 1994].

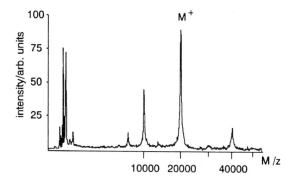

Figure 11. IR-MALDI spectrum of trypsin inhibitor obtained directly from a PVDF blot membrane. Succinic acid was used as matrix.

5.3 Combination of MALDI with Capillary Zone Electrophoresis

Mass spectrometry can be used as ideal detector for capillary zone electrophoresis (CZE) [Keough 1992; Van Veelen 1993; Castors 1992]. Proteins in the sub-pmol range were collected in 1–2 µL volumes in microcentrifuge tubes and than mixed with matrix solution and transferred to the sample probe of the mass spectrometer [Keough 1992]. Another approach has been reported where the effluent from the capillary is deposited on a moving sample stage together with a sheath flow of matrix solution [Van Veelen 1993]. Different peptides and proteins with concentrations in the lower femtomole range could be analyzed.

6 References

Annan, R. S., Köchling, H. J., Hill, J. A., Biemann, K. (1992) *Rapid Commun Mass Spectrom* 6: 298–302. Matrix-assisted Laser Desorption using a Fast Atom Bombardment Ion Source with a Magnetic Mass Spectrometer.

Bahr, U., Stahl-Zeng, J., Gleitsmann, E., Karas, M. (1997) *J Mass Spectrom* 32: 1111. Delayed Extraction Time-of flight MAKDI Mass Spectrometry of Proteins above 25 000 da.

Barber, M., Bordoli, R. S., Sedgwick, R. S., Tyler, A. N. (1981) *J Chem Soc Chem Commun, p* 325. Fast Atom Bombardment of Solids (FAB): A new Source for Mass Spectrometry.

Beavis, R. C., Chait, B. T. (1989a) *Rapid Commun Mass Spectrom* 3: 233. Factors effecting the Ultraviolet Laser Desorption of Proteins.

Beavis, R. C., Chait, B. T. (1989b) *Rapid Commun Mass Spectrom* 3: 436–439. Matrix assisted Laser Desorption Mass Spectrometry using 355 nm Radiation.

Beavis, R. C., Chait, B. T. (1990a) *Proc Natl Acad Sci USA* 87: 6873. Rapid, Sensitive Analysis of Protein Mixtures by Mass Spectrometry.

Beavis, R. C., Chait, B. T. (1990b) *Anal Chem* 62: 1836. High Accuracy Molecular Mass Determination of Proteins Using Matrix-assisted Laser Desorption Mass Spectrometry.

Beavis, R. C., Bridson, J. N. (1993) *J Phys D App Phys* 26, 442. Epitaxial Protein Inclusionin Sinapinic Acid Cristals.

Beckey, H. D. (1977) *Principles of field desorption mass spectrometry.* Pergamon Press, Oxford.

Billeci, T. M., Stults, J. T. (1993) *Anal Chem* 65: 1709–1716. Tryptic Mapping of Recombinant Proteins by Matrix-assisted Laser Desorption/Ionization Mass Spectrometry.

Blakely, C. R., Carmody, J. J., Vestal, M. L. (1980) *Anal Chem* 52: 1636.

Brown, R. S., Lennon, J. J. (1995) *Anal Chem* 67, 1998–2003. Mass Resolution Improvement by Incorporation of Pulsed Ion Extraction in a Matrix-Assisted Laser Desorption/ionization Linear Time-of-Floght Mass Spectrometer.

Buchanan, M. V., Hettich, R. L. (1993) *Anal Chem* 65: 245A–259A. Fourier Transform Mass Spectrometry of High Mass Biomolecules.

Caprioli, R. M., Whaley, B., Mock, K. K., Cotrel,l J. S. (1991) In: Villafranca, J. J. (Ed) *Techniques in protein chemistry II,* Academic Press, San Diego pp, 479–510. Sequence-ordered Peptide Mapping by Time-xourse Analysis of Protease Digests Using Laser Desorption Mass Spectrometry.

Castoro, J. A., Chiu, R. W., Monnig, C. A., Wilkins, C. L. (1992) *J Am Chem Soc* 114: 7571–7572. Matrix-assisted Laser Desorption-Ionization of Capillary Electrophoresis Effluents by Fourier Transform Mass Spectrometry.

Castoro, J. A., Köster, C., Wilkins, C. L. (1993) *Anal Chem* 65: 784–788. Investigation of a "screened" Electrostatic Ion Trap for Analysis of High-mass Molecules by Fourier Transform Mass Spectrometry.

Chambers, D. M., Goeringer, D. E., McLuckey, S. A., Glish, G. L. (1993) *Anal Chem* 65: 14–20. Matrix-assisted Laser Desorption of Biological Molecules in the Quadrupole Ion Trap Mass Spectrometer.

Colby, S. M., King, T. B., Reilly, J. P. (1994) *Rapid Commun Mass Spectrom* 8: 865. Improving the Resolution of Matrix-assisted Laser Desorption/Ionization Time-of-flight Mass Spectrometry by Exploiting the Correlation between Ion Position and Velocity.

Cotter, R. J., Yergey, A. L. (1981) *Anal Chem* 53: 1306–1307.

Cotter, J. R. (1987) *Anal Chim Acta* 195: 45–59. Laser Mass Spectrometry: An Overview of Techniques, Instruments and Application.

Eckerskorn, C., Strupat, K., Karas, M., Hillenkamp, F., Lottspeich, F. (1992) *Electrophoresis* 13: 664–665. Mass Spectrometric Analysis of Blotted Proteins after Gel-Electrophoretic Separation by Matrix-assisted Laser Desorption Ionization Mass Spectrometry.

Edmondson, R. D., Russell, D. H. (1996) *J Am Soc Mass Spectrom* 7: 995.

Egge, H., Peter-Katalinic, J., Karas, M., Stahl, B. (1991) *Pure Appl Chem* 63: 491–498. The Use of Fast Atom Bombardment and Laser Desorption Mass Spectrometry in the Analysis of Complex Carbohydrates.

Feigl, P., Schueler, B., Hillenkamp, F. (1983) *Int J Mass Spectrom Ion Phys* 47: 15. LAMMA 1000: A new Instrument for Bulk Microprobe Mass Analysis by Pulse Laser Irradiance.

Fenn, J. B., Mann, M., Meng, C. K., Wong, S. F., Whitehouse, C. M. (1990) *Mass Spectrom Rev* 9, 37–70. Electrospray Ionization – Principles and Practice.

Hardin, E. D., Vestal, M. L. (1981) *Anal Chem* 53: 1492–1497.

Harvey, D. J. (1993) *Rapid Commun Mass Spectrom* 7, 614–619. Quantitative Aspects of the Matrix-assisted Laser Desorption Mass Spectrometry of complex Oligosaccharides.

Hedrich, H. C., Isobe, K., Stahl, B., Nokihara, K., Kordel, M., Schmid, R. D., Karas, M., Hillenkamp, F., Spener, F. (1993) *Anal Biochem* 211: 288–292. Matrix-assisted Ultroviolet Laser Desorption/Ionization Mass Spectrometry Applied to Multiple Forms of Lipases.

Hill, J. A., Annan, R. S., Biemann, K. (1991) *Rapid Commun Mass Spectrom* 5: 395–399. Matrix-assisted Laser Desorption Ionization with a Magnetic Mass Spectrometer.

Hillenkamp, F., Ehring, H. (1992) In: Gross, M. L. (ed) *Mass spectrometry in the biological sciences: a tutorial.* Kluwer Academic Publishers, Dordrecht, Boston, London. Pp 165–179. Laser Desorption Mass Spectrometry. Part I – Basic Mechanisms and Techniques.

Hillenkamp, F., Karas, M. (1989) *Proceedings of the 37th Annual Conf. of the ASMS,* Miami Beach, May 21–26, pp. 1168–1169. Ultraviolet Laser Desorption/Ionization of Biomolecules in the High Mass Range.

Hillenkamp, F., Karas, M., Beavis, R. C., Chait, B. T. (1991) *Anal Chem* 63: 1196A–1203A. Matrix-Assisted Laser Desorption/Ionization Mass Spectrometry.

Hillenkamp, F., Karas, M., Ingendoh, A., Stahl, B. (1990) In: Burlingame, A. L., McCloskey, J. A. (eds) *Biological mass Spectrometry,* Elsevier, Amsterdam, pp 49–60. Matrix-Assisted Laser Desorption/Ionization: A new Approach to Mass Spectrometry of large Biomolecules.

Hillenkamp, F., Unsöld, E., Kaufmann, R., Nitsche, R. (1975) *Appl Phys* 8: 341–348. A High Sensitivity Laser Microprobe Mass Analyzer.

Jespersen, S., Niessen, W. M., Tjaden, U. R., van der Greef, J. (1994) *Rapid Commun Mass Spectrom* 8: 581. Attomole Detection of Proteins by Matrix-assisted Laser Desorption/Ionization Mass Spectrometry with the Use of Picolitre Vials.

Johnson, R. S., Martin, S. A., Biemann, K. (1988) *Int J Mass Spectrom* 86: 137.

Johnson, R. E., Banerjee, S., Hedin, A., Fenyo, D., Sundquist, B. U. R. (1991) In: Standing KG, Ens W (eds) *Methods and mechanisms for proceeding ions from large molecules.* Plenum Press, New York.

Jonscher, K., Currie, G., McCormack, A. L., Yates, J. R. (1993) *Rapid Commun Mass Spectrom* 7: 20–26. Matrix-assisted Laser Desorption of Peptides and Proteins on a Quadrupole Ion Trap Mass Spectrometer.

Juhasz, P., Costello, C. E. (1992) *J Am Soc Mass Spectrom* 3: 785–796. Matrix-Assisted Laser Desorption Ionization Time-of-Flight Mass Spectrometry of underivatized and permethylated Gangliosides.

Karas, M., Bachmann, D., Bahr, U., Hillenkamp, F. (1987) *Int J Mass Spectrom Ion Proc* 78: 53–68. Matrix-assisted Ultraviolet Laser Desorption of non-volatile Compounds.

Karas, M., Hillenkamp, F. (1988) *Anal Chem* 60: 2299–2301. Laser Desorption/Ionization of Proteins with Molecular Masses Exceeding 10 000 Da.

Karas, M., Bahr, U., Hillenkamp, F. (1989a) *Int J Mass Spectrom Ion Processes* 92: 231–242. UV-Laser Matrix Desorption/Ionization Mass Spectrometry of Proteins in the 100 000 Da Range.

Karas, M., Ingendoh, A., Bahr, U., Hillenkamp, F. (1989b) *Biomed Environm Mass Spectrom* 18: 841–843. Ultraviolet Laser Desorption/Ionization Mass Spectrometry of Femtomolar Amounts of Large Proteins.

Karas, M., Bahr, U., Ehring, H., Strupat, K., Hillenkamp, F. (1994) *Proc, 42th ASMS Conf. Mass Spectrom.*, Chicago, p. 7. What makes a Matrix Work for UV-MALDI Mass Spectrometry?

Karas, M., Bahr, U., Strupat, K., Hillenkamp, F., Tsarbopoulos, A., Pramanik, B. N. (1995) *Anal Chem* 67: 675. Matrix Dependence of metastable Fragmentation of Glycoproteins in MALDI TOF Mass Spectrometry.

Karas, M., Bahr, U., Stahl-Zeng, J. (1996) in: *Large Ions: Their Vaporization, Detection and Structural Analysis,* Bear, T., Ng, Cy., Powis, I. (Eds) Wiley and Sons Ldt, pp 27–47.

Keough, T., Takigiku, R., Lacey, M. P., Purdon, M. (1992) *Anal Chem* 64: 1594–1600. Matrix-assisted Laser Desorption Mass Spectrometry of Proteins isolated by Capillary Zone Electrophoresis.

Keough, T., Baker, T. R., Dobsen, R. L. M., Lacey, M. P., Riley, T. A., Hasselfield, J. A., Hesselberth, R. E. (1993) *Rapid Commun Mass Spectrom* 7: 195–200. Antisense DNA Oligonucleotides II: Use of Matrix-assisted Laser Desorption/Ionization Mass Spectrometry for the Sequence Verification of Methylphosphonate Oligodeoxyribonucleotides.

Kirpekar, F., Nordhoff, E., Kristiansen, K., Roepsdorff, P., Lezius, A., Hahner, S., Karas, M., Hillenkamp, F. (1994) *Nucl Acid Res.* 22, 3866. Matrix-assisted Laser Desorption/Ionization Mass Spectrometry of Enzymatically Synthesized RNA up to 150 kDa.

Köster, C., Kahr, M. S., Castoro, J. A., Wilkins, C. L. (1992) *Mass Spectrom Reviews* 11: 495–512. Fourier Transform Mass Spectrometry.

Liao, P.-C., Allison, J. (1995) *J Mass Spectrom* 30, 408–423. Ionization processes in Matrix-Assisted Laser Desorption/Ionization Mass Spectrometry: Matrix-dependant Formation of $[M+H]^+$ va $[M+Na]^+$ Ions of Small peptides and some Mechanistic Comments.

Macfarlane, R. D., Torgerson, F. D. (1976) *Science* 109: 920. Californium-252 Plasma Desorption Mass Spectrometry.

Mann, M., Hojrup, P., Roepsdorff, P. (1993) *Biol. Mass Spectrom.* 22, 338–345. Use of Mass Spectrometric Molecular Weight Information to Identify Proteins in Sequence Databases Mass Spectrometric Analysis of Blotted Proteins after Gel Electrophoretic Separation by Matrix-assisted Laser Desorption Ionization Mass Spectrometry.

Mock, K. K., Davey, M., Cotrell, J. S. (1991) *Biochem Biophys Res Commun* 177: 644. The Analysis of underivatised Oligosaccharides by Matrix-assisted Laser Desorption Mass Spectrometry.

Nordhoff, E., Ingendoh, A., Cramer, R., Overberg, A., Stahl, B., Karas, M., Hillenkamp, F., Crain, P. F. (1992) *Rapid Commun Mass Spectrom* 6: 771–776. Matrix-assisted Laser Desorption/Ionization Mass Spectrometry of Nucleic Acids with Wavelengths in the Ultraviolet and Infrared.

Nordhoff, E., Cramer, R., Karas, M., Hillenkamp, F., Kirpekar, F.. Kristiansen, K., Roepsdorff, P. (1993) *Nucl Acid Res* 21: 2247. Ion Stability of Nucleic Acids in Infrared Matrix-assisted Laser Desorption Ionization Mass Spectrometry.

Nordhoff, E., Kirpekar, F., Roepstorff, P. (1996) *Mass Spectrom Reviews* 15: 67. Mass Spectrometry of Nucleic Acids.

Parr, G. R., Fitzgerald, M. C., Smith, L. M. (1992) *Rapid Commun Mass Spectrom* 6: 369. Matrix-assisted Laser Desorption Mass Spectrometry of Synthetic Oligodeoxyribonucleotides.

Pieles, U., Zürcher, W., Schär, M., Moser, H. E. (1993) *Nucl Acid Res* 21 3191. Matrix-assisted Laser Desorption Ionization Time-of-Flight Mass Spectrometry: a Powerful Tool for the Mass and Sequence Analysis of Natural and Modified Oligonucleotides.

Posthumus, M. A., Kistemaker, P. G., Meuzelaar, H. L. C., deBrauw, M. C. (1978) *Anal Chem* 50: 985. Laser Desorption Mass Spectrometry of Polar Non-Volatile Bio-Organic Molecules.

Roepsdorff, P., Fohlmann, J. (1985) *Biomed Mass Spectrom* 12: 631. Proposed Nomenclature for Sequence Ions.

Russell, D. H., Edmondson, R. D. (1997) *J Mass Spectrom* 32: 263. High-resolution Mass Spectrometry and Accurate Mass Measurements with Emphasis on the Characterization of Peptides and Proteins by Matrix-assisted Laser Desorption/Ionization Time-of-flight Mass Spectrometry.

Salehpour, M., Perera, I., Kjellberg, J., Hedin, A., Islamian, M. A., Hakansson, P., Sundquist, B. U. R. (1989) *Rapid Commun Mass Spectrom* 3: 259. Laser-induced Desorption of Proteins.

Schär, M., Börnsen, K. O., Gassmann, E. (1991) *Rapid Commun Mass Spectrom* 5: 319–326. Fast Protein Sequencing Determination with Matrix-assisted Laser Desorption and Ionization Mass Spectrometry.

Schwartz, J. C., Bier, M. E. (1993) *Rapid Commun Mass Spectrom* 7: 27–32. Matrix-assisted Laser Desorption of Peptides and Proteins using a Quadrupole Ion Trap Mass Spectrometer.

Siegel, M. M., Hollander, I. J., Hamann, P. R., James, J. P., Karas, M., Ingendoh, A., Hillenkamp, F. (1991) *Anal Chem* 63: 2470. Matrix-assisted UV-Laser Desorption/Ionization Mass Spectrometric Analysis of Monoclonal Antibodies for the Determination of Carbohydrate, Conjugated Chelator, and Conjugated Drug Content.

Siegel, M. M., Hollander, I. J., Phipps, A., Ingendoh, A. (1993) Finnigan MAT, Vision 2000 Apl. Data by Matrix-assisted Laser Desorption/Ionization Mass Spectrometry Sheet No. 1. Molecular Weight Determination of Pure and Conjugated Monoclonal Antibodies.

Spengler, B., Cotter, R. J. (1990) *Anal Chem* 62: 793–796. Ultraviolet Laser Desorption/Ionization Mass Spectrometry of Proteins above 100 000 Dalton by Pulsed Ion Extraction Time-of-Flight Analysis.

Spengler, B., Kirsch, D., Kaufmann, R. (1991) *Rapid Commun Mass Spectrom* 5: 198. Metastable Decay of Peptides and Proteins in Matrix-assisted Laser Desorption Mass Spectrometry.

Spengler, B., Kirsch, D., Kaufmann, R. (1992) *Rapid Commun Mass Spectrom* 6: 105. Peptide Sequencing by Matrix-assisted Laser Desorption Mass Spectrometry.

Spengler, B., Kirsch, D., Kaufmann, R., Lemoine, J. (1994) *Org Mass Spectrom* 29: 782. Structure Analysis of Branched Oligosaccharides using Post-Source Decay in Matrix-Assisted Laser Desorption Ionization Mass Spectrometry.

Stahl, B., Steup, M., Karas, M., Hillenkamp, F. (1991) *Anal Chem* 63: 1463. Analysis of Neutral Oligosaccharides by Matrix-assisted Laser Desorption/Ionization Mass Spectrometry.

Stahl, B., Zeng, J. R., Karas, M., Hillenkamp, F., Steup, M., Sawatzki, G., Thurl, S. (1994) *Anal Biochem* 223, 218. Oligosaccharides from Human Milk as Revealed by Matrix-Assisted Laser Desorption/Ionization Mass Spectrometry.

Stoll, R., Röllgen, F. W (1979) *Org Mass Spectrom* 14: 642–645.

Strupat, K., Karas, M., Hillenkamp, F. (1991) *Int J Mass Spectrom Ion Process* 111: 89–101. 2,5 Dihydroxybenzoic Acid: A New Matrix for Laser Desorption/Ionization Mass Spectrometry.

Strupat, K., Karas, M., Hillenkamp, F., Eckerskorn, C., Lottspeich, F. (1994) *Anal Chem* 66, 464–470. Laser Desorption/Ionization Mass Spectrometry of Proteins Electroblotted after Polyacrylamide-Gelelectrophoresis.

Tanaka, K., Waki, H., Ido, Y., Akita, S., Yoshida, Y., Yoshida, T. (1988) *Rapid Commun Mass Spectrom* 2: 151–153. Protein and Polymer Analysis up to m/z 100 000 by Laser Ionization Time-of-Flight Mass Spectrometry.

Tang, K., Allman, S. L., Jones, R. B., Chen, C. H., Araghi, S. (1993) *Rapid Commun Mass Spectrom* 7: 63–66. Laser Mass Spectrometry of Polydeoxyribothymidylic Acid Mixtures.

Tang, K., Taranenko, N., Allman, S. L., Chang, L. Y., Chen, C. H. (1994) *Rapid Commun Mass Spectrom* 8, 727. Detection of 500 Nucleotide DNA by Laser Desorption Mass Spectrometry.

Thiede, B., Wittmann-Liebold, B., Biemert, M., Krause, E. (1995) *FEBS Lett* 357: 65. MALDI-MS for C-terminal Sequence Determination of Peptides and Proteins degraded by Carboxypeptidase Y and P.

Van Breemen, R. B., Snow, M., Cotter, R. J. (1983) *Int J Mass Spectrom Ion Phys* 49, 35–50. Time-resolved Laser Desorption Mass Spectrometry. I Desorption of Preformed Ions.

Van Veelen, P. A., Tjaden, U. R., van der Greef, J., Ingendoh, A., Hillenkamp, F. (1993) *Int J Chromatogr* 64: 367–374. Off-line coupling of Capillary Electrophoresis with Matrix-assisted Laser Desorption Mass Spectrometry.

Vastola, F. J., Mumma, R. O., Pirone, A. J. (1970) *J Org Mass Spectrom* 3: 101. Analysis of Organic Salts by Laser Ionization.

Vertes, A., Gijbels, R., Levine, R. D. (1990) *Rapid Commun Mass Spectrom* 4: 228. Homogeneous Bottleneck Model of Matrix-assisted Ultraviolet Laser Desorption of Large Molecules.

Vestal, M. L., Juhasz, P., Martin, S. A. (1995) *Rapid Commun Mass Spectrom* 9, 1044–1050. Delayed Extraction Matrix-Assisted Laser Desorption Time-of-Flight Mass Spectrometry.

Vorm, O., Roepsdorff, P., Mann, M. (1994a) *Anal Chem* 66, 3281. Improved resolution and very High Sensitivity in MALDI-TOF of Matrix Surfaces made by Fast Evaporation.

Vorm, O., Mann, M. (1994b) *Am Soc Mass Spectrom* 5, 955–958. Improved Mass Accuracy in Matrix-Assisted Laser Desorption/Ionization Time-of-flight Mass Spectrometry of Peptides.

Vorm, O., Roepsdorff, P. (1994c) *Biolog Mass Spectrom* 23: 734. Peptide Sequence Information derived by partial acid Hydrolysis and Matrix-assisted Laser desorption/Ionization Mass Spectrometry.

Weller, R. R., MacMahon, T. J., Freiser, B. S. (1990) In: Lubman DM (Ed) *Lasers and Mass Spectrometry* Oxford University, New York, pp 249–270.

Whittal, R. M., Li, L. (1995) *Anal Chem* 67: 1950–1954. High-Resolution Matrix-Assisted Laser Desorption/Ionization in a Linear Time-of Flight Mass Spectrometer.

Wu, K. J., Steding, A., Becker, C. H. (1993) *Rapid Commun Mass Spectrom* 7: 142–146. Matrix-assisted Laser Desorption Time-of-Flight Mass Spectrometry of Oligonucleotides using 3-Hydroxypicolinic acidas an ultraviolet-sensitive Matrix York, pp 89–99.

MALDI Postsource Decay Mass Analysis

Bernhard Spengler, Pierre Chaurand and Peter Bold

1 Introduction

Postsource Decay (PSD) analysis as an extension of MALDI mass spectrometry [Spengler 1992a, Spengler 1992b] became widely accepted for primary structure analysis of all kinds of biomolecules and other substance classes soon after its development [Yu 1993; Huberty 1993; Kaufmann 1993; Critchley 1994; Griffin 1994; Cornish 1994; Kaufmann 1994; Stahl-Zeng 1996; Medzihradszky 1996]. Especially in the field of peptide sequence analysis the high sensitivity of the method, which is in the range of 30 to 100 fmol of prepared sample, as well as the high tolerance of sample impurities and sample inhomogeneity (both properties inherited from the MALDI method) has made PSD analysis an attractive solid-phase alternative to liquid-phase electrospray ionization mass spectrometry [Biemann 1987; Biemann 1992; Wilm 1996a].

Mass spectrometrical methods, namely Electrospray Ionization and MALDI-PSD, now have started to replace classical analytical approaches such as Edman Degradation in certain cases. Their main advantages over automated Edman Degradation are the possibility to analyze chemically modified peptides, the higher sensitivity and the possibility to analyze multi-component samples. MALDI-PSD in comparison with Electrospray Ionization on triple-quadrupole or ion-trap mass spectrometers, offers a generally higher tolerance of sample impurity and, as a solid-phase method, is ideally suited for modern approaches of 2-D separation, 2-D immobilization and 2-D sample handling and storage. Electrospray Ionization, on the other hand, is the method of choice for on-line coupling in liquid-phase analyses. The simplicity of instrumentation necessary for MALDI-PSD as compared to other MS/MS methods as well as the simplicity of operation has made the method a success as a powerful analytical method in many fields of research, such as molecular biology [Machold 1995; Veelaert 1996; Wilm 1996b, Harvey 1996; Moormann 1997; Nilsson 1997; Yamagaki 1997; Schlueter 1997].

2 Methodology and Principle Mechanisms of Postsource Decay

2.1 Internal Energy Uptake and Ion Stability

Postsource decay (PSD) analysis is a method that allows to mass analyze fragment ions which are formed in the field-free region following the ion source of a time-of-flight mass spectrometer. These fragment ions are produced from the decay of MALDI generated ions. Mass determination of these PSD ions is based on using an electrostatical ion reflector as an energy analyzer.

The internal energy necessary for inducing fragmentation can be transferred to the molecular ion by three different mechanisms. The first energy uptake takes place in the solid phase during laser bombardment, mediated by several processes such as direct photon absorption of analyte molecules, solid state activation with phonon or exciton coupling to the analyte, temperature effects or excess energy uptake from ionization. The second, more efficient pathway of energy uptake is due to multiple low- to medium-energy collisions between analyte ions accelerated by the extracting field and neutral matrix molecules in a distance from the sample surface of up to a few hundred micrometers [Bökelmann 1995]. In state-of-the-art PSD instruments which use delayed ion extraction [Brown 1995; Kaufmann 1996] the influence of these in-source collisions on ion activation is strongly reduced since analyte ions are not immediately drawn through the early dense cloud of neutral molecules but only when this cloud has already become considerably rarefied. Among the alternative pathways of ion activation, however, collisions with the remaining neutral molecules during delayed analyte ion acceleration can still be taken as a considerable contribution to internal energy uptake. The third possible mechanism of energy uptake is due to high energy collisions taking place in the first field free region of the spectrometer. These collisions can be the result of controlled (using gas collision cells) or random (due to collisions with residual gas molecules) processes. Table 1 summarizes the three main mechanisms of ion activation in MALDI mass spectrometry.

Table 1. Activation mechanisms in MALDI leading to internal energy uptake and postsource decay.

Mode	1	2	3
activation mechanism	sample activation	in-source activation	post-source activation
activation region	surface/sample	selvedge	vacuum
activation processes	direct photon-molecule interactions, solid state activations, temperature effects, excess energy from ionization	multiple low-to-medium-energy collisions	high-energy collisions
time scale	ps .. ns	ns .. μs	μs .. ms
distance from surface	0	< 200 μm	> acceleration zone

All three ways of energy uptake have the same effect of increasing the probability for subsequent ion decay into structurally informative fragment ions. As a result of the larger number of degrees of freedom biomolecules (peptides, oligosaccharides, oligonucleotides etc.) usually have lower rate constants of fragmentation than smaller molecules. Ion decay thus takes place over a longer time range of several microseconds and is therefore in time-of-flight instruments located predominantly in the field-free region after leaving the ion source.

Using delayed ion extraction for PSD mass analysis has been found to have no considerable effect on reduction of analytical sensitivity even if ion activation and thus postsource decay is reduced. This is because the lower absolute number of fragment ions is compensated by an increased signal-to-noise ratio due to increased mass resolving power and due to reduced background noise [Kaufmann 1996].

In case of insufficient fragmentation which sometimes is observed for very stable molecules, high-energy collisional activation can be employed in order to enhance postsource decay. Using either the complete instrument as a "gas collision cell" (by increasing the residual gas pressure) or using a discrete collision cell permits to induce strong fragmentation of any kind of analyte ion.

2.2 Instrumentation for MALDI-PSD Analysis

The principle setup of a mass spectrometer for MALDI Postsource Decay analysis is shown in Figure 1.

The sample is irradiated by a pulsed ($\sim$ 3 ns) UV laser beam of typically 337 nm wavelength (N_2 laser). Ions formed are accelerated in a two-stage acceleration system, usually employing pulsed (delayed) ion extraction. After leaving the ion source all ions have the same nominal ion kinetic energy. Most of them are still unfragmented precursor molecular ions which have gained internal energy by the mechanisms described above (gas-phase collisions, laser irradiation, thermal mechanisms etc.). During the subsequent flight time (tens of microseconds) through the field-free drift region a certain number of analyte ions undergo postsource decay into product ions. These product ions still have basically the same velocity as their precursor ions. In linear instruments PSD ions are therefore detected at the same time as their precursors and cannot be mass analyzed. Due to their lower mass, however, PSD ions have a much lower kinetic energy than their precursor ions. The kinetic energy of the PSD product ions therefore is a measure of their mass. The ion reflector, in classical time-of-flight instruments used as a device for flight time compensation of initial energy distributions, in PSD instruments is used as an energy filter and thus as a mass analyzer for PSD ions. Due to their mass-dependent kinetic energies PSD ions are reflected at different positions within the reflector (at different equipotential surfaces) leading to mass dependent total flight times trough the instrument (Fig. 1).

An ion reflector as shown in Fig. 1 is capable of analyzing energies (and by that of PSD ion masses) within a certain range only. A complete PSD spectrum is therefore acquired in several steps of reflector potentials in typical instruments. Finally all sections of the PSD spectrum are compiled and mass calibrated by the computer.

Besides two-stage gridded reflectors (Fig. 1) [Spengler 1991], single-stage reflectors (for Secondary Ion Mass Spectrometry described in [Tang 1988]), gridless reflec-

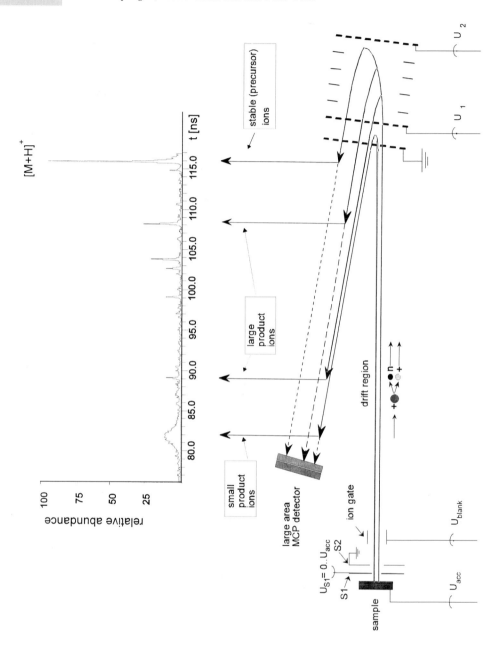

Figure 1. Scheme of a MALDI-PSD mass spectrometer, using a tilted two-stage gridded ion reflector. PSD ions reflected in the second stage of the reflector are observed as well-resolved ion signals in the spectrum. Smaller PSD ions reflected in the first stage of the reflector are detected as a broad unresolved signal. Mass analysis of these ions is done in a series of consecutive steps by lowering the potentials U_1 and U_2 of the reflector grids.

tors [Stahl-Zeng 1996] and non-linear reflectors [Cordero 1995] have been used as alternative approches for PSD analysis. The non-linear, "curved-field" reflectors allows to acquire PSD spectra in a single step but, so far, has not been demonstrated to provide for highest PSD performance with respect to sensitivity, mass revolving power etc.

2.3 Precursor Ion Selection

An important feature of MALDI-PSD time-of-flight mass spectrometry is its MS-MS capability for preselection of a certain precursor ion out of a mixture of multiple components. Precursor ion selection is realized by electrostatic "beam blanking" or "ion gating". All ions passing the beam blanking device (Fig. 1) are electrostatically deflected off the ion detector except for a well-defined mass window (i. e. flight time window) which is transmitted without deflection by turning off the deflection pulse. The ion gate device is typically built of small plates, wires or strips [Vlasak 1996]. The resolution of typical ion gate devices is currently in the range of $m/\Delta m$ = 100 to 200 (Full width at half maximum).

The selectivity of a high performance ion gate at 10 keV ion kinetic energy is demonstrated in Figure 2 showing the selection of a peptide at mass 1270.6 u out of a mixture with another peptide at mass 1281.6 u. As can be seen the signal at mass 1281.6 u is dropped to much less than 10 % of its original intensity while the peptide signal at 1270.6 u is not affected in intensity, mass resolution or flight time. Even the 5th isotope at mass 1274.6 u is still unaffected by the gating pulse. Figure 2 clearly demonstrates that a high performance ion gate is capable of discriminating ion signals of peptides that differ in mass by only a few units.

3 Applications and Spectra Interpretation

The majority of applications employing postsource decay MALDI mass spectrometry has concentrated on peptide characterization and to a similar extent on carbohydrate analysis. Only few investigations of oligonucleotides are reported, driven by a lower analytical demand (for mass spectrometry) rather than by methodological limitations. The sensitivity of the method is in the range of 30 to 100 fmol prepared sample for peptide analysis. Mass resolving power is in the range of 6000 to 10 000 for precursor ions and in the range of 1500 to 3000 for PSD ions. The mass accuracy of PSD ions can be as high as ± 10 to 100 ppm depending on the quality of PSD mass calibration and system stability.

The high sensitivity of MALDI-PSD is in part the result of the long time window available for ion decay between leaving the ion source and entering the ion reflector, which is typically in the range of 50 to 100 μs. One effect of the long time frame is that not only the fastest fragmentation reactions are observed as product ions but slower fragmentation reactions are observed to a similar extent. Another effect is that not only single-step fragmentations are observed but multiple-step fragmentations as well. Internal ions, which contain neither end of the molecule chain, are the result of at least two cleavage reactions. The prerequisite for multiple fragmentation reac-

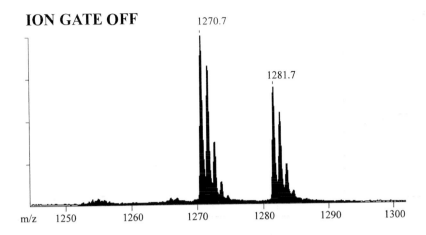

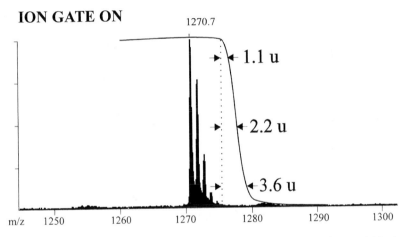

Figure 2. Demonstration of the performance of an ion gate for suppressing a neighboring ion. An estimated rise mass of 2.2 mass units (intensity drop from 90 % to 10 %) is observed corresponding to a selectivity M/ΔM of 512 for a peptide at mass 1270.6 u.

tions to be detected in PSD spectra is not so much the reaction rate constants but more the internal energy required for the reactions to take place. Especially oligosaccharides tend to undergo multiple cleavage reactions [Spengler 1995] but for peptides the formation of internal ions is pronounced as well.

The nomenclature for the typical C-terminal and N-terminal ion types observed from peptides in PSD analysis is described in Figure 3 [Roepstorff 1984; Johnson 1988]. All of these ion types can be accompanied by satellites due to loss of ammonia (–17 u) or water (–18 u). The proposed structures of a-, b- and y-type ions are shown in Figure 4.

Formation of internal ions (not described in Figure 3) is strongly sequence-specific. Peptides containing proline, for example, never fail to form a series of internal fragment ions starting with a proline into the C-terminal direction. This "proline

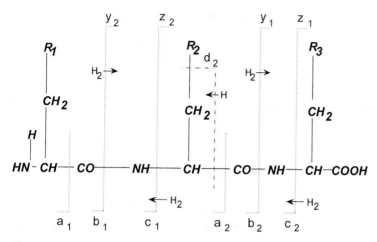

Figure 3. Nomenclature of peptide fragment ions typically observed in postsource decay analysis. Ions of type a, b, c and d are observed as N-terminal ions while those of type y and z are observed as C-terminal ions. All types of fragment ions might be accompanied by satellites due to loss of ammonia (–17 u) or water –18 u).

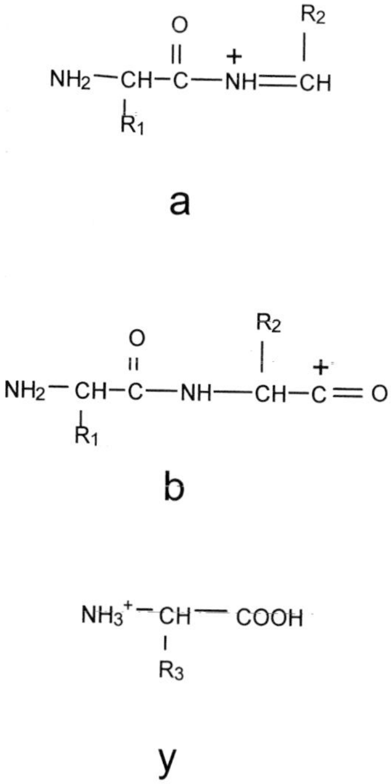

NH_2—CH–C—$\overset{+}{NH}$=CH with O (double bond) above C, R_1 below CH, R_2 above CH

a

NH_2—CH–C—NH—CH–$\overset{+}{C}$=O with O above C, R_1 below first CH, R_2 above second CH

b

NH_3^+—CH—COOH with R_3 below CH

y

Figure 4. Proposed structures of a-, b- and y-type ions.

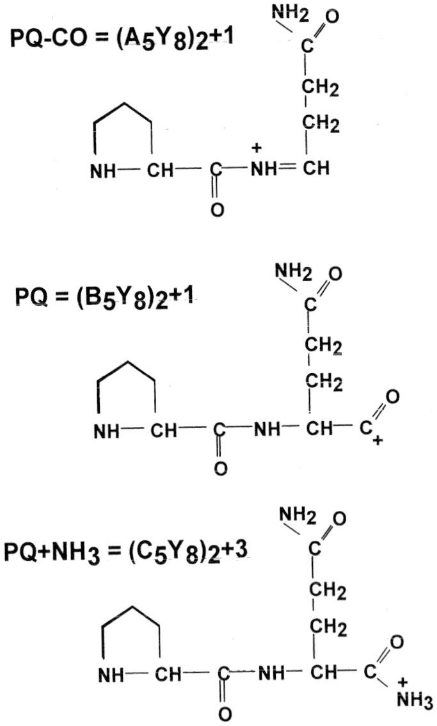

PQ-CO = $(A_5Y_8)_{2}{+1}$

PQ = $(B_5Y_8)_{2}{+1}$

PQ+NH$_3$ = $(C_5Y_8)_{2}{+3}$

Figure 5. Proposed structure of internal fragment ions containing the amino acids P and Q. Their principal structures are analogous to N-terminal ions of type a, b and c.

directed internal fragmentation" contains valuable additional information for confirmation of proposed amino acid sequences. Figure 5 describes the structure and nomenclature of internal fragment ions. These three types of internal fragment ions again can be accompanied by satellites due to loss of ammonia or water.

Internal ions that contain only one amino acid residue are typically observed as immonium ions. These immonium ions are of high analytical value since they indicate the presence or absence of certain amino acids in unknown peptides. The general structures of the most commonly observed immonium ions can be found in [Spengler 1993].

3.1 Sequence Analysis of Peptides

Peptide sequencing by mass spectrometry is a task of pattern interpretation rather than of direct reading of the amino acid sequence from the mass signals. Unequivocal sequence analysis of unknown peptides therefore is not necessarily possible in all cases and additional techniques that provide more structural information are sometimes required to solve the puzzle. A general strategy for sequence analysis of unknown peptides by MALDI-PSD is described in Figure 6, employing on-target derivatization reactions.

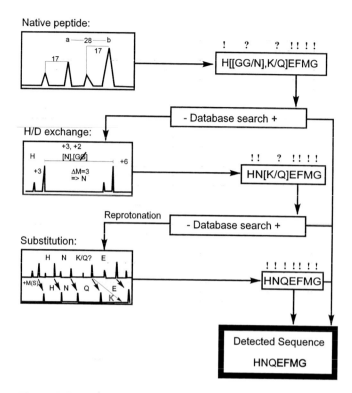

Figure 6. Strategy to determine the complete sequence of unknown peptides. After interpretation of the PSD spectrum of the native peptide by e.g. pattern identification, a database search might already lead to a listed protein and by that to the complete sequence of the peptide. If a database search is unsuccessful hydrogen-deuterium exchange and subsequent PSD analysis helps to further interpret the fragmentation pattern, in this example by differentiating between Asn (N) and Gly-Gly (GG). N-terminal acetylation as an example of on-target derivatization finally help to differentiate between N- and C-terminal ions, to direct the fragmentation pathways or (as here) to differentiate between Lys (K) and Gln (Q).

Certain derivatization methods are able to simplify mass spectral information, such as N-terminal charge-derivatization [Liao 1997; Spengler 1998] and arginine blocking [Spengler 1998]. The easiest and most general approach for obtaining additional structural information in mass spectrometrical peptide sequencing, however, certainly is the method of on-target hydrogen-deuterium exchange [Spengler 1993]. Both, the information about the total number of exchangeable hydrogens of the precursor ion and on that of the characteristic product ions decreases the number of possible interpretations of a PSD spectrum by orders of magnitude in general. Hydrogen-deuterium exchange can be performed with femtomolar amounts of sample on target within a few minutes.

The peptide chosen as an example for sequence interpretation is a human MHC class I restricted peptide [Flad 1998] isolated from a cancer cell line by a multiple step procedure. Before mass spectrometrical analysis the peptide pool of 300 to 3000 different peptides was fractionated by reversed-phase HPLC. The collected frac-

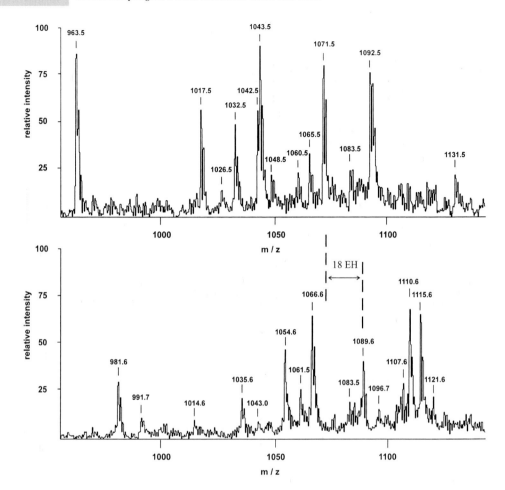

Figure 7. MALDI spectrum of an HPLC fraction from an MHC peptide pool. The number of exchangeable protons (EH) was determined by hydrogen-deuterium exchange. Native sample (top), fully deuterium exchanged sample (bottom).

tions were mass analysed by MALDI TOF-MS and it was observed that each fraction contained from 10 to 50 different MHC class I restricted peptides. The total amount of sample was found to be in the range of 100 femtomole up to 1.5 pmol for each peptide. The mass spectrum of the HPLC fraction sampled after 35.79 min is presented in Figure 7. The peptide detected at mass [M+H]$^+$ = 1071.5 was selected using the ion gate and its PSD fragment ion spectrum was acquired (Fig. 8a).

The procedure to elucidate the sequence of the MHC I peptide at mass [M+H]$^+$ = 1071.5 u is described in the following. As for most MHC class I peptides the expected number of amino acids in the peptide chain is between 8 and 10. The low mass end of the fragment ion spectrum exhibits mass signals clearly indicating the presence of the amino acids V (72 u), K or Q (84 and 129 u), L (86 u), W (159 u) in the peptide. Because of the absence of the usually strong immonium ion signals of H (110 u),

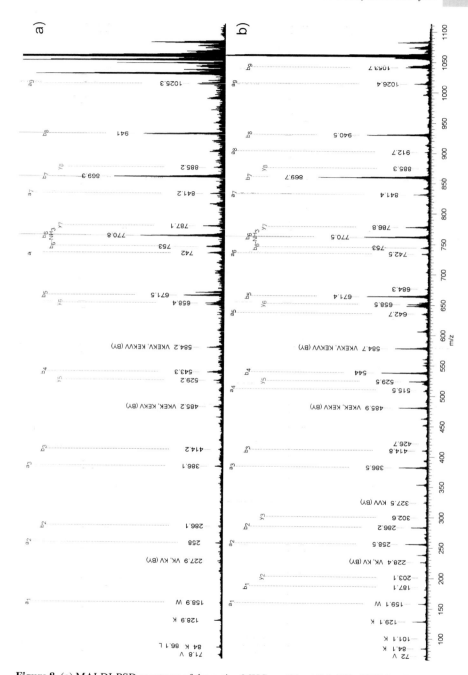

Figure 8. (a) MALDI-PSD spectrum of the native MHC peptide at $[M+H]^+=1071.5$ u. Interpretation of the spectrum and database search led to the sequence WVKEKVVAL. The corresponding fragment ions are labelled in the spectrum. (b) For comparison and for confirmation the peptide WVKEKVVAL was synthesized and PSD analyzed. The high degree of similarity is a strong indicator for the correctness of the sequence proposition.

F (120 u) and Y (136 u) these amino acids can be ruled out as constituents of the peptide sequence. Absence or presence of the other amino acids cannot be decided from the data.

Correspondences between N-terminal and C-terminal ions according to the relation:

$$m_b + m_y = m_{precursor} + 1$$

(m_b, m_y and $m_{precursor}$ being the masses of the b- and y-type fragment and precursor ions respectively) were found for the mass signal pairs at 286.1 u/787.1 u, 414.2 u/ 658.4 u, and 543.3 u/529.2 u. It can be assumed that one of each pairs is a b-type ion and the other is a y-type ion. Neighboring signals resulting from a-type ions or a- or b-type satellites indicating "b" as the correct fragment ion type were found for the mass signals 286.1 u and 414.2 u. With this information 2 positions within the peptide chain can be interpreted forming the sequence tag 286.1[(K/Q)E]543.3. The method of describing partial sequence information as "sequence tags" [Mann 1994] is useful for searching databases as shown later.

Because of their satellites, mass signals at 770.8 u and 869.3 u can also be proposed as being b-type fragment ion. This assumption is favored for 2 reasons. First the mass difference between the 2 mass signals corresponds to the amino acid valine (identified as being a constituent of the sequence) and second the mass difference between the mass signals at 770.8 u and 543.3 u corresponds to the sequences (K/Q)V or V(K/Q). Because of the presence of the ion signal at 671.5 u the first proposition is favored giving the sequence tag 286[(K/Q)E(K/Q)VV]869. This sequence tag is further confirmed by the presence of ion signals at 227.9, 485.2 and 584.2 u which can be attributed as being B/Y internal fragment ions of sequence (K/Q)V, (K/Q)E(K/Q)V and (K/Q)E(K/Q)VV respectively.

With this proposed partial sequence it is useful to question protein databases in order to find the source protein of the investigated peptide. Through database searches only one known protein in agreement with the above sequence tag has been found which is the human natural killer cells protein 4 precursor NK4_HUMAN (SwissProt P24001, MW 26.7 kDa), containing the peptide sequence WVKEKVVAL.

Confirmation of the proposition is at first hand done by comparison of the expected fragmentation pattern with the actual spectrum (Fig. 8a). Unexplained prominent peaks in the PSD spectrum are always a strong hint for misinterpretations. Figure 8a displays the observed spectrum labelled with the expected C-terminal and N-terminal fragment ion descriptions. Besides database searching of uninterpreted data [Griffin 1995; Eng 1994] this "sequence tag approach" [Mann 1994] has been found to be a very powerful method for peptides expected to be derived from database-listed proteins.

A final confirmation of the correctness of a sequence proposition (for a completely unknown or a database-listed peptide) can be obtained by synthesizing the peptide and comparing the PSD spectra of the native and the synthetic peptide. Correct propositions are usually characterized by a very high similarity of the two spectra, both in signal intensities and signal presence even if the homogeneity and purity of the two samples and the instrumental or preparational conditions are rather different. Figure 8b displays the fragment ion spectrum of the synthetic peptide WVKEKVVAL obtained under the same experimental conditions as that of the native sample. It shows a high degree of pattern similarity even in low intensity areas. The validity of the proposition WVKEKVVAL can therefore been taken as approved.

3.2 Primary Structure Analysis of Unknown Peptides

The method of elucidating sequence information as described above is not restricted to database-listed peptides by employing the sequence tag approach. With the same strategy and eventually the help of derivatization methods a considerable part of unknown (not database-listed) peptides can be characterized. This has been done, for example, for a number of MHC peptides which could not be assigned to any data-base-listed protein but could be confirmed by comparison with the synthetic peptides [Flad 1998] and has been described for a completely unknown peptide earlier [Spengler 1997a]. For the case of the peptide WVKEKVVAL it can be assumed that the sequence would have been elucidated without the need of databases as well, except for the principal uncertainties K/Q and I/L. Hydrogen-deuterium exchange of the fraction performed on target with the already analyzed sample gave the additional information that the peptide contains 18 exchangeable hydrogen atoms. Exchangeable hydrogens are all hydrogens linked to non-carbon atoms. With that information and the partial information that was extractable from the PSD spectrum of the hydrogen-deuterium exchanged sample there was not much possibility left for alternative sequence propositions.

3.3 Primary Structure Analysis of Modified Peptides and Other Biomolecules

Postsource decay analysis is not restricted to linear, unmodified peptides. Successful structure analyses have been described for oligonucleotides [Critchley 1994], oligo-saccharides [Harvey 1996; Spengler 1995; Lemoine 1996; Harvey 1995], branched peptides [Kaufmann 1995; Chaurand 1997], conjugates [Huberty 1993] and several other compound classes. Spectra interpretation, however, sometimes requires dedicated strategies which are still to be developed. In the case of characterizing a branched peptide from a crosslinking experiment [Chaurand 1997] the interpretation appeared to be very successful.

4 Potential and Perspectives

Extrapolating the current characteristics and properties of MALDI-PSD on the one hand and the current analytical demands in the biomedical field on the other hand it appears that solid-phase mass spectrometry could obtain a rather prominent position among other modern analytical techniques in the future. Exploring the proteome for example (the set of proteins produced in a cell under the actual biochemical conditions) as one of the next fundamental tasks after deciphering the human genome, will ask for completely new strategies of large scale peptide and protein analysis. The dynamical character of the proteome, in contrast to the static nature of the genome, is especially demanding a very fast analytical screening technology. Sophisticated robotics and computerized data mining and data warehousing will probably be essential in these future approaches. MALDI-PSD is well prepared for supporting new developments into this direction, allowing e. g. the readout of biochemical structure information from two-dimensionally separated samples with lateral resolutions in the submicrometer range. In this sense the offline character of solid-phase mass spectrometry turns out to be a strong analytical advantage over the online approach of liquid-phase ionization methods.

MHC peptide analysis is one example where the feasibility of primary structure elucidation by mass spectrometry has been demonstrated [Hunt 1992; Spengler 1997b], and which is now asking for a logistical solution to answer specific questions addressed to an extremely complex information pool. It is likely that in the future MALDI-PSD will be among the key technologies aiming to answer these questions.

5 References

Biemann, K., Martin, S.A. (1987) *Mass Spectrom. Rev. 6*, 1–76.

Biemann, K. (1992) *Annu. Rev. Biochem. 61*, 977–1010.

Bökelmann, V., Spengler, B., Kaufmann, R. (1995) *Eur. Mass Spectom. 1*, 81–93.

Brown, R. S., Lennon, J.J. (1995) *Anal. Chem. 67*, 1998–2003.

Chaurand, P., Kaufmann, R., Lützenkirchen, F., Spengler, B., Fournier, I., Brunot, A., Tabet, J.-C., Bolbach, G. (1997) *Proceedings of the 45th Annual Conference on Mass Spectrometry and Allied Topics*, Palm Springs, CA, June 1–5.

Cordero, M. M., Cornish, T. J., Cotter, R. J., Lys, I. A. (1995) *Rapid Commun. Mass Spectrom. 9*, 1356–1361.

Cornish, T. J., Cotter, R. J. (1994) *Rapid Commun. Mass Spectrom. 8*, 781–785.

Critchley, W. G., Hoyes, J. B., Malsey, A. E., Wolter, M. A., Engels, J. W., Muth, J. (1994) *Proceedings of the 42nd Annual Conference on Mass Spectrometry and Allied Topics*, Chicago, Illinois, May 29–June 3.

Eng, J. K., McCormack, A. L., Yates III, J. R. (1994) *J. Am. Soc. Mass Spectrom. 5*, 976–989.

Flad, Th., Spengler, B., Kalbacher, H., Brossart, P., Baier, D., Kaufmann, R., Meyer, H. E., Kurz, B., Müller, C. A. (1998) *Canses Res.*, in press.

Griffin, R. R., MacCoss, M J., Eng, J. K., Blevins, R. A., Aaronson, J. S., Yates III, J. R. (1995) *Rapid Comm. Mass Spectrom. 9*, 1546–1551.

Harvey, D. J., David, J., Naven, T. J. P., Kuster, B., Bateman, R. H., Green, M. R., Critchley, G. (1995) *Rapid Commun. Mass Spectrom. 9*, 1556–1561.

Harvey, D. J. (1996) *J. Chromatogr. A 720*, 429–446.

Huberty, M. C., Vath, J. E., Yu, W., Martin, S. A. (1993) *Anal. Chem. 65*, 2791–2800.

Hunt, D. F., Henderson, R. A., Shabanowitz, J., Sakaguchi, K., Michel, H., Sevilir, N., Cox, A. L., Appella, E., Engelhard, V. H. (1992) *Science 255*, 1261–1263.

Johnson, R. S., Martin, S. A., Biemann, K. (1988) *Int. J. Mass Spectrom. Ion Proc. 86*, 137.

Kaufmann, R., Spengler, B., Lützenkirchen, F. (1993) *Rapid Commun. Mass Spectrom. 7*, 902–910.

Kaufmann, R., Kirsch, D., Spengler, B. (1994) *Int. J. Mass Spectrom. Ion Proc. 131*, 355–385.

Kaufmann, R., Kirsch, D., Tourmann, J. L., Machold, J., Hucho, F. (1995) *Eur. Mass Spectrom. 1*, 313–325.

Kaufmann, R., Chaurand, P., Kirsch, D., Spengler, B. (1996) *Rapid Commun. Mass Spectrom. 10*, 1199–1208.

Lemoine, J., Chirat, F., Domon, B. (1996) *J. Mass Spectrom. 31*, 908–912.

Liao, P.-C., Huang, Z.-H., Allison, J. (1997) *J. Am. Soc. Mass Spectrom. 8*, 501–509.

Machold, J., Utkin, Y., Kirsch, D., Kaufmann, R., Tsetlin, V., Hucho, F. (1995) *Proc. Natl. Acad. Sci. USA 92*, 7282–7286.

Mann, M., Wilm, M. (1994) *Anal. Chem. 66*, 4390–4399.

Medzihradszky, K. F., Adams, G. W., Burlingame, A. L., Bateman, R. H., Green, M. R. (1996) *J. Am. Soc. Mass Spectrom. 7*, 1–10.

Moormann, M., Zahringer, U., Moll, H., Kaufmann, R., Schmid, R., Altendorf, K. (1997) *J. Bio. Chem. 272*, 10729–10738.

Nilsson, C. L., Murphy, C. M., Ekman, R. (1997) *Rapid Commun. Mass Spectrom. 11*, 610–612.

Roepstorff, P., Fohlmann, J. (1984) *Biomed. Mass Spectrom. 11*, 601.

Schlueter, H., Mentrup, D., Gross, I., Meyer, H. E., Spengler, B., Kaufmann, R., Zidek, W. (1997) *Anal. Biochem. 246*, 15–19.

Spengler, B., Kirsch, D., Kaufmann, R. (1991) *Rapid Commun. Mass Spectrom. 5*, 198–202.

Spengler, B., Kirsch, D., Kaufmann, R., Jaeger, E. (1992a) *Rapid Comm. Mass Spectrom. 6*, 105–108.

Spengler, B., Kirsch D., Kaufmann, R. (1992b) *J. Phys. Chem. 96*, 9678–9684.

Spengler, B., Lützenkirchen, E., Kaufmann, R. (1993) *Org. Mass Spectrom. 28*, 1482–1490.

Spengler, B., Kirsch, D., Kaufmann, R., Lemoine, J. (1995) *J. Mass Spectrom. 30*, 782–787.

Spengler, B. (1997a) *J. Mass Spectrom. 32*, 1019–1036.

Spengler, B., Bold, P., Chaurand, P., Kaufmann, R., Meyer, H. E., Flad, Th., Müller, C. A., Kalbacher, H. (1997b) *Proceedings ot the 45th Annual Conference on Mass Spectrometry and Allied Topics*, Palm Springs, CA, June 1–5.

Spengler, B., Lützenkirchen, F., Metzger, S., Chaurand, P., Kaufmann, R., Jeffery, W., Bartlet-Jones, M., Pappin, D. J. C. (1998) *Int. J. Mass Spectrom. Ion Proc. (in print)*.

Stahl-Zeng, J., Hillenkamp, F., Karas, M. (1996) *Eur. Mass Spectrom. 2*, 23–32.

Tang, X., Beavis, R., Ens, W., Lafortune, F., Schueler, B., Standing, K. G. (1988) *Int. J. Mass Spectrom. Ion Processes 85*, 43–67.

Veelaert, D., Devreese, B., Schoofs, L., Van Beeumen, J., Van den Broeck, J., Tobe, S. S., Loof, A. (1996) *Mol. Cell. Endocrinol. 122*, 183–190.

Vlasak, P. R., Beussman, D. J., Davenport, M. R., Enke, C. G. (1996) *Rev Sci Instrum 67*, 68–72.

Wilm, M., Shevchenko, A., Houthaeve, T., Breit, S., Schweigerer, L., Fotsis, T., Mann, M. (1996a) *Nature 379*, 466–469.

Wilm, M., Houthaeve, T., Talbo, G., Kellner, R., Mortensen, P., Mann, M. (1996b) *Mass Spectrom. Biol. Sci., Editors: A. Burlingame, S. A. Carr, Humana, Totowa, N. J.*, pp 245–265.

Yamagaki, T., Ishizuka, Y., Kawabata, S. I., Nakanishi, H. (1997) *Rapid Commun. Mass Spectrom. 11*, 527–531.

Yu, W., Vath, J. E., Huberty, M. C., Martin, S. A. (1993) *Anal. Chem. 65*, 3015–3023.

Electrospray Mass Spectrometry

Jörg W. Metzger, Christoph Eckerskorn,
Christoph Kempter and Beate Behnke

1 Introduction

Estimating molecular masses has always been an important aspect of protein and peptide characterization. Molecular mass measurements have been used to prove homogeneity of the sample, establish identity, analyze quaternary structure (e.g. the presence of subunits) and to detect modifications such as glycosylation or proteolysis. Most molecular mass estimates are made using dodecyl sulphate polyacrylamide gel electrophoresis (SDS-PAGE) and size exclusion chromatography calibrated with known standards. These techniques are able to give useful indications of purity, relative molecular masses and approximate amounts of material present. However, the techniques are unable to provide the mass accuracy and resolution really needed to detect mass changes due to posttranslational modifications (Table 1) and amino acid exchanges (Table 2). Two recent developments in mass spectrometry, electrospray ionization (ESI) [Yamashita 1984 a,b] and matrix-assisted laser desorption/ionization (MALDI) [Karas 1988] have extended the application of mass spectrometry to proteins and other biopolymers. The accuracy, sensitivity and resolving power of these new techniques permit the detection of these minor, but biologically significant protein modifications. In this account the essential features of ESI mass spectrometry are described in an attempt to provide some perspectives on how it works and how it can be used.

2 Instrumentation

ESI-MS like all other mass spectrometric techniques is based on the principle of producing molecular ions for subsequent separation and analysis. In contrast to "conventional" ionization techniques, however, ESI produces ions directly from liquids at atmospheric pressure. For ESI measurements the samples are dissolved in a suitable solvent, e.g. a mixture of methanol or acetonitrile and water. The sample solution is then infused into a glass capillary (fused silica) at a constant flow rate and introduced to a "source", where intact ionized molecules in the gas phase, free of solvent or other solute molecules are produced. In the mass analyzer the molecular ions are individu-

III.6 *Jörg W. Metzger, Christoph Eckerskorn, Christoph Kempter and Beate Behnke*

Table 1. Mass shifts due to some post-translationally modified residues in proteins.

Modification	Mass change
Disulphide bond formation	− 2.0
Desamidation of Asn or Gln	+ 1.0
Methylation	+ 14.0
Hydroxylation	+ 16.0
Oxidation of Met	+ 16.0
Formylation	+ 28.0
Acetylation	+ 42.0
Phosphorylation	+ 80.0
Sulphation	+ 80.0
Cysteinylation	+ 119.1
Pentoses (ara, rib, xyl)	+ 132.1
Deoxyoses (fru, rha)	+ 146.1
Hexoses (fru, gal, glc, man)	+ 162.1
Hexosamines (GalN, GlcN)	+ 161.2
Lipoic acid (amide bond to Lys)	+ 188.3
N-Acetylhexose amines (GalNAc, GlcNAc)	+ 203.2
Farnesylation	+ 204.4
Myristoylation	+ 210.4
Biotinylation (amide bond to Lys-ε-NH$_2$)	+ 226.3
Pyridoxal phosphate (Schiff base with Lys-ε-NH$_2$)	+ 231.1
Palmitoylation	+ 238.4
Stearoylation	+ 266.5
Geranylgeranylation	+ 272.5
N-Acetylneuraminic acid	+ 291.3
Addition of glutathione	+ 305.3
N-glycolylneuraminic acid	+ 307.3
5′-Adenosylation	+ 329.2
4′-Phosphopantetheine	+ 339.3
ADP-ribosylation	+ 541.3

ally selected and separated on the basis of their mass and charge. The ions reach the detector which is connected to a data acquisition system, where the abundance of ions at any given mass-to-charge ratio (m/z) is recorded.

2.1 The Electrospray Source

The ESI phenomenon is a process that produces intact molecules in ionized form from an analyte solution. A spray of fine, highly charged droplets is created at atmospheric pressure in the presence of a strong electric field. The ESI source may be just a metal capillary (capillary tip) at elevated voltage relative to a counter electrode (interface

Table 2. Mass differences resulting from replacement of one amino acid residue (one letter code; select in the first column) in a protein by another one (select in the first line).

	G	A	S	P	V	T	C	I	L	N	D	Q	K	E	M	H	F	R	Y	W
G	0	14	30	40	42	44	46	56	56	57	58	71	71	72	74	80	90	99	106	129
A	-14	0	16	26	28	30	32	42	42	43	44	57	57	58	60	66	76	85	92	115
S	-30	-16	0	10	12	14	16	26	26	27	28	41	41	42	44	50	60	69	76	99
P	-40	-26	-10	0	2	4	6	16	16	17	18	31	31	32	34	40	50	59	66	89
V	-42	-28	-12	-2	0	2	4	14	14	15	16	29	29	30	32	38	48	57	64	87
T	-44	-30	-14	-4	-2	0	2	12	12	13	14	27	27	28	30	36	46	55	62	85
C	-46	-32	-16	-6	-4	-2	0	10	10	11	12	25	25	26	28	34	44	53	60	83
I	-56	-42	-26	-16	-14	-12	-10	0	0	1	2	15	15	16	18	24	34	43	50	73
L	-56	-42	-26	-16	-14	-12	-10	0	0	1	2	15	15	16	18	24	34	43	50	73
N	-57	-43	-27	-17	-15	-13	-11	-1	-1	0	1	14	14	15	17	23	33	42	49	72
D	-58	-44	-28	-18	-16	-14	-12	-2	-2	-1	0	13	13	14	16	22	32	41	48	71
Q	-71	-57	-41	-31	-29	-27	-25	-15	-15	-14	-13	0	0	1	3	9	19	28	35	58
K	-71	-57	-41	-31	-29	-27	-25	-15	-15	-14	-13	0	0	1	3	9	19	28	35	58
E	-72	-58	-42	-32	-30	-28	-26	-16	-16	-15	-14	-1	-1	0	2	8	18	27	34	57
M	-74	-60	-44	-34	-32	-30	-28	-18	-18	-17	-16	-3	-3	-2	0	6	16	25	32	55
H	-80	-66	-50	-40	-38	-36	-34	-24	-24	-23	-22	-9	-9	-8	-6	0	10	19	26	49
F	-90	-76	-60	-50	-48	-46	-44	-34	-34	-33	-32	-19	-19	-18	-16	-10	0	9	16	39
R	-99	-85	-69	-59	-57	-55	-53	-43	-43	-42	-41	-28	-28	-27	-25	-19	-9	0	7	30
Y	-106	-92	-76	-66	-64	-62	-60	-50	-50	-49	-48	-35	-35	-34	-32	-26	-16	-7	0	23
W	-129	-115	-99	-89	-87	-85	-83	-73	-73	-72	-71	-58	-58	-57	-55	-49	-39	-30	-23	0

plate) with an orifice where ions entrained in a flow of gas enter the mass spectrometer (Figure 1). Liquid flow is generated by infusion syringes, separation devices (HPLC, CE) or other liquid sources, at flow rates usually between one and a few microliters per minute. An ESI interface designed to operate at low nanoliter flow rates was described recently [Wilm 1994].

The resulting field between capillary tip and the interface plate charges the surface of the emerging liquid, dispersing it by Coulomb forces into a fine spray of charged droplets (Figure 2). The fine droplets formed in such way carry an excess of charge and are attracted to the inlet of the mass spectrometer, which is held at a lower potential. A countercurrent flow of dry gas ("nitrogen curtain") to the droplets is used in some instruments to assist evaporation of solvent from each droplet, decreasing its diameter. Consequently, the charge density on its surface increases until the so-called Rayleigh limit is reached, at which the Coulomb repulsion becomes of the same order as the surface tension. The resulting instability, sometimes called "Coulomb explosion", tears the droplet apart, producing charged daughter droplets that also evaporate. This sequence of events repeats and finally produces droplets so small (nm-range) that the combination of charge density and radius of curvature at the droplet surface produces an electric field strong enough to finally desorb ions from the droplets into the ambient gas phase (Figure 2, pathway A). This ion desorption from the condensed

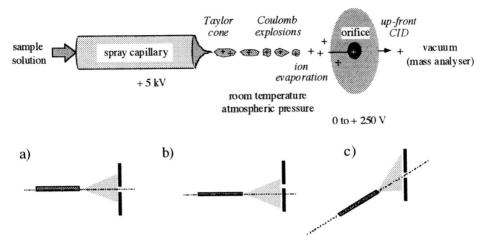

Figure 1. Electrospray ion source and different sprayer arrangements relative to the ion sampling orifice: a) on-axis, b) off-axis and c) diagonal.

phase ("ion-evaporation"), which was first proposed by Iribarne and Thomson [Iribarne 1976; Thomson 1979; Fenn 1990], and additionally competing and partially unknown processes produce finally so-called "quasi-molecular" (e.g. protonated) ions suitable for mass analysis. On the basis of energy calculations, however, this process was critisized [Röllgen 1987, 1989; Schmelzeisen-Redeker 1989]. Like Dole and coworkers [Dole 1968], who performed first ESI experiments already more than 25 years ago, Röllgen assumed that the droplets explode until a single analyte ion remains (Figure 2, pathway B: charged residue model). The formation of gas phase ions seems to occur at a very early stage in the ESI process. It is most likely that the majority of ions already desorb at the so-called "Taylor cone" directly at the solution/air interface at the tip of the capillary [Siu 1993].

In spite of quite a number of systematic investigations, in particular by the groups of Kebarle [Kebarle 1993] and Smith [Smith 1993 b], so far no satisfactory quantitative description exists of the formation of gas-phase ions from solute species in charged liquid droplets. Even so, the described concept for ion evaporation is widely accepted and has been a useful working hypothesis, although a lot of questions still remain about the details.

Several variations of the electrospray experiment exist (Figure 3), but all contain the essential elements described above. An advantageous improvement was introduced by Bruins in 1987. A concentrically applied nebulizer gas (compressed air) at the capillary tip is used to assist the formation of fine droplets (Figure 3b). This tech-

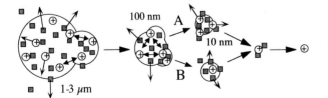

Figure 2. Formation of gas phase ions from highly charged droplets. A: ion evaporation model. B: charged residue model (dark squares: solvent molecules).

216

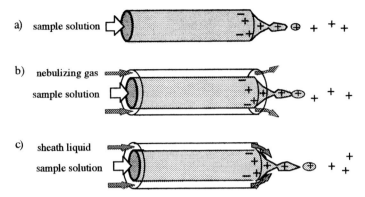

a) sample solution

b) nebulizing gas

sample solution

c) sheath liquid

sample solution

Figure 3. a) Electrospray b) pneumatically assisted electrospray ("ion spray") and c) electrospray with sheath liquid (frequently used for CE-MS).

nique was called "ion spray" by its originators to distinguish it from electrospray, however, it is based on the same ionization mechanism. The substantial difference between ion spray and electrospray lies in the mechanism of droplet formation: whereas in electrospray droplets are formed by charge-shearing of a liquid column as it exits a narrow tube, in ion spray additionally a jet of air shears the liquid, resulting in small (μm-range) droplets. With this pneumatic nebulization of liquid, ion spray produces stable ion currents over a wide range of flow rates, from 1 μL/min to 200 μL/min and during gradient flows from 100 % water to 100 % organic modifiers. This feature facilitates direct coupling of liquid separation techniques (e.g. HPLC) to mass spectrometry. Other possibilities supporting the droplet formation of ESI are ultrasound ("ultraspray", *Analytica of Branford*) or heating ("Turbo-Ion Spray", *PE-Sciex*).

Most electrospray interfaces nowadays commercially available are usually robust and rugged, easy to use, and the parameters for operation are relatively easy to optimize.

2.2 The Mass Analyzer

Because ESI produces ions continuously, it is compatible with scanning instruments. Electrospray was first demonstrated and also first commercialized on quadrupole mass spectrometers. The quadrupole is still the most common analyser used routinely in ESI.

A quadrupole consists of four parallel rods, in which opposite electrodes are electrically connected (Figure 4). To one pair of rods is applied a potential of $U+V \cos \omega t$, where U is a DC voltage and V is the peak amplitude of an RF (radio frequency) voltage at the frequency $\omega = 2\pi\nu$. To the other pair of rods a potential of the same amplitude is applied, but the polarity of the DC voltage is reversed and the RF voltage is shifted in phase by 180°. Ions injected parallel to the rods in the quadrupole undergo transverse oscillation caused by the perpendicular DC and RF voltages applied to the rods.

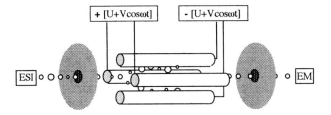

Figure 4. Schematic of a quadrupole mass analyzer (ESI, electrospray ion source; EM, electron multiplier).

For a proper selection of U and V, ions of a given mass-to-charge (m/z) ratio will have stable trajectories and will ultimately emerge the quadrupole towards the detector. Ions with other values of m/z will have unstable oscillations which increase in amplitude until they collide with the rods, thus not being transmitted.

Nowadays several other mass analyzers in combination with electrospray are commercially available (Figure 5). These mass spectrometers differ in mass accuracy, mass resolution, sensitivity, ease of handling and last but not least in price. In addition, the ion trap, the triple quadrupole, the quadrupole-TOF and the FT-ICR mass spectrometer allow tandem MS experiments.

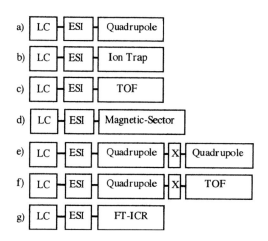

Figure 5. Commercially available mass spectrometers with ESI sources suitable for HPLC-MS (LC, liquid chromatography; ESI, electrospray ion source; TOF, time of flight analyzer; FT-ICR, Fourier-transform ion cyclotron resonance; X collision cell).

2.3 The Detector

Electron multipliers and channel electron multipliers are used universally for detecting the ions. Transmitted ions of the selected mass-to-charge ratio are deflected into the collector of the detector. An ion striking the collector causes an electron cascade and the resulting signal after amplification is sent to the data aquisition of a computer. Electron multipliers gradually lose sensitivity and have to be replaced after some months of operation.

3 Mass Spectra of Proteins

In the electrospray ionization process polyelectrolytes (= molecules containing several positively or negatively charged moieties, e.g. proteins) form multiply charged molecules $(M + nH)^{n+}$ (in positive ion mode) or $(M - nH)^{n-}$ (in negative ion mode). This gives rise to a series of consecutive, non-equidistant (!) peaks at $(M + n)/n$ and $(M - n)/n$, respectively, along the m/z scale of the spectrum (Figure 6). For most proteins the average charge state observed is proportional to the molecular mass. A typical positive mode mass spectrum of a protein with its characteristic slightly unsymmetrical bell shaped intensity distribution is shown in Figure 7. Multiple charging allows the determination of molecules of very high mass, because the m/z ratios of the molecular ions (z > 1) appear as a proportional fraction (1/z) of its molecular mass. A quadrupole mass analyzer of limited range (usually < m/z 4000; cf. Figure 6) suffices to measure masses up to 50 times larger [Hemling 1990].

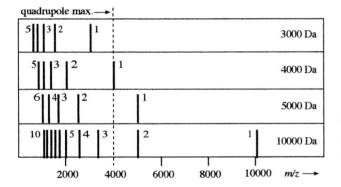

Figure 6. Calculated m/z values and number of charges of multiply charged molecular ions of peptides with different mass.

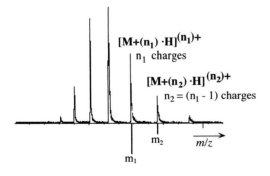

Figure 7. Typical electrospray spectrum of a protein. Each molecular peak allows the individual determination of the protein mass if the corresponding charge state is known.

For a single compound, each adjacent pair of peaks differs by one charge. The unknown charge state of the ion can be derived from any such pair.

If two adjacent peaks represent the adduct ions of the neutral molecule plus a proton:

$$m_1 = (M + n_1) / n_1 \qquad\qquad (1)$$

$$m_2 = (M + n_2) / n_2 \qquad\qquad\qquad (2) \text{ with } m_2 > m_1, n_2 < n_1$$

where M is the molecular weight of the protein, m_1 and m_2 are the measured mass-to-charge ratios, n_1 and n_2 are the numbers of additional protons on the protein molecule, respectively.

Then, if $n_2 = n_1 - 1$

$$m_1 = (M + (n_2 +1)) / (n_2 + 1) \qquad\qquad\qquad (3)$$

n_2 is given by equations (2) and (3):

$$n_2 = (m_1 -1) / (m_2 -m_1) \qquad\qquad\qquad (4)$$

and the molecular weight can be determined:

$$M = n_2 \cdot (m_2 - 1) \qquad\qquad\qquad (5)$$

This calculation, called deconvolution, is usually performed by a computer program. The relatively high precision (0.01 %) of the mass determination is chiefly the result of averaging individual measurements obtained in the same spectrum. In the mass spectrum of equine myoglobin (Figure 8) each of the mass peaks contribute to the molecular mass. The experimental error of the determination is largely dependent on the care with which the mass scale of the instrument is calibrated (usually polyethylene or -propylene glycol is used for calibration) and the stability of the calibration (dependent on the quality of the quadrupoles).

The net number of charged sites of a protein or peptide under solvation conditions is an important factor affecting the maximum extent of charging obtained in ESI mass spectra. For many proteins in aqueous solutions (pH < 4) an approximately linear cor-

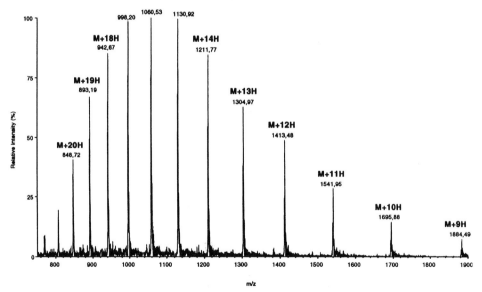

Figure 8. Electrospray mass spectrum of equine myoglobin, M = 16 951.6. The ion at m/z 1884.49 corresponds to the molecule with 9 protons attached. From the 12 ions ranging from 9+ (m/z 1884.49) to 20+ (m/z 848.72), a molecular mass of 16 951.34 ± 1.5 is calculated.

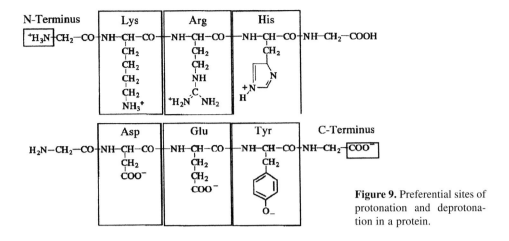

Figure 9. Preferential sites of protonation and deprotonation in a protein.

relation is observed between the maximum number of positive charges and the number of basic amino acid residues (e.g. arginine, lysine, histidine) plus the N-terminal amino group, unless it is not post-translationally modified [Smith 1990]. The same is true for acidic amino acids in the case of negative ion mass spectra. Deprotonation, which allows for recording negative ion ESI spectra, usually occurs at the side chains of Asp, Glu, Tyr and the carboxy-terminus (Figure 9).

The accessibility of these sites and the distribution of charge states depends, apart from the pH, on the temperature and any denaturing agents present in the solution. The pH effect can be used to probe conformational changes in proteins. For example, bovine cytochrome C, the most abundant ion, has 10 positive charges when electrospraying a solution at pH 5.2, but 16 charges at pH 2.6 and an intermediate bimodal distribution at pH 3.0 [Chowdhury 1990]. A similar effect is observed upon reduction of disulphide bonds. Lysozyme with four disulphide bonds shows a charge distribution centering at 12 charges, and after reduction with DTT a new cluster centering around 15 charges appears [Loo 1990].

4 Coupling of Chromatographic Methods to the Mass Spectrometer

4.1. On-Line HPLC-MS

A mass spectrometer constitutes a highly specific on-line detector for separation techniques such as RP-HPLC, capillary electrophoresis (CE) or capillary electrochromatography (CEC). Atmospheric pressure ion sources (API = electrospray and AP-CI) are optimal LC-MS interfaces [Covey 1986 a,b; Lee 1989], because they have the following features:

- The main portion of the eluent is removed at atmospheric pressure (therefore vacuum problems are rare and only minimal purification of the ion source – in most cases a cleaning of the orifice – is necessary).

- Solvents, gradient elution and volatile buffers commonly used for reversed-phase separations can be used within the whole range of concentration.
- Since no heat is applied (this is true for "pure" electrospray) the ionization is very soft (only the orifice plate is kept at ca. 50–60 °C to prevent its plugging by freezing of solvent).
- Optimization and operation of the ESI source are easy.
- No back pressure is built up (in contrast, for example, to thermospray MS).

The composition of the eluent and the flow rate have to be adjusted to the requirements of both HPLC and MS [Niessen 1992]. Because the optimal conditions for chromatography and ionization are often different, it can be of advantage to modify the mobile phase post-column. This can be done in order to reduce the surface tension (resulting in a more efficient ion evaporation process), to change the pH (for a better MS sensitivity) [Smith 1988] or to adjust the flow rate to the requirements of the interface (also for a better MS sensitivity) [Smith 1988 a,b; 1990]. Typical reversed-phase LC additives frequently used for peptide separations, e.g. phosphate, are not suitable for electrospray MS (high intensity of background ions, plugging of the sampling orifice, reduced MS sensitivity) and have to be replaced with volatile buffers, such as ammonium acetate. For negative ion operation the eluent should be adjusted to a lower pH, e.g. with ammonium hydroxide; in order to avoid corona discharge the use of isopropanol instead of methanol or acetonitrile / water mixtures can be effective.

Whereas pure ESI gives best results in the low µL/min range [Wong 1988], variants of ESI in which droplet formation is assisted by compressed air [Bruins 1987; Hopfgartner 1993], by temperature (e.g. "Turbo Ion Spray") or by ultrasound are able to handle flow rates up to 1–2 mL/min. Thus, the choice of HPLC column very much depends on the kind of interface being used, the scope being enlarged by using a pre- or post-column split (see Figure 10). Since ESI and nebulizer-assisted ESI constitute concentration and not mass flow-sensitive detectors, the MS sensitivity is not influenced by splitting the eluent [Hopfgartner 1993]. Generally, however, it is better to use smaller i.d. columns rather than a post-column split, because T-splitters often are the source of problems such as loss of sample or sensitivity due to possible dead volumes.

For HPLC-MS one end of a fused-silica capillary (50-100 µm i.d.) is placed into the metal capillary of the ESI interface and the other end is connected to the outlet of the HPLC column (Figure 10). The analytes of the sample are separated by chromato-

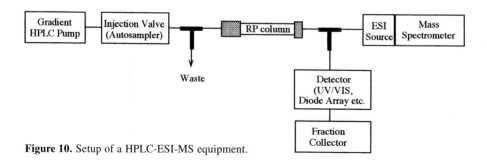

Figure 10. Setup of a HPLC-ESI-MS equipment.

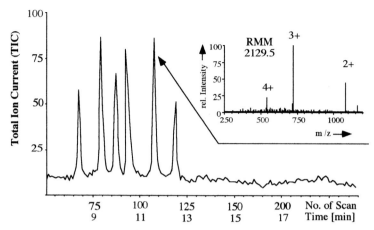

Figure 11. RP-HPLC-MS of a mixture of six peptides. The mass spectrum (inset) corresponding to one of the peaks in the TIC (indicated by the arrow) was obtained as sum of five scans. A relative molecular mass of 2129.5 was calculated.

graphy, reach the ion source at different time and are ionized. The m/z values of the formed ions are registered in a total ion current (TIC) *versus* time chromatogram (Figure 11). Usually more than one scan can be recorded during elution of an individual component. For the mass spectrum shown in Figure 11 (inset) five consecutive scans were accumulated. A fast chromatography requires a high scan speed, i.e. the time required to obtain a spectrum in a defined m/z range should be short. If it is too slow there is a possibility that components are registered in only one mass spectrum, although they are chromatographically separated.

In order to get a maximum of information on the analyte and also a higher specificity it is useful to connect further detectors, such as UV/VIS or fluorescence detectors, in series or parallel to the mass spectrometer (Figure 10) [Heath 1993]. This allows, for example, quantitation via the UV/VIS trace. Usually the software that comes with the mass spectrometer offers the possibility to register and compare simultaneously the TIC and UV/VIS signal.

An advantage of the mass spectrometric detection for HPLC is that peak purity can be determined very easily. The mass spectrum often unambiguously shows that several compounds coelute in one peak.

HPLC-MS is helpful for mixtures containing compounds with identical nominal mass (so-called isobaric compounds). Structural differences of isobaric peptides can be determined by on-line HPLC-MS-MS using collision-induced dissociation (CID). For the investigation of labile compounds, if only small sample amounts are available (pre-concentration on the column), or if the sample is contaminated with salts or other impurities on-line HPLC is essential.

For structure elucidations modification reactions (alkylations, acylations etc.) are helpful, e.g. for counting the number of residues with free SH or NH_2 -groups. Usually the modification reagent is used in high excess and further components like bases, acids or buffers are present, which makes direct measurement of the reaction cocktail with ESI-MS difficult. These reagents can be easily removed by HPLC-MS.

Mass spectrometric peptide mapping is of particular interest for protein structure analysis. For peptide mapping proteins are first cleaved enzymatically or chemically. The resultant smaller peptides are then characterized by HPLC-MS [Ling 1991; Hess 1993; Guzetta 1993] and HPLC-MS-MS.

4.2 Coupling of a Protein Sequencer to an ESI Mass Spectrometer

In a modern protein sequencer the protein is stepwise degraded automatically from the N- to the C-terminus by cleaving off the N-terminal residue as PTH-amino acid. For identification the cleaved PTH-amino acid is analyzed by RP-HPLC and detected by UV absorbance. Many peptidic natural compounds contain modified or unusual amino acids. Usually these residues are not available as standards for a qualitative and quantitative analysis. Thus, it can be of advantage to use the more specific mass spectrometer rather than the commonly used UV detector for identification after Edman degradation.

In order to couple a protein sequencer to an ESI mass spectrometer, the HPLC-column of the sequencer is linked to the ESI source as described above for "standard" HPLC-MS experiments. The involatile phosphate buffer of the protein sequencer has to be replaced by the ESI compatible ammonium acetate buffer (1 mM; pH 4,3). In Figure 12 the UV trace and the TIC of an analysis of an amino acid standard (100 pMol) are compared. Under these conditions not all of the 20 PTH-amino acids are baseline-separated. However, the higher selectivity of the mass spectrometer compensates for this disadvantage. Coeluting PTH residues like PTH-Val and PTH-Pro can be unequivocally differentiated based on their different masses (Figure 13).

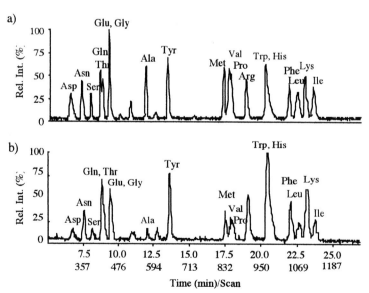

Figure 12. Automated Edman degradation coupled on-line to an ESI mass spectrometer. The standard buffer of the protein sequencer 476A (ABI Weiterstadt, Germany) was replaced with ammonium acetate buffer (1 mM; pH 4,3). 100 pMol of an amino acid standard were analyzed. a) UV trace. b) TIC.

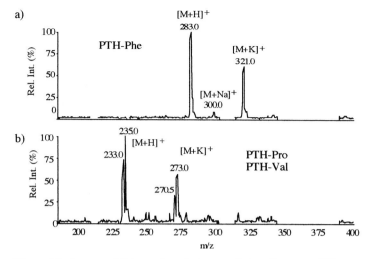

Figure 13. Automated Edman degradation coupled on-line to an ESI mass spectrometer (cf. Figure 12). a) ESI mass spectrum of PTH-Phe, b) ESI mass spectrum of PTH-Pro and PTH-Val, which are not separated under the applied chromatographic conditions.

Edman degradation with MS identification was successfully applied in the structure elucidation of peptide antibiotics of the CDA group, which contain several unusual amino acids such as 4-hydroxyphenylglycine, 3-phosphohydroxyasparagine and 3-methylglutamic acid [Kempter 1997].

4.3 Microcapillary LC Coupled to Mass Spectrometry

The use of capillary columns is especially suitable for small sample volumes and low amounts (10 pmol) [Huang 1991; Lewis 1994; Chowdhury 1995]. The signal intensity of the mass spectrometer increases with increasing concentration of the analytes. Consequently the sample consumption is significantly reduced compared to conventional 2–4 mm i.d. columns. This makes micro LC particularly interesting for the analysis of biological and biochemical samples when only miniature amounts are available. With columns of 150–500 μm i.d. the flow rate is in the range of 1–20 μL/min which is an ideal flow rate range for stable spray formation in nebulizer-assisted electrospray ionization. Furthermore the small flow rates and the resulting reduced solvent consumption make the use of expensive eluents such as completely deuterated solvents feasible [Karlson 1993]. Conventional HPLC systems can easily be modified for capillary LC by using a flow splitter consisting of a T-piece and a resistor capillary. For additional UV detection a capillary detection cell is required. Packed capillary columns are nowadays commercially available with a wide range of column dimensions and stationary phases.

Figure 14 shows the TIC of a capillary LC-MS analysis of 12 picomole tryptic digest of cytochrome C. Rapid peptide mapping was achieved with a 300 μm i.d. column packed with small reversed phase particles of 1.5 μm particle diameter. The cor-

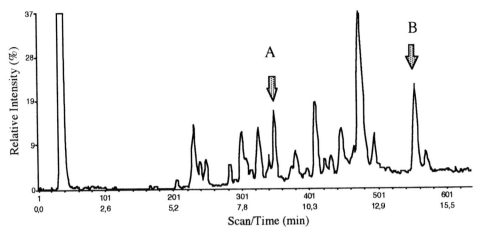

Figure 14. Micro LC-MS of a tryptic map of cytochrome C (12 pmol); column: 300 μm × 5 cm; Gromsil ODS-2; dp = 1.5 μm; A = 0.1 % Tfa in H₂O, B = acetonitrile; 0–50 % B in 15 min; TIC: m/z 300–1700.

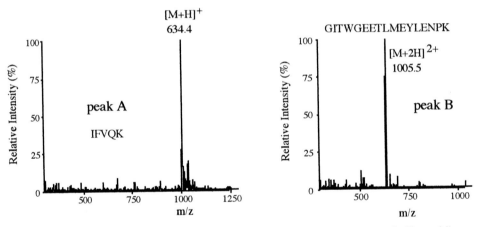

Figure 15. ESI-mass spectra of two peptide fractions A and B (indicated by arrows in Figure 14).

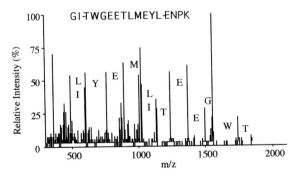

Figure 16. Product ion spectrum of the peptide corresponding to peak B in Figure 14 allowing a partial sequence determination (CID of [M+2H]²⁺).

responding mass spectra have a very good signal/noise ratio (Figure 15). In a second experiment sequence information of individual peaks was obtained using on-line LC-MS-MS and collision induced dissociation (CID) of a doubly protonated molecular ion (Figure16).

4.4 Capillary Electrophoresis Coupled to Mass Spectrometry

Capillary electrophoresis has gained special attraction because of its separation power, fast analysis times and low consumption of sample and solvent. Whereas HPLC-ESI-MS can be used routinely, the coupling of CE with the mass spectrometer requires more optimization and more time [Smith 1993a]. This is due to the low flow rates for CE (nL/min), the low sample amount (pmol range) and especially the buffers (like citric acid or borate) necessary for effective separation. Organic buffers are ionized very easily and give interfering peaks over the whole mass range. The presence of non-volatile buffers like phosphate or borate leads to a high conductibility in the droplet, causing an instable spray [Mann 1990]. One way of solving the problem of the low electroosmotic flow is to add a second flow ("make up" or "liquid sheath" flow with several microliters per minute; see Figure 3 c). This can be achieved either by an additional coaxial capillary, which surrounds the CE capillary [Smith 1988b] or by mixing in a dead-volume free mixing chamber behind the exit of the column [Lee 1988; Garcia 1992].

In both cases the analyte is diluted resulting in a lower MS sensitivity. In addition to the sheath flow also interfaces have been used for CE-MS, which offer the possibility to introduce an additional nebulizing gas for assisting droplet formation (Figure 17. Despite all problems, CE-MS has considerable analytical potency, especially for highly polar compounds, which cannot be separated on RP-columns [Olivares 1987; Smith 1988a, 1989; Edmonds 1989].

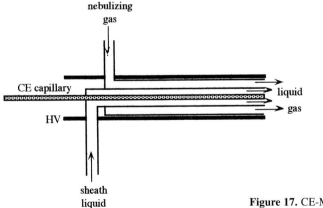

Figure 17. CE-MS interface (HV, high voltage)

5 Purity Control of Synthetic Peptides

5.1 Sample Introduction with an Autosampler

ESI-MS is a very powerful and meaningful method for checking the purity of crude synthetic peptides and determination of by-products [Metzger 1991, 1993]. A high sample throughput is often desired for such routine ESI-MS measurements. This can be achieved with an autosampler (ideally capable of handling 96-well titer plates) in combination with a HPLC pump using the same set-up as for HPLC-MS, but without a column (Figure 7). With this instrumentation, ca. 50–60 samples per hour can be investigated. The individual samples (injection volume ca. 5 μL) are injected into a continuous flow of solvent (e.g. methanol/ 0.1 % formic acid; flow rate ca. 80–100 μL/min). The time required for the analysis of the ESI mass spectra of crude synthetic peptides is about 1 min per sample.

The nonapeptide SNKLYLKNI with a monoisotopic relative molecular mass (RMM) of 1091.6 was obtained by solid-phase peptide synthesis using Fmoc/tBu strategy. The side chain protecting group for serine and tyrosine was tert.-butyl and for lysine tert.-butyloxycarbonyl; the asparagine side chain was not protected.

After cleavage of the peptide from the resin with trifluoroacetic acid the crude peptide was precipitated from ether, lyophilized and an ESI mass spectrum was recorded (Figure 18 a). The spectrum showed the expected singly and doubly protonated ions of the peptide at m/z 1092.5 and 546.8. However, a closer look showed that in the

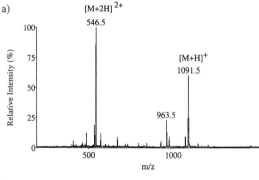

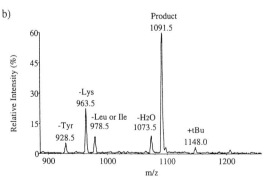

Figure 18. a) ESI mass spectrum of crude SNKLYLKNI (RMM 1091.6). b) Enlarged view of the [M+H]+ range indicating the presence of by-products.

[M+H]$^+$ range additional ions at m/z 1149, 1074, 980, 965, 1073 and 930 were found, which indicated the presence of by-products (Figure 18 b); these masses found are in agreement with a tert.-butylated peptide ([M+H+56]$^+$), a dehydrated peptide ([M+H-18]$^+$), and truncated peptides lacking leucine or isoleucine ([M+H-113]$^+$), lacking lysine ([M+H-128]$^+$) or lacking tyrosine ([M+H-163]$^+$), respectively. The formation of these by-products can be explained by assuming incomplete deprotection of the Ser and Thr side chains, dehydration of the side chain amide group of Asn forming a nitrile function [Metzger 1993] and incomplete coupling steps [Metzger 1994].

5.2 Off-Line HPLC-MS

Although on-line HPLC-MS using an electrospray interface is a straightforward technique nowadays (see above), it can be of advantage to analyze routine samples such as synthetic peptides off-line, especially if the mass spectrometer is in full operation and analysis is urgent. For this purpose and to speed up analysis fractions of an analytical HPLC run after separation by a routinely used analytical column (e.g. 4.6 mm I.D.) are first collected and then individually investigated by ESI-MS. ESI is a highly sensitive method, which allows the determination of molecular mass of peptides in the low picomole to femtomole range, in some cases even in the attomole range. Thus, the concentration of such diluted HPLC samples is generally sufficiently high to obtain a mass spectrum with a reasonable signal/noise ratio. For the analysis of small sample volumes of ca. 1 μl "nanoelectrospray" was developed, which operates at ca. 25 nL/min [Wilm 1994]. The outside of the capillary of this miniaturized source is coated with gold.

5.3 Characterization of Combinatorial Compound Collections

Mixtures consisting of equimolar mixtures of defined organic compounds obtained by combinatorial chemistry can be directly used in biological screening assays. Peptide mixtures ("peptide libraries"), for example, were used for studying antigen/antibody, receptor/ligand or enzyme/ substrate interaction or for the detection of antibiotic activities [Geysen 1993].

For the determination of the composition and purity of peptide mixtures ESI mass spectrometry [Stevanovic 1993; Metzger 1993, 1994 a,b] is well suited. These techniques allow fast optimization of the synthesis of even complex mixtures.

The ESI mass spectra of a mixture which contains each peptide in equimolar amounts shows the protonated molecular ions of all peptides present in the mixture. The ion intensities reflect the mass distribution, which can be easily calculated by a computer program. The lightest and heaviest peptide of a mixture mark the range within which the protonated molecular ions should be found. The presence of mass peaks at lower or higher m/z values indicates the presence of by-products (e.g. deletion peptides or peptides with protecting groups). Usually peptide libraries contain many peptides with the same mass. Therefore not only one but several isobaric molecular ions give rise to a particular mass peak in the ESI mass spectrum. Often it is possible to differentiate the isobaric peptides of peptide libraries by HPLC-MS-MS.

Isobaric peptides in less complex peptide mixtures can be differentiated by HPLC-MS as shown for the octapeptide mixture $LNYRFX_1X_2X_3$. All 48 peptides of this mixture have the same five N-terminal residues but differ in the three C-terminal residues X_1, X_2 and X_3. In position 6 (X_1), they contain Ser, Thr, Ile or Glu, in position 7 (X_2), Asn, Lys or Gln and in position 8 (X_3), Val, Leu, Ile or Met. In the ESI mass spectrum the $[M+H]^+$ ions of the peptides of this mixture are expected at m/z 1013 (1x), 1027 (5x), 1039 (1x), 1041 (8x), 1045 (1x), 1053 (4x), 1055 (5x), 1059 (3x), 1067 (4x), 1069 (4x), 1071 (1x), 1073 (2x), 1083 (4x), 1085 (2), 1087 (1x) and 1100 (2x).

HPLC-MS of this mixture was performed using a C-18 column at a flow rate of 200 µL/min and a linear gradient (5–20 % B in 20 min; A = 0.1 % aqueous trifluoroacetic acid; B = 0.1 % trifluoroacetic acid in acetonitrile). A post-column split (1:5) allowed ca. 40 µL/min to reach the ESI interface. The TIC showed four distinct peak groups with maxima at retention times of 15.2, 16.1, 17.2 and 18.2 min (Figure 19).

Analysis of this TIC was facilitated by using a two-dimensional display of the mass-analyzed chromatogram, in which the mass signals observed are plotted against a given retention time or mass scan (Figure 20). It can be seen that most of the isobaric peptides of the mixture were separated from each other. However, due to the complexity of the mixture most of the components are coeluting with further peptides of different mass.

The two-dimensional display also shows that peak groups III and IV (cf. Figure 19) clearly correspond to six and four different peptides, respectively.

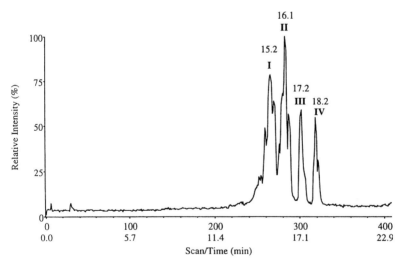

Figure 19. HPLC-ESI-MS of the 48 component octapeptide mixture LNYRF $X_1X_2X_3$ with X_1 = S,T,I,E, X_2 = N,K,Q and X_3 = V,L,I,M. Reconstructed total ion current chromatogram (m/z 800–1200) showing four distinct peak groups I-IV.

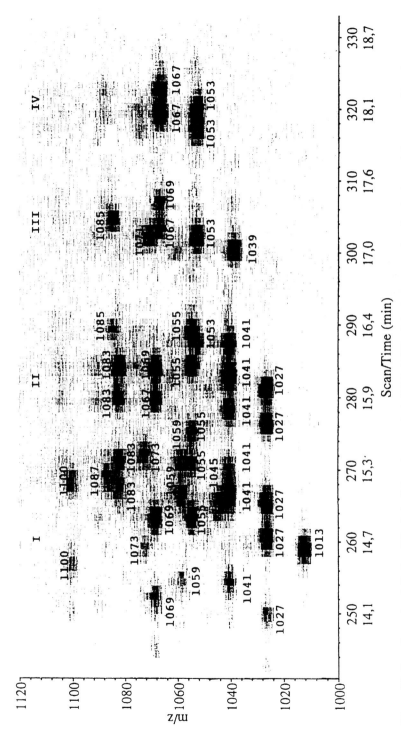

Figure 20. Two dimensional display (2D-plot) of the m/z range (m/z 1000–1120) of singly protonated molecular ions of the octapeptide mixture.

6 References

Bruins, A. P., T. R. Covey, J. D. Henion (1987) *Anal. Chem.* 59, 2642–2646. Ion spray interface for combined liquid chromatography/atmospheric pressure ionization mass spectrometry.

Chowdhury, S. K., Katta, V., Chait, B. T. (1990) *J. Am. Chem. Soc.* 112, 9012–9013. Probing conformational changes in proteins by mass spectrometry.

Chowdhury, S. K., Eshraghi, J., Wolfe, H., Forde, D., Hlavac, A. G., D. Johnston (1995) *Anal. Chem.* 67, 390–398. Mass spectrometric identification of amino acid transformations during oxidation of peptides and proteins: modifications of methionine and tyrosine.

Covey, T. R., E. D. Lee, A. P. Bruins, J. D. Henion (1986a) *Anal. Chem.* 58, 1451A–1461A. Liquid chromatography/mass spectrometry.

Covey, T. R., E. D. Lee, J. D. Henion (1986b) *Anal. Chem.* 58, 2453–2460. High-speed liquid chromatography/ tandem mass spectrometry for the determination of drugs in biological samples.

Covey, T. R., A. P. Bruins, J. D. Henion (1988) *Org. Mass Spectrom.* 23, 178–186. Comparison of thermospray and ion spray mass spectrometry in an atmospheric pressure ion source.

Ding, J., P. Vouros (1997) *Anal. Chem.* 69, 379–384. Capillary electrochromatography and capillary electrochromatography-mass spectrometry for the analysis of DNA adduct mixtures.

Dole, M., L. L. Mack, R. L. Hines, R. C. Mobley, L. D. Ferguson, M. B. Alice (1968) *J. Chem. Phys.* 49, 2240–2249. Molecular Beams of Macroions.

Edmonds, C. G., J. A. Loo, C. J. Barinaga, H. R. Udseth, R. D. Smith (1989) *J. Chromatogr.* 474, 21–37. Capillary electrophoresis-electrospray ionization-mass spectrometry.

Fenn, J. B., M. Mann, C. K. Meng, S. F. Wong, C. M. Whitehouse (1990) *Mass Spectrom. Rev.* 9, 37–70. Electrospray ionization – principles and practice.

Geysen, H. M., T. J. Mason (1993) *Bioorg. Med. Chem. Lett.* 3, 397–404. Screening chemically synthesized peptide libraries for biologically-relevant molecules.

Guzetta, A. W., L. J. Basa, W. S. Hancock, B. A. Key, W. F. Bennett (1993) *Anal. Chem.* 65, 2953–2962. Identification of carbohydrate structures in glycoprotein peptide maps by the use of LC/MS with selected ion extraction with special reference to tissue plasminogen activator and a glycosylation variant produced by site directed mutagenesis.

Hess, D., T. C. Covey, R. Winz, R. W. Brownsey, R. Aebersold (1993) *Protein Sci.* 2, 1342–1351. Analytical and micropreparative peptide mapping by high performance liquid chromatography / electrospray mass spectrometry of proteins purified by gel electrophoresis.

Heath, T., A. B. Giordani (1993) *J. Chromatogr.* 638, 9–19. Reversed-phase capillary high performance liquid chromatography with online UV, fluorescence and electrospray ionization mass spectrometric detection in the analysis of peptides and proteins.

Hemling, M. E., G. D. Roberts, W. Johnson, S. A. Carr, T. R. Covey (1990) *Biomed. Environ. Mass Spectrom.* 19, 677–691. Analysis of proteins and glycoproteins at the picomole level by on-line coupling of microbore high-performance liquid chromatography with flow fast atom bombardment and electrospray mass spectrometry: a comparative evaluation.

Hopfgartner, G., T. Wachs, K. Bean, J. Henion (1993) *Anal. Chem.* 65, 439–446. Highflow ion spray liquid chromatography / mass spectrometry.

Huang, E. C., J. Henion (1991) *Anal. Chem.* 63, 732–739. Packed-capillary chromatography / ionspray tandem mass spectrometry. Determination of biomolecules.

Iribarne, J. V., B. A. Thomson (1976) *J. Phys. Chem.* 64, 2287-2293. On the evaporation of small ions from charged droplets.

Karas, M., F. Hillenlamp (1988) *Anal. Chem.* 60, 2299–2301. Laser desorption ionization of proteins with molecular masses exceeding 10 000 daltons.

Karlson, K.-E. (1993) *J. Chromatogr.* 647, 31–38. Deuterium oxide as a reagent for the modification of mass spectra in electrospray microcolumn liquid chromatography mass spectrometry.

Kebarle, P., L. Tang (1993) *Anal. Chem.* 63, 2709–2715. From ions in solution to ions in the gas phase.

Kempter, C., Kaiser, D., Haag, S., Nicholson, G., Gnau, V., Walk, T., Gierling, K. H., Decker, H., Zähner, H. (1997) *Angew. Chem. Int. Ed. Engl. 36*, 498–501. CDA: calcium-dependent peptide antibiotics from *Streptomyces coelicolor* A3(2) containing unusual residues; *Angew. Chem. 109*, 510–513. Calcium-abhängige Peptidantibiotika mit ungewöhnlichen Bausteinen aus *Streptomyces coelicolor* A3(2).

Lee, E. D., J. Henion, T. R. Covey (1989) *J. Microcolumn Sep.* 1, 1–48. Microbore high performance liquid chromatography – ion spray mass spectrometry for the determination of peptides.

Lee, E. D., J. D. Henion (1992) *Rapid Commun. Mass Spectrom.* 6, 727–733. Thermally assisted electrospray interface for liquid chromatography / mass spectrometry.

Lewis, D. A., Guzzetta, A. W., W. S. Hancock (1994) *Anal. Chem.* 66, 585–595. Characterization of humanized anti-TAC, an antibody directed against the interleukin 2 receptor, using electrospray ionization mass spectrometry by direct infusion, LC/MS, and MS/MS.

Ling, V., A. W. Guzzetta, E. Canova-Davis, J. T. Stults, W. S. Hancock, T. R. Covey, B. I. Shushan (1991) *Anal. Chem.* 63, 2909–2915. Characterization of the tryptic map of recombinant DNA derived tissue plasminogen activator by high performance liquid chromatography – electrospray ionization mass spectrometry.

Loo, J. A., Edmonds, C. G., Udseth, H. R., Smith, R. D. (1990) *Anal. Chem.* 62, 693–698. Effect of reducing disulfide-containing proteins an electrospray ionization mass spectra.

Mann, M. (1990) *Org. Mass Spectrom.* 25, 575–587. Electrospray: its potential and limitations as an ionization method for biomolecules.

Metzger, J., G. Jung (1991) API-MS is a powerful new tool for on-line peptide and protein analysis with HPLC and CZE. In: *Peptides 1990, Proceedings of the 21st European Peptide Symposium* (Giralt, E., Andreu, D.; eds.) p. 341–342, Escom, Leiden.

Metzger, J. W., Jung, G. (1993a) Peptide and protein analysis with ion spray mass spectrometry. In: *Chemistry of peptides and proteins* (Brandenburg, D., Ivanov, V., Voelter, W.; eds.) pp. 171–180, Vol. 5/6, DWI Reports 112 A, Verlag Mainz.

Metzger, J. W., K.-H. Wiesmüller, V. Gnau, J. Brünjes, G. Jung (1993b) *Angew. Chem. Int. Ed. Engl.* 32, 894–896. Ion-spray mass spectrometry and high-performance liquid chromatography-mass spectrometry of synthetic peptide libraries.

Metzger, J. W., C. Kempter, K.-H. Wiesmüller G. Jung (1994a) *Anal. Biochem* 219, 261–277. Electrospray mass spectrometry and tandem mass spectrometry of multi-component peptide mixtures: determination of composition and purity.

Metzger, J. W., S. Stevanovic, J. Brünjes, K.-H. Wiesmüller, G. Jung (1994b) *Methods Enzymol.* 6, 425–431. Electrospray mass spectrometry and multiple sequence analysis of synthetic peptide libraries.

Niessen, W. M. A., R. A. M. Van Der Hoeven, J. Van Der Greef (1992) *Org. Mass Spectrom.* 27, 341–342. Analysis of intact oligosaccharides by liquid chromatography / mass spectrometry.

Olivares, J. A., N T. Nguyen, C. R. Yonker, R. D. Smith (1987) *Anal. Chem.* 59, 1230–1232. On-line mass spectrometric detection for capillary zone electrophoresis.

Röllgen, F. W., E. Bramer-Weger, L. Buetfering (1987) *J. Phys. Colloq.* (Paris) 48, C6-253/C6-256. Field ion emission from liquid solutions: ion evaporation against electrohydrodynamic disintegration.

Röllgen, F. W., H. Nehring, U. Giessmann (1990) Mechanisms of fieldinduced desolvation of ions from liquids. In: *Ion Formation from Organic Solids* (Hedin, A., B.U.R. Sundqvist, A. Benninghoven; eds.) Proceedings of the Fifth International Conference, Lovaanger, Sweden, June 18–21, 1989. Wiley, Chichester.

Schmelzeisen-Redeker, G., L. Bütfering, F. W. Röllgen (1989) *Int. J. Mass Spectrom. Ion Proc.* 90, 139–150. Desolvationof ions andmolecules inthermospray mass spectrometry.

Schmeer, K., Behnke, B., E. Bayer (1995) *Anal. Chem.* 67, 3656–3658. Capillary electrochromatography-electrospray mass spectrometry: a microanalysis technique.

Siu, K. W. M., R. Guevremont, J. C. Y. Le Blanc, R. T. O'Brian, S. S. Berman (1993) *Org. Mass Spectrom.* 28, 579–584. Is droplet evaporation crucial in the mechanism of electrospray mass spectrometry?

Smith, R. D., J. A. Olivares, N. T. Nguyen, H. R. Udseth (1988a) *Anal. Chem.* 60, 436–441. Capillary zone electrophoresis – mass spectrometry using an electrospray ionization interface.

Smith, R. D., C. J. Barinaga, H. R. Udseth (1988b) *Anal. Chem.* 60, 1948–1952. Improved electrospray ionization interface for capillary zone electrophoresis – mass spectrometry.

Smith, R. D., J. A. Loo, C. J. Barinaga, C. G. Edmonds, H. R. Udseth (1989) *J. Chromatogr.* 480, 211–232. Capillary zone electrophoresis and isotachophoresis mass spectrometry of polypeptides and proteins based upon an electrospray ionization interface.

Smith, R. D., J. A. Loo, C. G. Edmonds, C. J. Barinaga, H. R. Udseth (1990) *J. Chromatogr.* 516, 157–165. Sensitivity considerations for large molecule detection by capillary electrophoresis-electrospray ionization mass spectrometry.

Smith, R. D., K. J. Light-Wahl (1993a) *Biol. Mass Spectrom.* 22, 493–501. The observation of non-covalent interactions in solution by electrospray ionization mass spectrometry: promise, pitfalls and prognosis.

Smith, R. D., J. H. Wahl, D. R. Goodlett, S. A. Hofstadler (1993b) *Anal. Chem.* 65, 574A–584A. Capillary electrophoresis / mass spectrometry.

Stevanovic, S., K.-H. Wiesmüller, J. Metzger, A. G. Beck-Sickinger, G. Jung (1993a) *Bioorg. Med. Chem. Lett.* 3, 431–436. Natural and synthetic peptide pools: characterization by sequencing and electrospray mass spectrometry.

Stevanovic, S., G. Jung (1993b) *Anal. Biochem.* 212, 212–220. Multiple sequence analysis: pool sequencing of synthetic and natural peptide libraries.

Thomson, B. A., J. V. Iribarne (1979) *J. Chem. Phys.* 71, 4451–4463. Field induced ion evaporation from liquid surfaces at atmospheric pressure.

Valascovic, G. A., Kelleher, N. L., F. W. McLafferty (1996) *Science* 273, 1199–1202. Attomole Protein Characterization by Capillary Electrophoresis-Mass Spectrometry.

Wilm, M. W., Mann, M. (1994) *Int. J. Mass Spectrom. Ion Processes* 136, 167.

Wong, S. F., C. K. Meng, J. B. Fenn (1988) *J. Phys. Chem.* 92, 546–550. Multiple charging in electrospray ionization of poly(ethylene glycols).

Yamashita, M., J. B. Fenn (1984a) *J. Phys. Chem.* 88, 4451–4459. Electrospray ion source. Another variation on the free jet theme.

Yamashita, M., J. B. Fenn (1984b) *J. Phys. Chem.* 88, 4671–4675. Negative ion production with the electrospray ion source.

Fourier-Transform Ion Cyclotron Resonance Mass Spectrometry (FT-ICR-MS)

Jörg W. Metzger

1 Introduction

Fourier transform mass spectrometry (FT-MS) [Asamoto 1991] has received considerable attention for its ability to perform mass measurements with a very high resolution (Table 1) and accuracy. The first FT-MS instrument was built over twenty years ago by Comisarow and Marshall [Comisarow 1974]. Interest in FT-MS for peptide and protein analysis, however, has arosen only since electrospray (ESI) and matrix-assisted laser desorption (MALDI) have been used for ionization [Buchanan 1993; McIver 1994]. Three companies offer FT-MS commercially: Bruker-Spectrospin, Finnigan-Extrel FT-MS and IonSpec. All instruments consist of four main components: a magnet (generally a superconducting magnet with field strengths between 3 and 9 Tesla), an analyzer cell, an ultra-high vacuum system and a sophisticated data system. For bioanalytical applications the spectrometers are equipped with an external ion source (ESI or MALDI) [Amster 1996].

Table 1. Resolution and upper m/z value obtained in routine operation for various mass analyzers.

Mass analyzer	Resolution	Upper m/z
Quadrupole	unit mass	4,000
Ion Trap	unit mass or lower	2,500
Sector (double focusing)	50,000	10,000
Time-of-Flight	500 (5,000)*	1,000,000
FT-ICR-MS	> 50,000 (1,000,000)*,**	50,000**

  * has been demonstrated under pristine conditions
** depends on m/z

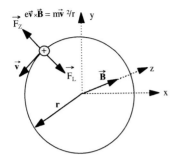

Figure 1. Ion movement in a homogeneous magnetic field: counterbalance of Lorentz force by centrifugal force.

2 Principle

FT-MS is based on the ion cyclotron resonance (ICR) principle. After ionization the ions are trapped in an ICR cell, which is situated in the homogeneous region of a magnet (either a permanent magnet, electromagent or superconducting magnet). In the strong magnetic field **B** the ions of charge e and of a velocity **v** are constrained to move in circular orbits (Figure 1). They experience the Lorentz force $F_L = e \cdot v \times B$, that is perpendicular to both the direction of their velocity and the magnetic field. This force is directed towards the center of the cyclotron orbit which is counterbalanced by the outward directed centrifugal force $F_Z = mv^2/r$. With $v/r = \omega = 2\pi v$ the cyclotron frequency can be written as v [Hz] $= B \cdot e/2\pi m$ and for an ion with charge state z (e/πu = 15357 C/g with elementary charge e = $1.602177 \cdot 10^{-19}$ C, atomic mass unit u = $1.66054 \cdot 10^{-24}$ g and π = 3.141...):

$$v \text{ [kHz]} = 15357 \cdot B \cdot z/m$$

Cyclotron frequencies fall in the range of tens of kilohertz to megahertz. A singly charged ion of a relative molecular mass of 2000 in a magnetic field of 3 Tesla would have an ion cyclotron frequency of 23.036 kHz and the doubly charged ion of the same molecule 46.071 kHz, i.e. double the frequency (Table 2). The cyclotron frequency v increases with increasing magnetic field strength but decreases with increasing mass. Higher frequencies can be measured with a better signal-to-noise

Table 2. Cyclotron frequencies of multiply charged ions in a magnetic field and physical limitation of mass resolution.

Rel. Molecular Mass	Charge	Magnetic Field [Tesla]	Cyclotron Frequency [kHz]	Duration of Transient [s]	Resolution
1000	1	3	46,071	2	46071
2000	1	3	23,036	2	23036
2000	2	3	46,071	2	46071
2000	3	3	69,107	2	69107
2000	1	3	23,036	4	46071
2000	1	7	53,750	2	53750
10000	1	7	10,750	2	10750
10000	10	7	107,499	2	107499

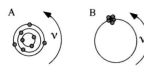

Figure 2. Movement of ions with the same m/z in a homogeneous magnetic field (the field vector **B** is perpendicular to the drawing plane) (A) ions of different velocity before excitation and (B) ions forming an ion packet after excitation (ion cyclotron resonance).

ratio, so that high magnetic field strengths of the spectrometer lead to better results (like in NMR). The ICR frequency is independent of the velocity of the ions, which is one of the fundamental reasons why FT-ICR mass spectrometers are able to achieve ultra-high resolution.

If the magnetic field strength **B** is constant, all ions of the same m/z ratio have the same frequency, but may have different velocities. An ion with higher velocity has a cyclotron orbit with a larger radius r = **v** (m/e·**B**) = const.·**v** than one with lower velocity (Figure 2 A).

Ions that move parallel to the magnetic field (Figure 3 a) are not influenced by the field. In order to trap the ion in an ICR cell a small voltage (usually ca. + or − 1–2 V, whatever positive or negative ions are to be trapped) is applied to the two plates (the "trapping plates") perpendicular to the magnetic field (Figure 3 b). This potential well causes the ions to undergo harmonic oscillations between the trapping plates and stores an ion for seconds, minutes or even hours. The movement of ions is influenced both by the magnetic field (cyclotron motion) and the electric field (trapping motion); these two fields give rise to the magnetron motion of the ion.

Common ICR cells are cubic, orthorhombic or cylindrical with measurements of several cm. The field strenghts of magnets used for FT-MS appear to be low compared to those used for NMR spectroscopy. However, it has to be considered that the bore of the superconducting magnet of the instrument is relatively wide to accomodate the ICR cell.

In the cubic and orthorhombic cell one opposing pair of plates is oriented orthogonal to the direction of the magnetic field lines and two pair of plates lie parallel to the field. The plates that are perpendicular to the field are the trapping plates (Figure 3 b).

One opposing pair oriented parallel to the field is used for ion excitation (transmitter plates) and the other pair is used for ion detection (receiver plates) (Figure 3 b). For

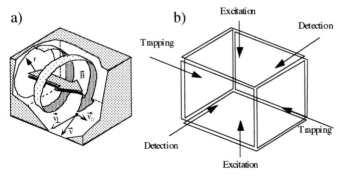

Figure 3. a) Movement of an ion in an ICR cell in the presence of a homogeneous magnetic field. b) Arrangement of trapping plates, excitation (transmitter) plates and receiver (detection) plates of an orthorhombic ICR cell.

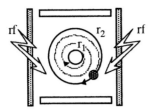

Figure 4. Excitation of an ion in the ICR cell (r_1 and r_2 orbit radius before and after excitation).

excitation an external oscillating electric field (excite pulse) is applied, which is transmitted by a sine wave signal generator via the two transmitter plates (Figure 4). If the frequency of the external oscillating field equals the cyclotron frequency (ion cyclotron resonance, ICR), all ions with a particular m/z are steadily and coherently accelerated to a larger orbit radius. After excitation these ions move as a single "ion packet" on an orbit with a radius, which is independent of the original velocity of the ions (Figure 2 B). The radius $r = A_{rf} \cdot t / \mathbf{B}$ increases with increasing time t and increasing amplitude A_{rf} of the excite pulse and decreases with increasing magnetic field strength.

Each "ion packet" emits a rf signal to the receiver plates at its characteristic cyclotron frequency. A model for the signal formation is shown in Figure 5 [Comisarow 1978]. A positive ion packet coming closer to the receiver plates attracts electrons in the external circuit and induces a so-called image current. This signal causes a small ac voltage to develop, which is amplified and registered as transient. The amplitude of the transient is proportional to the number of ions in the cell. Usually 100–1000 ions generate a detectable image current. Collisions of the ion packet in the cell with neutral molecules reduce the velocity of the ions, which decreases the radius and leads to the dampening of the signal. A very good vacuum in the ICR cell is therefore required ($< 10^{-9}$ Torr) to achieve high resolution. If an external ion source such as ESI is used several stages of differential pumping are required to preserve the high vacuum in the analyzer cell. The ions remain in the analyzer cell after detection (non-destructive detection) and can be analyzed in further experiments (e.g. by tandem mass spectrometry).

In FT-MS usually all ions of different m/z are excited and detected simultaneously (multi-channel analysis). A composite transite signal is obtained that represents a "time-domain" spectrum, i.e. the signal intensity is recorded *versus* time (Figure 6). Usually several transients are accumulated to improve the signal-to-noise ratio.

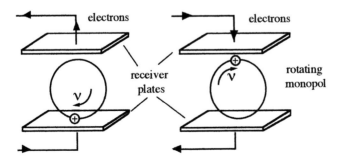

Figure 5. Principle of detection of a signal (image current), which is induced by a coherently moving ion packet.

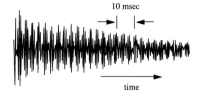

Figure 6. Transient ion image current signal induced by a mixture of ions of different m/z.

In order to get the frequency components representing each m/z value of the ions in the ICR cell from this spectrum, it is converted into a "frequency-domain" spectrum by a mathematical algorithm called Fourier transform (Jean-Baptiste Fourier 1768–1830). The following example will illustrate the principle of Fourier transform: if the G string of a violin is bowed the pure G frequency is mixed with other frequencies (overtones, harmonics) typical of this musical instrument. If a microphone is used to transmit the signal to an oscilloscope a complex interferogram, the time domain spectrum (in units of seconds) is obtained. Fourier transform converts this spectrum into the much easier to interpret frequency spectrum (in units of reciprocal time s^{-1} = Hz), that contains the relative intensities of all frequencies emitted by the violin. Calculating the Fourier transform requires a computer program, which first digitizes the time domain interferogram, performs the FT (usually within seconds) and than allows to plot the frequency domain spectrum on a printer. From the cyclotron frequency at a constant magnetic field the m/z of an ion is then calculated.

In contrast to other mass spectrometers ionization, mass analysis and ion detection occcur in the same space, the analyzer cell. In FT-MS these function are spread out in time. A simple experimental sequence for getting a spectrum with FT-MS is shown in

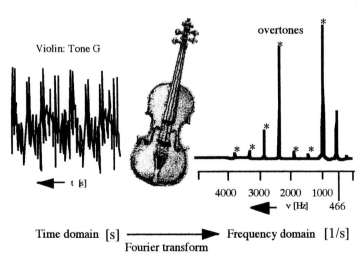

Figure 7. The G string of a violin is bowed. Left side: signal recorded by a microphone and transmitted to an oscilloscope as changes in intensity *versus* time. Right side: signal after Fourier transform reveals each frequency with its the relative intensity.

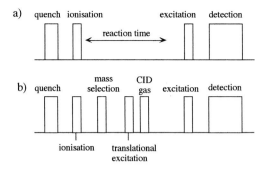

Figure 8. Experimental sequences in FT-MS experiments. a) simple experiment b) sequence for tandem mass spectrometry.

Figure 8a. The ICR cell is first completely emptied (quench) by applying a positive voltage to one trapping plate and a negative voltage to the other one. Ions are then formed inside (by "classical" electron impact or chemical ionization) or outside (by ESI or MALDI) the ICR cell. After external ionization the ions have to be guided into the cell by a rf ion guide (quadrupole) or electrostatic lenses. For a certain period of time the ions in the cell can be left alone to undergo reactions, e.g. with gas molecules (collisional damping). To excite simultaneously the ions in the cell that have different m/z many different frequencies are applied. Several techniques are available for this purpose, e.g. frequency sweep (rf chirp), impulse excitation and stored waveform inverse Fourier transform (SWIFT) [Wang 1986; Chen 1987]. The last step of the experimental sequence includes the simultaneous detection of various cyclotron frequencies of the coherent ion packets in the cell.

2.1 Resolution, Mass Accuracy, Mass Range and Sensitivity

Mass resolution R (full width at half height, FWHH) in FT-MS is directly proportional to the cyclotron frequency v and the time duration T of the acquired transient signal: $R = 1/2 \, v \cdot T$ (Table 2). Long transients and therefore high resolution can only be obtained when the pressure in the ICR cell is a factor of 1000 lower (10^{-8} to 10^{-9} Torr) than needed for most other mass analyzers. Besides the pressure also the amount of memory in the transient digitizer for recording the data and the amount of computer memory for performing the Fourier transform limits the length of the transient (usually $< 10^6$ data points). Ions with low cyclotron frequency (high mass ions) are principly measured with lower resolution.

For recording ultra-high resolution spectra (R = several hundred thousand and higher) often so-called heterodyne detection (narrow-band detection, mixer mode) is used, in which the signal is multiplied by a reference signal to reduce the sampling rate. Mass accuracy not necessarily improves with increased resolution in FT-MS. With a careful calibration, however, even for ions with m/z >1000 accuracies in the 1–10 ppm range can be obtained.

For biopolymer analysis it is important that there is a theoretical upper mass limit determined by the trapping potential and the magnetic field (Figure 9). The trapping potential causes a radial electric field which can lead to ejection of ions from the

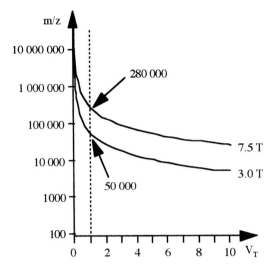

Figure 9. Fundamental upper limit of the mass range of an FT mass spectrometer: dependence on trapping potential V_t and magnetic field.

cell. The critical mass beyond which the ions escape radially from the cell is proportional to the magnetic field and to the reciprocal trapping potential. For high mass analysis therefore a high magnetic field is required and the trapping potential should be as low as possible (Figure 9).

The sensitivity of the instrument is increased by increasing the magnetic field strength (approximately linear dependence), prolongation of the aquisition time, accumulation of several transients n (improves with $n^{1/2}$), reduction of pressure (smaller line width), choice of A/D converter (digitalization noise).

2.2 Tandem Mass Spectrometry

Tandem mass spectrometry can be performed with FT-MS by adding additional steps in the experimental sequence (Figure 8b). After the quench and ionization step all ions but the precursor ions are ejected from the cell by an excitation pulse with appropriate frequencies and amplitudes. The selected precursor ion population is then accelerated into a larger orbit radius to increase the kinetic energy. For the dissociation of multiply charged ions sustained off-resonance irradiation (SORI) is particularly suitable [Hofstadler 1994]. A gas is introduced into the cell and the precursor ion dissociates by collision with the gas. Finally, the formed fragment ions are excited and detected as described above. The whole experiment can be repeated after isolation of a fragment ion as new precursor ion. This high order mass spectrometry (MS^n) is only possible with the FT-ICR-analyzer or an ion trap.

Due to the charge repulsion large multiply charged proteins formed by ESI are more susceptible to fragmentation using CID than singly charged ions [Senko 1994]. If CID of multiply charged ions of large proteins is performed, fragment ions having various charge states are obtained. The big advantage of FT-ICR-MS/MS is that – due to its high resolution – it allows the determination of the respective charge state as the reciprocal of the isotope spacing (a spacing of 0.2 correlates to a charge state of 5)

241

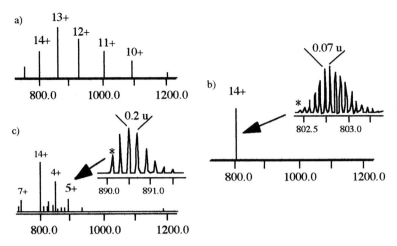

Figure 10. FT-ICR-MS/MS a) ESI mass spectrum of a protein, b) selection of the [M+14H]$^{14+}$ precursor ion (inset: isotope pattern) and c) CID of the precursor ion (inset: isotope pattern of a fragment ion indicating a charge state of five).

(Figure 10) [Chen 1994]. Therefore, CID can reveal sequence information of protein ions even of several 10 000 mass units.

3 Electrospray and FT-ICR-MS

Based on its ultra-high resolution and mass accuracy FT-ICR-MS is capable of generating extraordinary mass spectra. In the past, however, it often has been questioned whether there is any application that can exploit this performance. In the meantime ions produced by ESI can be analyzed and the results of quite a number of intriguing experiments have been reported [Williams 1994]. The m/z-values of protein ions formed by ESI usually fall in a range of 500–2000 for which FT-MS provides very good performance [Henry 1991; Buchanan 1993; Winger 1993; Winger 1994; Li 1994]. Due to the long storage times in the ICR cell FT-ICR-MS can be used to study the high-order structure of ESI ions (real time monitoring!) or chemical reactions (after introduction of a reactant gas into the ICR cell). H/D exchange experiments can be performed to investigate gas-phase ion conformation [Suckau 1995]. High resolution tandem mass spectrometry is capable of revealing structural information of large proteins [Qinyuan 1995; Senko 1994]. The technique is also a powerful tool for mixture analysis, e.g. the characterization of combinatorial libraries [Nawrocki 1996]. Capillary electrophoresis has been coupled successfully to FT-ICR-MS [Hofstadler 1993; Hofstadler 1994; Hofstadler 1995; Valaskovic 1996]. Recently, it has been shown that it is even possible to trap and detect one single ion by FT-ICR-MS [Bruce 1994; Smith 1994].

4 References

Asamoto, B. (Ed.) (1991) VCH, New York. *FT-ICR/MS: analytical applications of Fourier transform ion cyclotron resonance mass spectrometry.*

Amster, I. J. (1996) *J. Mass Spectrom. 31*, 1325–37. Fourier Transform Mass Spectrometry.

Bruce, J. E., Cheng, X., Bakhtiar, R., Wu, Q., Hofstadler, S. A., Anderson, G. A., Smith, R.D. (1994) *J. Am. Chem. Soc. 116*, 7839–47. Trapping, detection, and mass measurement of individual ions in a Fourier transform ion cyclotron resonance mass spectrometer.

Buchanan, M. V., Hettich, R. L. (1993) *Anal. Chem. 65*, 245A–259A. Fourier transform mass spectrometry of high-mass biomolecules.

Chen, L., Marshall, A. G. (1987) *Int. J. Mass Spectrom. Ion Proc. 79*, 115–125. Stored Waveform Simultaneous Mass-Selective Ejection/Excitation for Fourier Transform Ion Cyclotron Resonance Mass Spectrometry.

Chen, R., Wu, Q., Mitchell, D. W., Hofstadtler, S. A., Rockwood, A. L., Smith, R. D. (1994) *Anal. Chem. 66*, 3964–69. Direct charge number and molecular weight determination of large individual ions by electrospray ionization Fourier transform ion cyclotron resonance mass spectrometry.

Comisarow, M. B. (1978) *J. Chem. Phys. 69*, 4097–4104. Signal modeling for ion cyclotron resonance.

Comisarow, M. B., Marshall, A. G. (1974) *Chem. Phys. Letters 25*, 282–283. Fourier transform ion cyclotron resonance spectroscopy.

Henry, K. D., Quinn, J. P., McLafferty, F. W. (1991) *J. Am. Chem. Soc. 113*, 5447–49. High-Resolution Electrospray Mass Spectra of Large Molecules.

Hofstadler, S. A., Swanek, F. D., Gale, D. C., Ewing, A., Smith, R. D. (1995) *Anal. Chem. 67*, 1477–80. Capillary Electrophoresis-Electrospray Ionization Fourier Transform Ion Cyclotron Resonance Mass Spectrometry for the Direkt Analysis of Cellular Proteins.

Hofstadler, S. A., Wahl, J. H., Bakhtiar, R., Anderson, G. A., Bruce, J. E., Smith, R. D. (1994) *J. Am. Mass Spectrom. 5*, 1–6. Capillary Electrophoresis Fourier Transform Ion Cyclotron Resonance Mass Spectrometry with Sustained Off-Resonance Irradiation for the Characterization of Protein and Peptide Mixtures.

Hofstadler, S. A., Wahl, J. H., Bruce, J. E., Smith, R. D. (1993) *J. Am. Chem. Soc. 115*, 6983–84. On-Line Capillary Electrophoresis with Fourier Transform Ion Cyclotron Resonance Mass Spectrometry.

Li, Y., McIver Jr., R. T., Hunter, R. L. (1994) *Anal. Chem. 66*, 2077–83. High-Accuracy Molecular Mass Determination for Peptides and Proteins by Fourier Transform Mass Spectrometry.

McIver, J., Robert T., Li, Y., Hunter, R. L. (1994) *Proc. Natl. Acad. Sci. USA 91*, 4801–05. High-resolution laser desorption mass spectrometry of peptides and small proteins.

Nawrocki, J., Wigger, M., Watson, C. H., Hayes, T. W., Senko, M. W., Benner, S. A., Eyler, J. R. (1996) *Rapid Commun. Mass Spectrom. 10*, 1860–64. Analysis of Combinatorial Libraries Using Electrospray Fourier Transform Ion Cyclotron Resonance Mass Spectrometry.

Qinyuan, W., Van Orden, S., Cheng, X., Bakhtiar, R., Smith, R. D. (1995) *Anal. Chem. 67*, 2498–2509. Characterization of Cytochrome c Variants with High-Resolution FTICR Mass Spectrometry: Correlation of Fragmentation and Structure.

Senko, M. W., Beu, S. C., McLafferty, F. W. (1994) *Anal. Chem. 66*, 415–417. High-Resolution Tandem Mass Spectrometry of Carbonic Anhydrase.

Smith, R. D., Cheng, X., Bruce, J. E., Hofstadler, S. A., Anderson, G. A. (1994) *Nature 369*, 137–139. Trapping, Detection and Reaction of Very Large Single Molecular Ions by Mass Spectrometry.

Suckau, D., Shi, Y., Beu, S. C., Senko, M. W., Quinn, J. P., Wampler, F. M., McLafferty, F. W., (1995) *Proc. Natl. Acad. Sci. USA 90*, 790–793. Coexisting stable conformations of gaseous protein ions.

Valaskovic, G. A., Kelleher, N. L., McLafferty, F. W. (1996) *Science 273*, 1199–1202. Attomole protein characterization by capillary electrophoresis-mass spectrometry.

Williams, E. R. (1994) *Trends Anal. Chem. 13*, 247–251. Electrospray ionization with Fourier transform mass spectrometry.

Wang, T.-C. L., Ricca, T. L., Marshall, A. G. (1986) *Anal. Chem. 58*, 2935–38. Extension of Dynamic Range in Fourier Transform Ion Cyclotron Resonance Mass Spectrometry via Stored Waveform Inverse Fourier Transform Excitation.

Winger, B. E., Hein, R. E., Becker, B. L., Campana, J. E. (1994) *Rapid Commun. Mass Spectrom. 8*, 495–497. High-resolution characterization of biomolecules by using an electrospray ionization Fourier-transform mass spectrometer.

Winger, B. E., Hofstadler, S. A., Bruce, J. E., Udseth, H. R., Smith, R. D. (1993) *J. Am. Soc. Mass Spectrom. 4*, 566–577. High-resolution accurate mass measurements of biomolecules using a new electrospray ionization ion cyclotron resonance mass spectrometer.

Sequence Analysis of Proteins
and Peptides by Mass Spectrometry

Christoph Siethoff, Christiane Lohaus
and Helmut E. Meyer

1 Introduction

A protein is characterized not only by its amino acid sequence but also by any post-translational modification, e.g. phosphorylation, carbohydrate attachment or lipidation. However, determination of the primary structure is often the first step in speculating on the shape and function of the analyzed protein. The classical method of sequencing proteins and peptides is Edman degradation (chapter III.2). For this, the purified proteins have to be digested enzymatically with different endoproteases or split chemically, e.g. with cyanogen bromide. After fractionation of the resulting peptides by reversed-phase high-performance liquid chromatography (HPLC), the separated peptides can be sequenced. Another method of determining the primary structure is sequencing the gene or cDNA encoding the desired protein. More often, partial amino acid sequences are determined to construct oligonucleotide probes allowing a cDNA or genomic DNA library to be screened for positive DNA clones. Sequencing of those DNA clones will result in identification of the amino acid sequence of the encoded protein. Both methods, Edman degradation and DNA sequencing, have the important limitation that they cannot identify posttranslational modifications. For example, modified amino acids or disulfide bridges need special treatment to be analyzed.

Mass spectrometry has been improved tremendously over the last few years and new ionization techniques have been developed, allowing peptides or even proteins to be analyzed. The most important ionization methods concerning peptides and proteins nowadays are electrospray ionization [Yamashita 1984a; Kebarle 1993; Banks 1996; Gaskell 1997] and matrix-assisted laser desorption ionization [Karas 1988; Beavis 1996] which will be used for sequencing either in collision-induced dissociation MS/MS or post-source-decay (MALDI-PSD), respectively. Mass spectrometry can overcome the above-mentioned drawbacks and is complementary to Edman degradation. It is more rapid and efficient in determining the nature of an N-terminal blocking group or in recognizing and localizing posttranslational modifications. By developing coupling interfaces between HPLC or capillary zone electrophoresis (CZE) and tandem mass spectrometry and by improving matrix-assisted laser desorption ionization (MALDI) to analyze metastable peptide ions, a particular peptide even in a complex peptide mixture can be selected and investigated.

MALDI-TOF-MS for analysis of large biomolecules was first described in 1988 [Karas 1988]. Laser light is used to irradiate a UV- or IR-absorbing matrix, e.g. sinapinic acid, while the peptide or protein of interest is embedded in a large excess of this matrix. Energy is transferred onto the target molecules and ionization and desorption takes place. Analyte ions are emitted and fly towards the detector. Different masses cause different flight times (time-of-flight (TOF) mass analyzer) [Guilhaus 1995] and ions can be detected separately (chapter III.4).

Another technique for analyzing proteins and peptides by mass spectrometry uses electrospray ionization (ESI). Two decades ago no method of ionizing proteins was available. In 1984 an ionization method was developed producing an electrospray which could be combined with mass spectrometry [Yamashita 1984a, 1984b]. Fenn et al. [Fenn 1989] have shown that this technique can also be used for mass spectrometric analyses of proteins. The protein or peptide solution passes with a flow rate of 1–10 µL/min through a fused silica capillary which ends in a stainless steel capillary held at a potential of ± 3–6 kV. The resulting electrostatic field generates positively or negatively charged droplets. These droplets shrink by solvent evaporation supported by heating or by passing a curtain of dry gas until the charge accumulation on the liquid surface is high enough for ion evaporation. These ions enter the mass analyzer and their mass-to-charge (m/z) ratio can be determined [Edmonds 1990; Smith 1990]. Electrospray mass spectrometry generates multiply charged ions, thus direct molecular weight analysis of even large proteins (up to 133.000 Da [Loo 1989; Mann 1989]) with high sensitivity and precision is possible.

However, electrospray ionization delivers singly or multiply charged ions with little excess of energy. This results in very stable ions and little fragmentation. To obtain structural information the kinetic energy has to be converted into vibrational energy by collision of the ions with a neutral gas (for example helium or argon) inducing fragmentation of the analyzed ions. This is most promising done with triple quadrupole MS, ion trap MS or more recently combining quadrupole and time-of-flight MS (Q-TOF)-MS [Biemann 1990, 1992; Schwartz 1996; Shevchenko 1997].

The characteristics of electrospray mass spectrometry can be summarized as followed:

- Soft ionization method, therefore little structural information. But even without tandem mass spectrometry due to the interface construction structure elucidation is possible by in-source fragmentation.
- Low transmission rate concerning the transport of ions from the high-pressure region into vacuum.
- The generation of multiply charged ions, allowing molecular weight determination of molecules far in excess of the mass range of the analyzer used and which can be fragmented efficiently by collision with a neutral gas (tandem mass spectrometry [Smith 1990]).
- Excellent compatibility with HPLC and CZE since the ionization takes place at atmospheric pressure.
- Acceptance of aqueous and many organic mobile phases and volatile buffers.

This chapter will describe the forthcoming methods of sequencing proteins and peptides by mass spectrometry. First, the basics in peptide fragmentation will be regarded followed by a discussion about the most frequently used instruments and their characteristics in acquiring fragment ion spectra. Finally, the interpretation of product ion spectra will be explained either by data base search or de novo.

2 Basics in Peptide Fragmentation

For electrospray mass spectrometry doubly charged tryptic peptide ions are the most promising ions for collision-induced dissociation (CID). In that case the charges are located at the opposite ends of the peptide, one at the N-terminus, the other at the C-terminal lysine or arginine. Thus, fragmentation results in singly charged daughter ions (Figure 1). The cleavage leads to fragments of type a_n, b_n and c_n if the charge is located at the N-terminus and to fragments of type x_n, y_n and z_n if the charge is retained on the C-terminus. If a complete ion series is obtained, the analyzed peptide can be sequenced, because the adjacent signals differ by the mass of one amino acid residue (Table 1).

The isobaric amino acids leucine and isoleucine and the modified amino acid hydroxyproline cannot be differentiated using a MALDI-PSD instrument or low energy collision-induced dissociation for example with a triple stage quadrupole MS since all three amino acid residues have the same nominal mass of 113 Da. This is only feasible either with a magnetic instrument where the precursor ions have a kinetic energy of 5–10 keV resulting in high energy collisions or with MALDI-TOF-MS in combination with collision-induced dissociation. Typically a collision energy of 10 to 30 keV is used [Biemann 1992; Medzihradszky 1994, 1996; Stimson 1997].

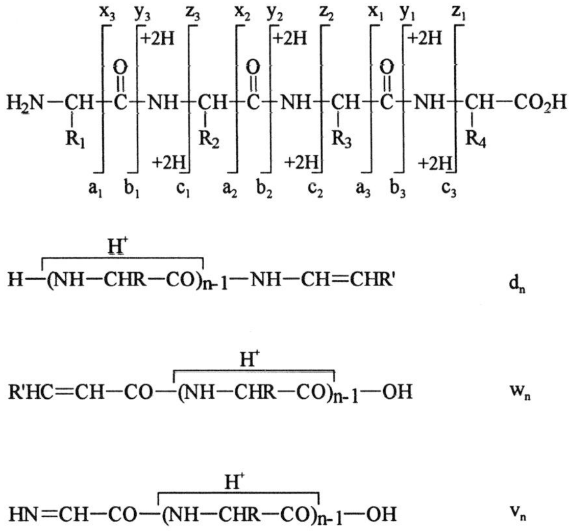

Figure 1. Notation of fragment ions [Biemann 1990]. The most prominent fragments observed after low energy collision-induced dissociation are the fragments belonging to the b_n- and y_n-series, reflecting cleavage of the peptide bond. These fragments are mainly used for deducing the primary sequence of an analyzed peptide. High-energy collision-induced dissociation yields also series of d-, w-, and v-ions due to fragmentation of the side chain.

Table 1. Monoisotopic and average masses of amino acids.

Amino acid	Letter code		Mass	
	Three	One	Monoisotopic	Average
Glycine	Gly	G	57.021	57.052
Alanine	Ala	A	71.037	71.079
Serine	Ser	S	87.032	87.078
Proline	Pro	P	97.053	97.117
Valine	Val	V	99.068	99.133
Threonine	Thr	T	101.048	101.105
Cysteine	Cys	C	103.009	103.139
Isoleucine	Ile	I	113.084	113.159
Leucine	Leu	L	113.084	113.159
Asparagine	Asn	N	114.043	114.104
Aspartic acid	Asp	D	115.027	115.089
Glutamine	Gln	Q	128.059	128.131
Lysine	Lys	K	128.095	128.174
Glutamic acid	Glu	E	129.043	129.116
Methionine	Met	M	131.040	131.193
Histidine	His	H	137.059	137.141
Phenylalanine	Phe	F	147.068	147.177
Arginine	Arg	R	156.101	156.188
Tyrosine	Tyr	Y	163.063	163.176
Tryptophan	Trp	W	186.079	186.213

This high energy collisions produce an additional pattern of fragments (d_n- and w_n-ions (Figure 2) allowing differentiation between leucine and isoleucine. As seen in Figure 2 (insert) isoleucine generates two side-chain fragment ions (d_n-ions) in a high energy CID mass spectrum with a mass difference of 14 Da. In contrast, leucine produces only one side-chain fragment. In Figure 2 the fourth amino acid can be assigned as isoleucine based on fragment ions d_{4a} and d_{4b}. However, the combination of low energy CID with sequence analysis and amino acid analysis may determine the presence of hydroxyproline, leucine or isoleucine.

The other pair of isobaric amino acid residues is lysine/glutamine. These two amino acids can be distinguished with high resolution instruments and accurate mass measurement of fragment ions since in contrast to leucine and isoleucine the exact masses of the amino acid residues are different (lysine 128.095, glutamine 128.059) or by acetylation of the peptides with acetic anhydride. The acetylation increases the mass by 42 Da for the free amino-terminus and for every ε-amino group of lysine.

Recently, the differentiation of Lys/Gln in peptide sequence analysis could be shown by using an ESI-ion trap [Bahr 1998].

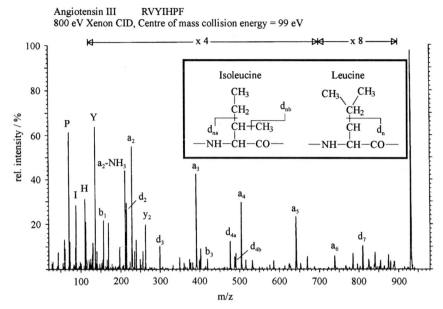

Figure 2. High energy MALDI-PSD mass spectrum of Angiotensin III RVYIHPF m/z 930.51 (Auto-Spec-Tof, Micromass). The fragment ions d_{4a} and d_{4b} indicate the isoleucine due to the side chain cleavage shown in the insert.

3 Instrumentation and Generation of MS/MS Data Sets

3.1 MALDI-PSD Time-of-Flight Mass Spectrometry

In most cases, MALDI is coupled directly with a time-of-flight mass spectrometer as a mass separation and detection device [Cotter 1992]. Indeed, nowadays it is the most important MS instrument for MALDI but other instruments are also used (ion trap, FT-ICR). In the beginning, it was reported as a characteristic feature of MALDI that any fragmentation of the analyte molecules hardly occurs. However, a metastable dissociation of peptides after the desorption process has been described [Spengler 1992; Kaufmann 1993; Kaufmann 1994; Kaufmann 1995]. Analyte ions gain additional activation energy and fragmentation occurs by collisions with matrix molecules and residual gas molecules during their flight in the field free drift path; this post-source decay creates metastable ions. By stepping the reflector voltage, the metastable ions can be brought into the scope of the detector and their flight times and, hence, fragment masses can be analyzed.

In Figure 3 an example of peptide sequencing with MALDI-PSD is shown. The peptide with the mass of $(M+H)^+ = 1133.4$ Da is mixed 1:1 with a 3,5-dihydroxy benzoic acid (DHB) matrix (10 g/L dissolved in 1:1 acetonitrile/water). The MALDI-TOF instrument employed in this measurements is house built and is used in conjunction with a high-voltage switch for delayed extraction (for details see [Kaufmann 1996]). The sequence of this peptide, shown at the top of the figure is

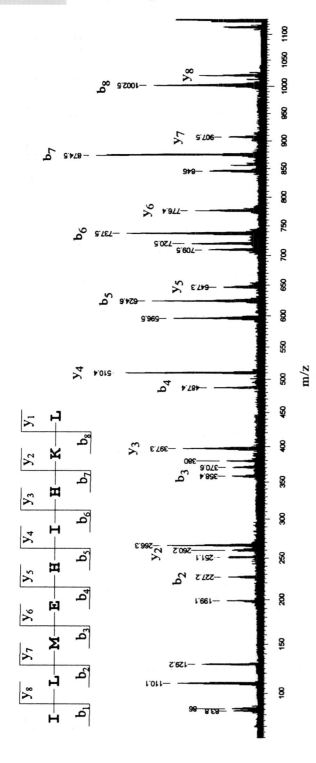

Figure 3. Low energy MALDI-PSD mass spectrum of ILMEHIHKL m/z 1133.4 (Aladim, Kaufmann). The peptide is mixed 1:1 with DHB matrix. At the top of the figure the sequence of the peptide is shown and the corresponding fragment ions of the b- and y-series are marked in the spectrum.

identified as a part of the ribosomal protein L19 using the *Sequest* algorithm [Immler 1997].

A detailed description of the MALDI-PSD technique is given in Chapter III.5.

3.2 ESI-Triple Stage Quadrupole Mass Spectrometry

In a quadrupole mass spectrometer ions of different mass-to-charge ratio will be dispersed using a high frequency voltage. The quadrupole field is formed by four hyperbolic poles (rods). The superposition of a high frequency and a variable dc-voltage applied to the four rods allows the separation of ions due to their different mass-to-charge ratio [Dawson 1976]. A triple stage mass spectrometer with electrospray ionization can be divided roughly into three parts (Figure 4). The first part consists of the ionization source and the electrospray interface. In this part of the instrument the pressure is reduced gradually from atmospheric pressure to the vacuum of the mass spectrometer. The second part consists of the mass analyzers namely the two mass filter (Q1 and Q3) separated by the collision cell (Q2). Performing a MS experiment either the quadrupole Q1 or Q3 is scanned and the other is switched to the so called rf-only mode for highest transmission of all ions. Finally, the detection of ions by a multiplier takes place in the third part.

If a MS/MS experiment is performed on a triple stage mass spectrometer the first quadrupole is used as a mass filter allowing only selected ions (parent or precursor ions) to reach the second quadrupole, where hexapoles or octapoles are employed for better ion transmission. Here, the collision of the parent ions with argon or xenon atoms (collision-induced dissociation (CID)) occurs, producing daughter or product ions. These daughter ions are analyzed in the third quadrupole [Busch 1988].

A triple quadrupole instrument produces ions with 20–50 eV or less energy delivering low energy collision; therefore, fragmentation occurs mainly in the peptide backbone as described before. The optimal collision energy for a good product ion yield depends on the primary structure, the charge and the mass of the parent ion. Additionally to the above mentioned most frequently used acquisition of **daughter ion** spectra two other MS/MS scan modes are available. In the **parent ion** mode all parent ions are detected which yield a certain product ion. During a **neutral loss** scan the elimination of a certain neutral molecule, e.g. water or phosphoric acid is monitored. These scan functions sometimes are useful for the solution of special analytical problems.

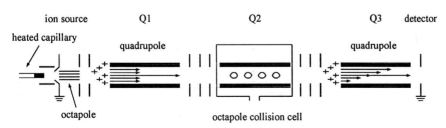

Figure 4. Schematic diagram of a triple stage quadrupole mass spectrometer.

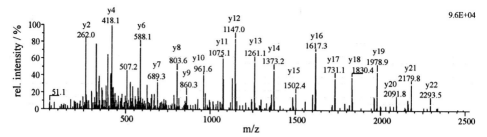

Figure 5. MS/MS spectrum of the peptide AASLISLFTLDELIANTGNTGLGVSR m/z 2633.4 (TSQ 7000, Finnigan MAT). The doubly charged ion m/z 1317.2 was fragmented with argon as collision gas (0.4 Pa), a collision energy of 43 eV and corresponding y-ions are marked.

In Figure 5 a MS/MS spectrum of a tryptic peptide of xylosidase from Aspergillus niger is shown. The mass spectrum was recorded during a HPLC analysis using a 75 µm inner diameter C18-reversed-phase column (see Figure 11). Mass spectrometry was done on a TSQ 7000 equipped with an electrospray ion source. The MS/MS spectrum was acquired in the positive ion mode in the mass range m/z 50 to 2500 with a scan duration of 2 s. Collision-induced dissociation was done with argon as collision gas (0.4 Pa) and a collision energy of 43 eV. The MS/MS spectra are interpreted with the *Sequest* algorithm and parts of the results are shown in Figures 14 to 16 [Sander 1998].

3.3 ESI-Ion Trap Mass Spectrometry

The latest technique for sequencing peptides uses a three-dimensional analog of the quadrupole mass filter, the ion trap [Johnson 1990; Cooks 1991; Jonscher 1997; March 1997]. It consists of three electrodes: two end-cap electrodes at ground potential and between them a ring electrode to which a radio frequency voltage is applied (Figure 6). The ion trap like the quadrupole mass filter employs an alternating electric field to stabilize and destabilize the passing ions. In the case of the ion trap the electric field is used to store ions and to release them in time to be analyzed. The mass separation can be done by several methods. The most important for analytical purposes are:

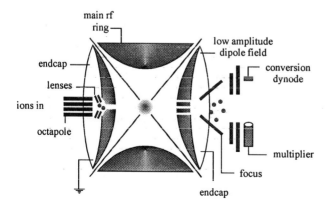

Figure 6. Schematic diagram of an ion trap mass spectrometer.

- Stafford's mass-selective instability mode scans the fundamental rf-voltage amplitude applied to the ring electrode. Thus, ions of increasing mass-to-charge ratio adopt unstable flight paths and leave the trap towards the detector.
- Resonance ejection utilizes again the fundamental rf-voltage amplitude to bring the ions into resonance with the supplementary rf-voltage at the end cap electrodes. Thus they absorb sufficient power to leave the ion trap. In this way the mass range of the ion trap can be extended.

The resonance ejection or axial modulation provides several advantages. The most important is the order of magnitude improvement in sensitivity achieved while maintaining unit resolution. The use of helium damping gas within the trapping volume also improves the resolution and the overall trapping efficiencies of the instrument. The ion trap mass analyzer applied to peptide sequencing generally are combined with an ESI ion source. The externally generated ions must be pulsed using gating lenses to repel ions after the injection and trapping of ions.

A MS/MS experiment starts with ionization and injection of the generated ions into the ion trap. Here the parent ion is isolated and collision-induced dissociation takes place as a result of collisions in the helium damping gas due to the resonance excitation of the ions. The resulting fragment ion spectra are similar to those acquired with a triple quadrupole mass spectrometer. After fragmentation, the CID spectrum of the selected precursor ion is recorded by sequentially ejecting the product ions. In contrast to triple stage quadrupole instruments, one limitation of the ion trap is visible: only masses higher than 28 % of the parent ion can be stabilized inside the ion trap, i.e. the immonium- and side-chain fragments and the low mass fragment ions cannot be detected. On the other hand the ion trap is about 50 times more sensitive than a triple quadrupole mass spectrometer and the resolution can be improved using a special scan mode. The mass resolution of the ion trap mass spectrometer can be increased by reducing the scan speed (normally 5500 amu/s for the LCQ, Finnigan MAT). Due to the longer scan time the mass range is limited to 10 amu around a center mass (Zoom-Scan). In future another advantage of the ion trap in opposite to a triple quadrupole mass spectrometer may yield more sequence information in shorter time: the possibility to do a number of sequential MS/MS operations. Thus, one can perform multiple stages of fragmentation separated in time, i.e. one can select the precursor ion for CID resulting in a daughter ion spectrum, followed by isolation and fragmentation of one selected daughter ion to generate a MS^3 spectrum. In this way up to 10 sequential MS/MS operations can be done, theoretically. However, they are not routine and difficult to interpret.

In Figure 7A a product ion spectrum generated with an ion trap mass spectrometer is depicted. Here the peptide LIQPIQDKIKNE with the mass of $[M+H]^+ = 1438.8$ Da was introduced into the mass spectrometer with a concentration of 2 pmol/µL via a nano-electrospray ion source. The spectrum was recorded with the LCQ ion trap mass spectrometer (Finnigan MAT). In a first experiment the doubly charged peptide ion was fragmented followed by a further collision-induced dissociation of the fragment ion m/z 598.2 in a MS^3-experiment (Fig. 7B). For the determination of the charge state of the ion m/z 598.2 the resolution was increased in a Zoom-Scan (Fig. 7C).

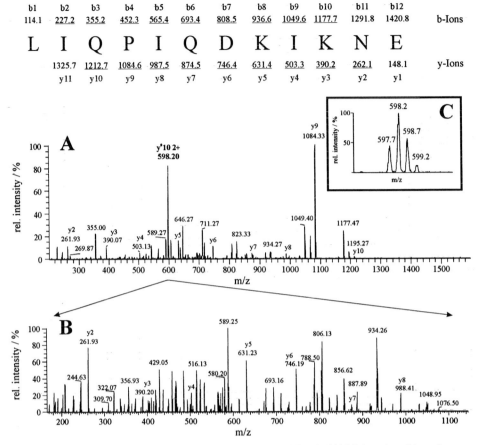

	b1	b2	b3	b4	b5	b6	b7	b8	b9	b10	b11	b12	
	114.1	227.2	355.2	452.3	565.4	693.4	808.5	936.6	1049.6	1177.7	1291.8	1420.8	b-Ions

L   I   Q   P   I   Q   D   K   I   K   N   E

	1325.7	1212.7	1084.6	987.5	874.5	746.4	631.4	503.3	390.2	262.1	148.1	y-Ions
	y11	y10	y9	y8	y7	y6	y5	y4	y3	y2	y1	

Figure 7. Product ion spectra of the peptide LIQPIQDKIKNE m/z 1438.8 introduced into the mass spectrometer via a nano-electrospray ion source. The concentration of the peptide was 2 pmol/µL. **A)** Fragmentation of the doubly charged peptide ion m/z 719.9 (MS^2).) **B** Further collision-induced dissociation of the fragment m/z 598.2 (MS^3). **C)** Zoom-Scan of m/z 598.2 for determination of the charge state (LCQ, Finnigan MAT). At the top of the figure the sequence of the peptide is presented and the y- and b-ions found in the fragment ion spectrum are underlined.

The MS^3-experiment may help to verify the proposed sequence of the product ion. The ion m/z 598.2 is a doubly charged y$^0_{10}$ ion, as can be deduced from the Zoom-Scan (Fig. 7C).

3.4 ESI-Quadrupole-TOF Mass Spectrometry

The high sensitivity of a time-of-flight mass spectrometer is attributed to the simultaneous transmission of all masses to the detector which is typically a fiftyfold over that of scanning mass spectrometers [Guilhaus 1997]. The combination of a quadrupole mass spectrometer and a time-of-flight instrument with orthogonal

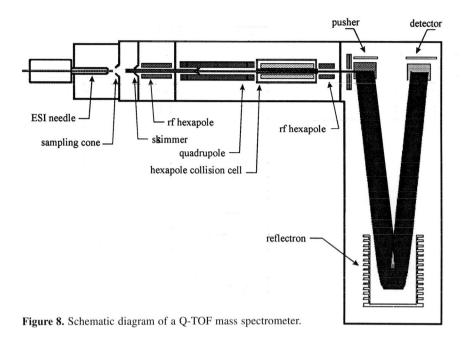

Figure 8. Schematic diagram of a Q-TOF mass spectrometer.

acceleration (Q-TOF) allows the coupling of continuous ion sources like electrospray ionization to a TOF mass analyzer. The use of such an orthogonal acceleration system offers many desirable features: high sensitivity, nearly unlimited mass range, efficient duty cycle, high speed, MS/MS capability and an increased mass resolution when a reflectron is used [Dawson 1989]. Figure 8 shows a schematic assembly of an ESI-Q-TOF.

By switching the quadrupole in the rf-only mode, mass spectra are directly acquired using the time-of-flight mass spectrometer. The acquisition of MS/MS spectra is accomplished by switching the quadrupole in the narrow bandpass mode transmitting only a selected parent ion. The first hexapole is an rf-only hexapole increasing the transmission of ions. The second hexapole is used in the MS/MS mode as a collision cell. The TOF mass spectrometer itself is a discontinuous mass analyzer whereas a pusher is used to generate ion packages which are analyzed due to the different flight time of ions with varying mass-to-charge ratio. The moderate to high resolution facilitates the interpretation of MS and MS/MS spectra due to the straightforward assignment of the charge state of multiply charged ions.

In Figure 9 a MS/MS spectrum of the doubly charged ion of the peptide GDHFA-PAVTLYGK with the mass of $[M+H]^+ = 1375.7$ Da is shown. The sample was introduced into the mass spectrometer via a nano-electrospray ion source with a concentration of 2 pmol/µL. The measurement was done on a Q-TOF mass spectrometer (Q-TOF, Micromass).

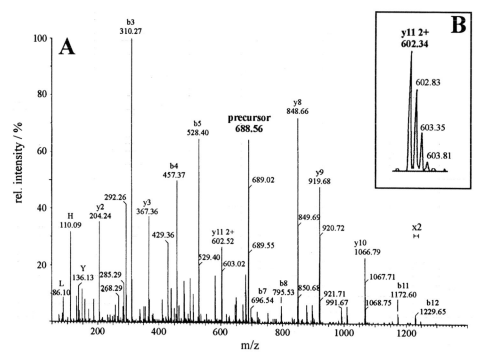

Figure 9. A) MS/MS spectrum of the peptide GDHFAPAVTLYGK m/z 688.4 introduced into the mass spectrometer via a nano-electrospray ion source with a concentration of 2 pmol/μL. **B)** Resolution of the doubly charged ion m/z 602.34 (Q-TOF, Micromass).

4 Coupling Methods for Mass Spectrometry

The high sensitivity of the now available mass spectrometer allows the analysis of subpicomole amounts of protein or peptide. To handle these small amounts of material the transfer steps and the sample volumes have to be reduced because of unspecific adsorption of peptides to the surface. So on-line coupling methods for mass spectrometry are developed predominantly to analyze the sample without any loss of material. In many cases off-line techniques are imperative, but there are ways to minimize or avoid sample loss [Immler 1997]. In Figure 10 possible on-line and off-line techniques are presented, which are described in the following chapter.

4.1 HPLC Mass Spectrometry

The development of a reliable on-line LC-MS coupling has been accompanied with some compatibility problems namely the composition of mobile phase, the high flow rate previously used and the nature of the investigated analytes which are normally polar in HPLC analysis. In the past 20 years a variety of interfaces have been constructed [Linscheid 1994; Niessen 1995; Barcelo 1996]. However, electrospray

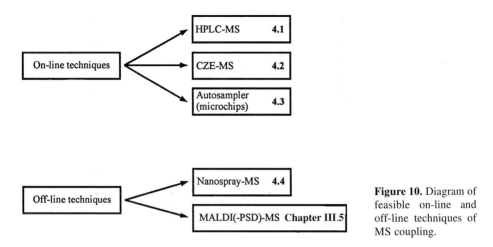

Figure 10. Diagram of feasible on-line and off-line techniques of MS coupling.

ionization is the most widely used interface for the on-line LC-MS analysis of polar ionic compounds therefore also for proteins and peptides [Wachs 1991]. Flow rates in the range of 0.1 µL/min to 1 mL/min are easily tolerated. The limitations in the choice of mobile phases which should be volatile are compensated through the mass detection capability of the detector.

For protein identification, first a LC-MS-analysis can be done with an aliquot of the protein digest separating the peptide mixture by reversed-phase HPLC. An advantage of this procedure is that residual proteolytic enzymes and buffer salts, which may disturb mass spectrometric measurements, are removed. The sample buffer should contain some organic solvent [Langen 1993] and trifluoroacetic acid (TFA) should be used at low concentrations (< 0.025 %) or avoided (it drastically quenches the ion yield) and replaced by acetic acid or formic acid, if compatible. Hydrophobic proteins and peptides which are not soluble in solutions usually used for electrospray MS may be analyzed in chloroform/methanol/water-mixtures as well [Schindler 1993].

The LC-MS analysis yields a peptide mass fingerprint (peptide map) which is characteristic for the digested protein. With this information a database search is possible and the protein can be identified if it is already known and in the database. Afterwards another part of the digest is committed to electrospray tandem mass spectrometry, whereby primary structure information can be obtained for every peptide by collision-induced dissociation (LC-MS/MS) and therefore the matched protein can be verified.

Two programs for the automated LC-MS/MS analysis of protein digests on a TSQ 7000 (Finnigan MAT) have been developed [Stahl 1992; Mylchreest 1995] at which the choice of collision parameters depending among other things on the mass-to-charge ratio and the collision gas pressure may be considered [Haller 1996; Moritz 1996]. They use the instrument control language (ICL) to fully automate the switching from MS- to MS/MS-mode and the acquisition of MS/MS spectra. At the end of each scan in MS-mode both programs determine the peak intensity and the signal-to-noise ratio of the base peak. If one of these two parameters is below a specified value no MS/MS analysis is done from this base peak ion and the next scan in MS-mode is done. If the parameters fit the specified values the instrument switches into MS/MS-mode. Here the two programs differ:

- The first program is used for MS/MS analysis of unknown compounds, where CID spectra are desired for all abundant ions found in a protein digest.
- The second program considers ions for MS/MS analysis only if their masses are present in a user list. Here, CID analysis is performed to specific ions expected from a protein digest.

When the criteria for MS/MS analysis are fulfilled several scans are accumulated before the instrument is again switched to MS-mode.

In Figure 11 an ESI-LC-MS chromatogram of a tryptic digest of xylosidase is shown. The peptide separation was done on a 75 µm inner diameter nano-HPLC column. First a peptide mass fingerprint was acquired followed by an automatic LC-MS/MS analysis of selected peptide masses; one of these MS/MS spectra is shown in Figure 5. The intensity of the reconstructed ion current in Figure 11B decreases dramatically due to the automatic switching into the MS/MS mode if the above described conditions are fulfilled. One limitation in performing MS/MS experiments during a HPLC-analysis on a triple stage mass spectrometer is the time-consuming process to

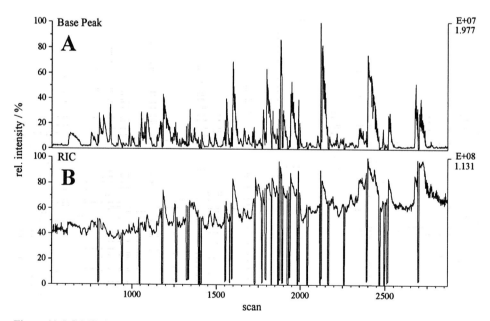

Figure 11. LC-MS chromatogram of tryptic digested xylosidase (TSQ 7000, Finnigan MAT). Peptides were separated on a Hypersil C18 BDS reversed-phase column (150 mm x 75 µm, LC Packings) with a flow rate of 200 nL/min. Elution of the peptides was done with a gradient starting at 95 % 0.02 % TFA (A). Content of solution B (acetonitrile/water/TFA (84/16/0.025)) was increased to 50 % in 91 min and from 50 % to 100 % in 15 min. **A)** Base peak chromatogram. **B)** Reconstructed ion chromatogram. Results shown in Figures 5, 14 and 16 belong also to this experiment.

Table 2. Parameters of capillary- and nano-HPLC columns.

Column	I. D. [mm]	Flow rate [μL/min]	Capacity [mg]	Gain in sensitivity
Conventional-LC	4.6	400–2000	1000	1
Narrowbore-LC	2.0	50–400	200	5
Microbore-LC	1.0	20–100	50	35
Capillary-LC	0.1–0.5	1–100	5	200
Nano-LC	0.05–0.1	0.1–1	0.2	3500

reach the desired collision gas pressure. A higher performance without filling and evacuating the collision cell can be realized with an ion trap mass spectrometer.

Since a mass spectrometer equipped with an electrospray interface acts like a concentration sensitive detector, the dimensions of the HPLC column can be decreased without a loss in sensitivity if the sample amount is reduced in the same way. Normally, the protein amount is limited and the capacity of columns with 4 mm or 2 mm inner diameter is not fully used. Therefore, if capillary- or nano-HPLC columns are used the injected sample amount can be increased without an overload of the column. Large volume injection of peptides in aquatic solutions can be performed using capillary- or nano-HPLC columns drastically increasing mass spectrometric sensitivity. The diameters and flow rates of capillary- and nano-HPLC columns are summarized in Table 2. The low flow rates are achieved by precolumn splitting of the HPLC eluent via a T-piece.

4.2 CZE Mass Spectrometry

Coupling of capillary electrophoresis with mass spectrometry (CE-MS) combines the high separation efficiency of CE with information about molecular masses and, in the case of MS/MS with structural information. During the last years several ionization methods for CE-MS have been developed: plasma desorption and MALDI work via fraction collection as **off-line** coupling methods, electrospray ionization, ion spray and continuous-flow fast atom bombardment (FAB) have been used for **on-line** coupling. Essentially three CE-MS interfaces have been reported: the widely used coaxial sheath-flow interface, the liquid junction interface and the sheathless interface [Smith 1993; Cai 1995]. But there are some limitations, which have prevented the widely accepted use of CE-MS:

- The limited sample volume, which can be analyzed without decreasing separation efficiency causes a high concentration detection limit.
- The analyte migration time tends to fluctuate.
- The reproducibility of CE-MS is not as good as of LC-MS.
- The limitation in electrolyte selection.

Concerning the poor concentration sensitivity several attempts for improvement are reported, e.g. on-line preconcentration, the use of mass spectrometer types with higher sensitivity and the use of coated capillaries with small inner diameters [Smith 1993; Cai 1995]. Nevertheless CE-MS is not the mostly used instrument for sequence analysis of proteins and peptides at the moment, although fragmentation of peptides in the attomole range is reported [Valaskovic 1996].

4.3 Microchips

Miniaturization of sample introduction systems shows some advantages over conventional systems such as low cost, high speed, compact size, small surface contact of the analyte and parallel analyses. One micro-instrumentation which possesses these advantages is the microchip. Microchips are low cost devices which are fabricated on glass, quartz or plastic substrates using standard micro-machining techniques such as thin-film deposition, photolithography or wet chemical etching [Paulus 1997]. Electrophoresis, electrochromatography or liquid chromatography with different packing materials can be done on microchips [Jacobson 1994a; Jacobson 1994b; Jacobson 1997]. The coupling of microchip devices with electrospray mass spectrometry was shown by Ramsey [1997]. If a high speed separation will be done on a microchip a fast scanning mass spectrometer such as a TOF-MS must be used. As a continuous infusion method for multiple samples such micro-fabricated devices are used for rapid protein identification using an ion trap mass spectrometer (Figure 12) [Figeys 1997].

4.4 Nanospray

Another strategy for acquiring MS/MS data is the use of the nanospray ion source without any chromatographic separation of peptides [Wilm 1996a]. Typically, electrospray sources operate at a flow rate of 1–10 µL/min generating droplets with an initial diameter of more than 1 µm. In this flow rate range only a small part of analyte molecules will be introduced into the mass spectrometer. The use of an electrospray device which yields flow rates below 30 nL/min is very efficient since smaller

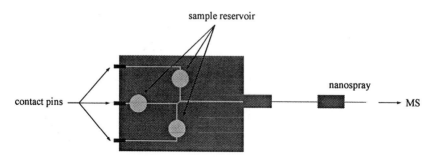

Figure 12. Schematic diagram of a microchip used as a micro-autosampler [Figeys 1997].

initial droplets are formed and the sample consumption is very low. The electrospray needle will be made from 0.5–1 mm i.d. glass capillary tubing which is heat softened and pulled down to form tips of 1–3 µm i.d., for electrical conductivity the tip is gold coated. Without additional pumping these device yields a stable electrospray with improved desolvation and ionization efficiency [Wilm 1996b]. Several groups [Wilm 1996c; Blackburn 1997] have shown that protein identification can be done with nanospray in the low femtomole range. McLafferty et al. [Valaskovic 1995] have used fused silica capillary tubing with an inner diameter of 5 to 20 µm which are pulled and etched to form tips of 2 µm i.d. with a flow rate of approximately 1 nL/min. For practical use the positioning of the nanospray needle in front of the orifice of the mass spectrometer is done using a microscope or a CCD camera. Mass spectrometric measurements can be done with 1 to 2 µL sample volume over 30 to 60 min. The low flow rate of the nanospray source provides enough time to optimize the collision energy for each peptide individually. Thus for peptide sequencing the best tandem mass spectrometric data can be obtained. Nano-electrospray mass spectrometry was also successfully employed in protein identification from proteins separated by polyacrylamide gel electrophoresis [Wilm 1996c; Immler 1998]. A protocol of the in-gel digestion procedure of a protein for use with nano-electrospray is given in Figure 13. This protocol works with volatile buffers and needs no prior desalting step. The solution volumes used should be as small as possible.

5 Identification of Posttranslational Modifications

5.1 Disulfide Bond Location

One problem in sequencing proteins and peptides is the localization of disulfide bonds. The principle is to digest the oxidized form of the protein with a specific endoprotease and to determine the molecular weight of the resulting peptides by mass spectrometry. The re-determination of the molecular weight of the peptides after reduction and alkylation gives information about the total number of disulfide bonds. Peptides with an internal -S-S- bridge increase in mass by 57 Da (carbamidomethylation), 58 Da (carboxymethylation) or 105 Da (ethylpyridylation) per cysteine, respectively. On the other hand the signal of two peptides linked by a disulfide bond will disappear after reduction and two new signals will arise. Through knowledge of the total number of disulfide bonds and the peptides where they appear the disulfide bonds can be identified [Biemann 1992].

Recently, a novel approach for assignment of disulfide bonds using MALDI mass spectrometry was published [Wu 1997]. The denatured protein undergoes partial reduction and subsequent cyanylation. The mixture of intact protein and partially reduced and cyanylated species is separated by reversed-phase HPLC. A following mass analysis identifies singly reduced isomers by a mass shift of 52 Da. These isomers are subjected to specific chemical cleavage and complete reduction. For localization of disulfide bonds the resulting peptides are again analyzed by MALDI mass spectrometry.

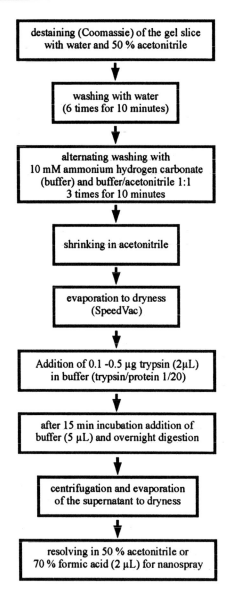

Figure 13. Protocol of in-gel digestion of a protein for use with nano-electrospray MS.

5.2 Other Posttranslational Modifications

After determination of the amino acid sequence of a protein, the next step is to search for posttranslational modifications, like deamidation, oxidation, glycosylation, phosphorylation, sulfation, methylation, blocked N-, C-termini or ragged C-/N-terminal ends [Covey 1991] which may give a hint to the function or localization of the protein in the cell. In this area mass spectrometry again renders good service as to see in the publication of Qin and Chait [Qin 1997].

Table 3. Some posttranslational modifications and their corresponding mass changes.

Modification	Monoisotopic mass change	Average mass change
Homoserine formed from Met by CNBr treatment	−29.9928	−30.0935
Pyroglutamic acid formed from Gln	−17.0265	−17.0306
Disulphide bond formation	−2.0157	−2.0159
C-Terminal amide formed from Gly	−0.9840	−0.9847
Deamidation of Asn and Gln	−0.9840	−0.9847
Methylation	14.0157	14.0269
Hydroxylation	15.9949	15.9994
Oxydation of Met	15.9949	15.9994
Proteolysis of a single peptide bond	18.0106	18.0153
Formylation	27.9949	28.0104
Acetylation	42.0106	42.0373
Carboxylation of Asp and Glu	43.9898	44.0098
Phosphorylation	79.9663	79.9799
Sulfation	79.9568	80.0642
Cysteinylation	119.0041	119.1442
Pentoses (Ara, Rib, Xyl)	132.0423	132.1161
Deoxyhexoses (Fuc, Rha)	146.0579	146.1430
Hexosamines (GalN, GlcN)	161.0688	161.1577
Hexoses (Fru, Gal, Glc, Man)	162.0528	162.1424
Lipoic acid (amide bond to lysine)	188.0330	188.3147
N-Acetylhexosamines (GalNAc, GlcNAc)	203.0794	203.1950
Farnesylation	204.1878	204.3556
Myristoylation	210.1984	210.3598
Biotinylation (amide bond to lysine)	226.0776	226.2994
Pyridoxal phosphate (Schiff base formed to lysine)	231.0297	231.1449
Palmitoylation	238.2297	238.4136
Stearoylation	266.2610	266.4674
Geranylgeranylation	272.2504	272.4741
N-Acetylneuraminic acid (Sialic acid, NeuAc, NANA, SA)	291.0954	291.2579
Glutathionylation	305.0682	305.3117
N-Glycolylneuraminic acid (NeuGc)	307.0903	307.2573
5′-Adenosylation	329.0525	329.2091
4′-Phsophopantetheine	339.0780	339.3294
ADP-ribosylation (from NAD)	541.0611	541.3052

If the amino acid sequence of a protein is known, differences between the expected molecular weight and the mass determined by mass spectrometry can be related to a posttranslational modification (Table 3). A more comprehensive list of mass changes observed for common chemical and enzymatic modifications is available for use on the World Wide Web [http://www.medstv.unimelb.edu.au/WWWDOCS/SVIMR-Docs/MassSpec/deltamassV2.html]. The accuracy of the mass measurement may enable identification of a posttranslational modification, perhaps in combination with other data. The expected modification is confirmed if it can be reversed by a specific process, e.g. the treatment of a glycoprotein with glycosidase, followed by a redetermination of the molecular weight, and a corresponding decrease in molecular mass may confirm the glycosylation of the analyzed protein.

One of the most abundant posttranslational modifications is the reversible phosphorylation of a protein. There exist different forms of amino acid phosphates: O-phosphates, N-phosphates, S-phosphates and acyl phosphates. Most common are O-phosphoamino acids like phosphoserine, -threonine and -tyrosine. Phosphoserine- and -threonine-containing peptides can be identified by the loss of their phosphate group (**neutral loss**) by increasing the orifice potential [Ding 1994], by increasing the collision energy in the collision cell [Hunter 1994] or in the octapole region in front of the first quadrupole in combination with a hybrid-scanning technique [Jedrzejewski 1995]. The phosphate group may be eliminated as H_3PO_4 (decrease in mass: 98 Da, or 49 Da for the doubly charged ion) or HPO_3 (decrease in mass: 80 Da, respectively 40 Da for the doubly charged ion). Thus, serine- and threonine-phosphorylated peptides can be identified by using the negative ion mode (detecting 79 Da for PO_3^-) or the **neutral loss** scan function of a quadrupole mass spectrometer. The latter means screening for all doubly charged peptides showing an additional signal with a mass difference of 49 or 40 Da between the phosphopeptide and the dephosphorylated peptide [Covey 1991; Meyer 1993]. Phosphotyrosine-containing peptides may be established by the difference between determined and expected molecular mass, because they do not undergo neutral loss. Identification of the specific phosphorylated serine, threonine or tyrosine in a phosphorylated peptide may be performed by tandem mass spectrometry or by other methods [Meyer 1993; Busman 1996].

6 Interpretation of Mass Spectrometric Data

6.1 Identification of Proteins by Database Search

The identification of protein components which are already listed in a database can be achieved by a peptide map according to the fragment masses [Pappin 1993]. Thus, the protein is digested with specific proteases and the masses of the resulting peptides will be determined by mass spectrometry. The obtained masses are screened against a database. Database search will be more and more successful with the continuation of genome sequencing. The different search programs mostly available in the World Wide Web will be mentioned in short since they are explained in detail in chapter IV.1 of this book. Some of these programs use only peptide masses for a database search, other ones need MS/MS data or additional informations. The first category includes the following programs:

- *MS-Fit* from the UCSF Mass Spectrometry Facility
  http://falcon.ludwig.ucl.ac.uk/msfit.htm
  http://prospector.ucsf.edu/htmlucsf/msfit.htm

- *PeptideSearch* from EMBL in Heidelberg, protein identification by peptide mass data
  http://www.mann.embl-heidelberg.de/Services/PeptideSearch/FR_PeptideSearch-Form.html

- *MOWSE* molecular weight search peptide-mass database from the SERC Daresbury Laboratory
  http://gserv1.dl.ac.uk/SEQNET/mowse.html

- *ProFound* from the Rockefeller University
  http://prowl.rockefeller.edu/cgi-bin/ProFound

- *PEPTIDE MASS SEARCH* from the Max-Delbrück Center for molecular medicine
  http://www.mdc-berlin.de/~emu/peptide_mass.html

The second category includes the programs:

- *MS-Tag* from the UCSF Mass Spectrometry Facility
  http://falcon.ludwig.ucl.ac.uk/mstag.htm
  http://prospector.ucsf.edu/htmlucsf/mstag.htm

- *PeptideSearch* from EMBL in Heidelberg, protein identification by peptide sequence tags [Mortz 1996]
  http://www.mann.embl-heidelberg.de/Services/PeptideSearch/FR_PeptidePattern-Form.html

The previously mentioned programs are available via the World Wide Web. In contrast another most promising algorithm for protein identification is distributed by Finnigan MAT :

- *Sequest* from John Yates and Jimmy Eng [Eng 1994], an automatic program for database search using uninterpreted MS/MS data. Descriptions of the routine are noted at:
  http://thompson.mbt.washington.edu/sequest

In the following *Sequest* is explained in detail. *Sequest* identifies sequences from MS/MS spectra by comparing a measured spectrum to calculated spectra of peptides which are taken from a protein or nucleic acid database. The minimum input for this algorithm is the peptide mass, the charge state and the spectrum. The program package consists of several files: *lcms_dta*, which extracts and pre-processes the acquired MS/MS spectra from the TSQ7000 LC-MS/MS data file which are in the ICIS format to the final input format (Figure 14).

The generated input file contains the m/z value of $[M+H]^+$ and the charge of the parent ion followed by the m/z values and intensities of the reduced mass spectrometric data. The determination of the charge and the molecular weight from the parent ion should be controlled since the automatic assignment is not always correct. Sequence tags or immonium ions can be added next to the charge state of the peptide (Fig. 14A). For optimal search results some parameters can be varied in the *sequest.-params* file before running *Sequest*. First, the database has to be chosen (Figure 15).

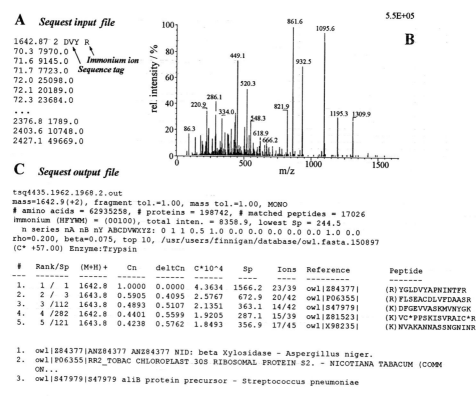

A *Sequest input file*

```
1642.87  2  DVY  R
70.3  7970.0
71.6  9145.0        Immonium ion
71.7  7723.0        Sequence tag
72.0  25098.0
72.1  20189.0
72.3  23684.0
...
2376.8  1789.0
2403.6  10748.0
2427.1  49669.0
```

C *Sequest output file*

```
tsq4435.1962.1968.2.out
mass=1642.9(+2), fragment tol.=1.00, mass tol.=1.00, MONO
# amino acids = 62935258, # proteins = 198742, # matched peptides = 17026
immonium (HFYWM) = (00100), total inten. = 8358.9, lowest Sp = 244.5
   n series nA nB nY ABCDVWXYZ: 0 1 1 0.5 1.0 0.0 0.0 0.0 0.0 0.0 1.0 0.0
rho=0.200, beta=0.075, top 10, /usr/users/finnigan/database/owl.fasta.150897
(C* +57.00) Enzyme:Trypsin
```

#	Rank/Sp	(M+H)+	Cn	deltCn	C*10^4	Sp	Ions	Reference	Peptide
1.	1 / 1	1642.8	1.0000	0.0000	4.3634	1566.2	23/39	owl|Z84377|	(R) YGLDVYAPNINTFR
2.	2 / 3	1643.8	0.5905	0.4095	2.5767	672.9	20/42	owl|P06355|	(R) FLSEACDLVFDAASR
3.	3 /112	1643.8	0.4893	0.5107	2.1351	363.1	14/42	owl|S47979|	(K) DFGEVVASKMVNYGK
4.	4 /282	1642.8	0.4401	0.5599	1.9205	287.1	15/39	owl|Z81523|	(K) VC*PPSKISVRAIC*R
5.	5 /121	1643.8	0.4238	0.5762	1.8493	356.9	17/45	owl|X98235|	(K) NVAKANNASSNGNINR

```
1.  owl|Z84377|ANZ84377 ANZ84377 NID: beta Xylosidase - Aspergillus niger.
2.  owl|P06355|RR2_TOBAC CHLOROPLAST 30S RIBOSOMAL PROTEIN S2. - NICOTIANA TABACUM (COMM
    ON...
3.  owl|S47979|S47979 aliB protein precursor - Streptococcus pneumoniae
```

Figure 14. A) *Sequest* input file. B) MS/MS spectrum of a tryptic xylosidase peptide. C) *Sequest* output file.

Database search can be performed using protein- or DNA-databases which can be downloaded via the World Wide Web [e.g. ftp://ncbi.nlm.nih.gov/repository/OWL, ftp://ncbi.nlm.nih.gov/repository/dbEST]. The *owl* database is a nonredundant database comprising protein sequences from GenBank (National Centre for Biotechnology Information, Washington, DC), SWISS-PROT (Swiss Institute of Technology), Protein Information Resource (National Biology Resource Foundation, Georgetown, Washington, DC) and National Research Laboratory (Brookhaven National Laboratory, Brookhaven, NY) databases. *EST* databases (**E**xpressed **S**equence **T**ags) are randomly sequenced cDNA libraries. Furthermore, the weighting of fragment ions must be chosen. In the case of high energy collision fragments from side-chain elimination should also be considered. Additional information like the specificity of the protease, modifications of cysteine etc. may be used to narrow down the search results. A highlight of this program is the possibility to look for modifications like phosphorylation of amino acids or acetylation of the N-terminus. It is not only possible to identify them but also to localize them exactly within a peptide [Yates 1995]. If a modification is always present at the amino acid the growth in amino acid mass should be added to the relevant amino acid below line six. If the modification is not necessarily present the information should be inserted in line four (e. g. carboxymethylation of cysteine).

```
/usr/users/finnigan/database/owl.fasta      | database file
0 1 0 1 0                                    | 0/1 = prot/nuc;long off/on;short off/on;avg/mono masses;print ions off/on
0 1 1 0.5 1.0 0.0 0.0 0.0 0.0 0.0 0.0 1.0 0.0 | (neutral-a)(neutral-b)(neutral-y) a b c d v w x y z
58 C  ←— Carboxymethylation of cysteine      | mass, amino acid for differential modification amino acid (up to 2 mods)
1 2                                          | fragment tolerance (use integers here), mass tolerance
10 3 1  ←— Trypsin                           | # of output lines (1-500), # description lines (1-500), enzyme # (below)
0.0   | added to C-terminus (peptide mass & all Y"-ions)
0.0   | added to N-terminus (B-ions)
0.0   | added to G - Glycine         avg.  57.0519,  mono.  57.02146
0.0   | added to A - Alanine         avg.  71.0788,  mono.  71.03711
0.0   | added to S - Serine          avg.  87.0782,  mono.  87.02303
0.0   | added to P - Proline         avg.  97.1167,  mono.  97.05276
0.0   | added to V - Valine          avg.  99.1326,  mono.  99.06841
0.0   | added to T - Threonine       avg. 101.1051,  mono. 101.04768
0.0   | added to C - Cysteine        avg. 103.1388,  mono. 103.00919
0.0   | added to L - Leucine         avg. 113.1594,  mono. 113.08406
0.0   | added to I - Isoleucine      avg. 113.1594,  mono. 113.08406
0.0   | added to X - L or I          avg. 113.1594,  mono. 113.08406
0.0   | added to N - Asparagine      avg. 114.1038,  mono. 114.04293
0.0   | added to O - Ornithine       avg. 114.1472,  mono. 114.07931
0.0   | added to B - avg. of N & D   avg. 114.5962,  mono. 114.53494
0.0   | added to D - Aspartic acid   avg. 115.0886,  mono. 115.02694
0.0   | added to Q - Glutamine       avg. 128.1307,  mono. 128.05858
0.0   | added to K - Lysine          avg. 128.1741,  mono. 128.09496
0.0   | added to Z - avg. of Q & E   avg. 128.6231,  mono. 128.55059
0.0   | added to E - Glutamic acid   avg. 129.1155,  mono. 129.04259
0.0   | added to M - Methionine      avg. 131.1926,  mono. 131.04049
0.0   | added to H - Histidine       avg. 137.1411,  mono. 137.05891
0.0   | added to F - Phenylalanine   avg. 147.1766,  mono. 147.06841
0.0   | added to R - Arginine        avg. 156.1875,  mono. 156.10111
0.0   | added to Y - Tyrosine        avg. 163.1760,  mono. 163.06333
0.0   | added to W - Tryptophan      avg. 186.2132,  mono. 186.07931
```

#	name	offset	sites	no-sites
0.	No_Enzyme	0	-	-
1.	Trypsin	1	KR	P
2.	Chymotrypsin	1	FWY	P
3.	Clostripain	1	R	-
4.	Cyanogen_Bromide	1	M	-
5.	IodosoBenzoate	1	W	-
6.	Proline_Endopept	1	P	-
7.	Staph_Protease	1	E	-
8.	Trypsin_K	1	K	P
9.	Trypsin_R	1	R	P
10.	AspN	0	D	-
11.	Cymotryp/Modified	1	FWYL	P
12.	Elastase	1	ALIV	P
13.	Elastase/Tryp/Chymo	1	ALIVKRWFY	P

Figure 15. *Sequest.params* file.

measured mass · *peptide mass* · *cross correlation value* · *preliminary score value* · *preliminary rank* · *matched ions/occurred ions* · *database entry* · *peptide sequence*

#	MassI	MassA	Xcorr*	DelCn*	Sp	RSp	Ions	Reference*	Sequence**
6	1664.3	1663.7	5.3142	0.543	2013.0	1	25/ 39	owl|Z84377|	(R)SHLIC*DESATPYDR
32	2633.4	2633.4	6.8643	0.506	2396.9	1	32/ 75	owl|Z84377| +1	(R)AASLISLFTLDELIANTGNTGLGVSR
25	1638.6	1637.9	5.4222	0.464	2586.4	1	26/ 42	owl|Z84377| +1	(R)TLIHQIASIISTQGR
29	2178.5	2177.1	3.2207	0.428	1007.6	1	22/ 57	owl|AB00347	(K)NGVFTATDNWAIDQIEALAK
20	1643.0	1642.8	4.6872	0.405	1148.8	1	24/ 39	owl|Z84377| +1	(R)YGLDVYAPNINTFR
28	2024.3	2025.1	4.2986	0.393	1758.6	1	47/102	owl|Z84377| +1	(R)LGLPAYQVWSEALHGLDR
7	1677.8	1677.7	3.8579	0.380	1261.2	2	21/ 39	owl|Z84377|	(R)SHLIC°DESATPYDR

. . .

1	owl|Z84377|	124 (12,0,0,1,0) {3 4 5 6 7 10 15 20 22 25 28 32} beta Xylosidase, Aspergillus niger
2	owl|AB00079	20 (1,0,1,1,0) {14} beta Xylosidase resp. Xylosidase with 2 exchanges
3	owl|P32138|	10 (1,0,0,0,0) {31}
4	owl|Z54174|	10 (1,0,0,0,0) {30}
5	owl|U23685|	10 (1,0,0,0,0) {19}

. . .

C* = carboxymethylated cysteine
C° = acrylamide modified cysteine

Figure 16. *Sequest_summary* file. About 30 peptide ions have been fragmented and 21 of them have been identified unambiguously as xylosidase peptides. Not identified peptide masses may contain amino acid changes in comparison to the database sequence or may have posttranslational modifications, so that the Sequest algorithm cannot match theoretical and found masses.

In Figure 14 the *Sequest* input and output file of one MS/MS spectrum of the xylosidase digest are shown together with the acquired MS/MS spectrum. The search results are easily evaluated using a WWW-browser.

When running the program each entry of the database is screened for peptides that match the given mass within a certain range. The best match is made via a combination of an ion intensity based preliminary score and a cross correlation analysis. The final score value is normalized to 1.0 and is termed C_n. The score value and the ΔC_n value allow to judge whether the similarity is significant or not. In the *sequest_summary* file (Figure 16) a concise overview of the search results of all MS/MS data sets is presented. Here, only the protein which has the highest score value in each output file is shown. Additionally, a ranking of the best matching proteins is given which allows the identification of the analyzed protein.

A previous selection of the database will drastically reduce the runtime of the search. The *select* program allows to generate a new database which contains for example only proteins from human or mouse. Naturally, only known sequences can be identified by this approach. However, as more and more sequences become available in databases, this approach can successfully be applied in many, if not most cases.

6.2 De novo Sequencing of Peptides

The programs described above have besides many advantages some disadvantages in the interpretation of product ion spectra. The *Sequest* algorithm is not tolerant of sequence variants because of the peptide mass prefilter and it is not possible to make non-identical matches between query peptides and database homologs. In this context an interesting method of labelling the C-terminal ion series x_n, y_n and z_n has to be mentioned which simplifies the de novo interpretation of MALDI-PSD- or ESI-product ion spectra [Schnölzer 1996]: the incorporation of ^{18}O into peptide fragments by usage of endoproteases. This results in C-terminal fragment ions with a mass shift of 2 or 4 Da, i.e. using a mixture of $H_2^{16}O/H_2^{18}O$ generates ion doublets. Together with accurate mass determination of peptide fragments this method increases the precision of sequence determination [Shevchenko 1997]. A drawback of this strategy is up to now the manual interpretation of MS/MS data which is a real time consuming process.

J. A. Taylor and R. S. Johnson [Taylor 1997] have described a computer program using sequence database searches via de novo peptide sequencing by tandem mass spectrometry. The *Lutefisk97* algorithm is based on graph theory and rescores the final list of completed sequences using a combination of an ion intensity-based score plus a cross correlation score. The ambiguity of such a de novo interpretation can be reduced if the de novo peptide sequencing program is used in conjunction with a public available search program (FASTA) which is implemented in the CIDentify. The program allows also the interpretation of modified peptides and the use of some enzymes. The source code for use on multiple platforms and a compiled application for the Macintosh PowerPC is available via World Wide Web [http://www.lsbc.com:70/Lutefisk97.html]. Here also an illustrated example of how the program works is shown.

However, at the moment no perfectly working program for the interpretation of completely unknown peptides is available. For this reason most mass spectrometric companies are working to improve their *de novo* interpretation programs which are implemented in the instrument software packages.

7 References

Bahr, U., Karas, M., Kellner, R. (1998) *Rapid Commun. Mass Spec.* **12**, 1382–88. Differentation of Lysine / Glutamine in Peptide Sequence Analysis by Electrospray Ionization Sequential Mass Spectrometry compled with a Quadrupole Ion Trap.

Banks, J. F. and Whitehouse, C. M. (1996) *Methods Enzymol.* **270**, Chap. 21, 486–519. Electrospray ionization mass spectrometry.

Barcelo, D. (ed.) (1996) *Journal of chromatography library* – volume 59. Applications of LC-MS in environmental chemistry. Elsevier Scientific Publishing Company, Amsterdam-Lausanne-New York-Oxford-Shannon-Tokyo.

Beavis, R. C. and Chait, B. T. (1996) *Methods Enzymol.* **270**, Chap. 22, 519–551. Matrix-assisted laser desorption ionization mass spectrometry of proteins.

Biemann, K. (1990) *Biological Mass Spectrometry*, ed. by Burlingame, A.L.; McCloskey, J.A., Elsevier, Amsterdam-Oxford-New York-Tokyo, 179–196. Applications of tandem mass spectrometry to peptide and protein structure.

Biemann, K. (1992) *Annu. Rev. Biochem.* **61**, 977–1010. Mass spectrometry of peptides and proteins.

Blackburn, R. K. and Anderegg, R. J. (1997) *J. Am. Soc. Mass Spectrom.* **8**, 483–494. Characterization of femtomole levels of proteins in solution using rapid proteolysis and nano-electrospray ionization mass spectrometry.

Busch, K. L.; Glish, G. L. and McLuckey, S. A. (1988) *Mass spectrometry/mass spectrometry: techniques and applications of tandem mass spectrometry.* VCH Verlagsgesellschaft mbH, Weinheim.

Busman, M.; Schey, K. L.; Oatis, J. E. and Knapp, D. R. (1996) *J. Am. Soc. Mass Spectrom.* **7**, 243–249. Identification of phosphorylation sites in phosphopeptides by positive and negative mode electrospray ionization-tandem mass spectrometry.

Cai, J. and Henion, J. (1995) *J. Chrom. A* **703**, 667–692. Capillary electrophoresis-mass spectrometry.

Cooks, R.G.; Glish, G.L.; McLuckey, S.A. and Kaiser, R.E. (1991) *C&EN*, 26–41. Ion trap mass spectrometry.

Cotter, R. J. (1992) *Anal. Chem.* **64**, 1027A–1039A. Time-of-flight mass spectrometry for the structural analysis of biological molecules.

Covey, T.; Shushan, B.; Bonner, R.; Schröder, W. and Hucho, F. (1991*) Methods in Protein Sequence Analysis*, ed. by Jörnvall, H.; Höög, J.O. and Gustavsson, A.M., Birkhäuser Verlag, Basel, Boston, Berlin, 249–256. LC/MS and LC/MS/MS screening for the sites of posttranslational modification in proteins.

Dawson, P. H. (ed.) (1976) *Quadrupole mass spectrometry and its applications.* Elsevier Scientific Publishing Company, Amsterdam-Oxford-New York.

Dawson, J. H. J. and Guilhaus, M. (1989) *Rapid Commun. Mass Spec.* **3**, 155–159. Orthogonal-acceleration time-of-flight mass spectrometer.

Ding, J.; Burkhart, W. and Kassel, D. B. (1994) *Rapid Commun. Mass Spec.* **8**, 94–98. Identification of phosphorylated peptides from complex mixtures using negative-ion orifice-potential stepping and capillary liquid chromatography/electrospray ionization mass spectrometry.

Edmonds, C. G. and Smith, R. D. (1990) *Methods in Enzymology* **193**, 412–431. Electrospray ionization mass spectrometry.

Eng, J. K.; McCormack A. L.; Yates III, J. R. (1994) *J. Am. Soc. Mass Spectrom.* **5**, 976–989. An approach to correlate tandem mass spectral data of peptides with amio acid sequences in a protein database.

Fenn, J. B.; Mann, M.; Meng, C. K.; Wong, S. F. and Whitehouse, C. M. (1989) *Science* **246**, 64–71. Electrospray ionization for mass spectrometry of large biomolecules.

Figeys, D.; Ning, Y. and Aebersold, R. (1997) *Anal. Chem.* **69**, 3153–3160. A microfabricated device for rapid protein identification by microelectrospray ion trap mass spectrometry.

Gaskell, S. J. (1997) *J. Mass Spectrom.* **32**, 677–688. Electrospray: principles and practice.

Guilhaus, M. (1995) *J. Mass Spectrom.* **30**, 1519–1532. Principles and instrumentation in time-of-flight mass spectrometry.

Guilhaus, M.; Mlynski, V. and Selby, D. (1997) *Rapid Commun. Mass Spec.* **11**, 951–962. Perfect timing: time-of-flight mass spectrometry.

Haller, I.; Mirza, U. A. and Chait, B. T. (1996) *J. Am. Soc. Mass Spectrom.* **7**, 677–681. Collision induced decomposition of peptides. Choice of collision parameters.

Hunter, A. P. and Games, D. E. (1994) *Rapid Commun. Mass Spec.* **8**, 559–570. Chromatographic and mass spectrometric methods for the identification of phosphorylation sites in phosphoproteins.

Immler, D.; Blüggel, M.; Hamscher, G.; Spengler, B.; Kaufmann, R. and Meyer, H. E. (1997) *Würzburger Kolloquium Fortschrittsberichte '97*, Bertsch Verlag, 165–175. Protein identification by on-line or off-line capillary LC and mass spectrometry.

Immler, D.; Gremm, D.; Kirsch, D.; Spengler, B.; Presek, P. and Meyer, H. E. (1998) *Electrophoresis* **19**, 1015–23. Identification of phosphorylated proteins from thrombin-activated human platelets isolated by 2-D-gel electrophoresis by nanospray-ESI-MS/MS and LC-ESI-MS.

Jacobson, S. C.; Hergenröder, R.; Koutny, L. B. and Ramsey, J. M. (1994a) *Anal. Chem.* **66**, 2369–2373. Open channel electrochromatography on a microchip.

Jacobson, S. C.; Hergenröder, R.; Koutny, L. B.; Warmack, R. J. and Ramsey, J. M. (1994b) *Anal. Chem.* **66**, 1107–1113. Effects of injection schemes and column geometry on the performance of microchip electrophoresis devices.

Jacobson, S. C. and Ramsey, J. M. (1997) *Anal. Chem.* **69**, 3212–3217. Electrokinetic focusing in microfabricated channel structures.

Jedrzejewski, P.; Schnölzer, M.; Lorenz, P. and Lehmann, W. D. (1995) 28. Diskussionstagung der Arbeitsgemeinschaft Massenspektrometrie, Tübingen, June 6-9 1995. Nachweis von sulfatierten und phosphorylierten Peptid-Fragmenten mit LC-ESI MS/MS und Hybrid-Scantechniken.

Johnson, J. V.; Yost, R. A.; Kelley, P. E. and Bradford, D. C. (1990) *Anal. Chem.* **62**, 2162–2172. Tandem-in-space and tandem-in-time mass spectrometry: triple quadrupoles and quadrupole ion traps.

Jonscher, K. R. and Yates, J. R. (1997) *Anal. Biochem.* **244**, 1–15. The quadrupole ion trap mass spectrometer- a small solution to a big challenge.

Karas, M. and Hillenkamp, F. (1988) *Anal. Chem.* **60**, 2299–2301. Laser desorption ionization of proteins with molecular masses exceeding 10.000 daltons.

Kaufmann, R. (1995) *J. Biotechnol.* **41**, 155–175. Matrix-assisted laser desorption ionization (MALDI) mass spectrometry: a novel tool in molecular biology and biotechnology.

Kaufmann, R.; Kirsch, D. and Spengler, B. (1994) *Int. J. Mass Spectrom. Ion Proc.* **131**, 355–385. Sequencing of peptides in a time-of-flight mass spectrometer: evaluation of postsource decay following matrix-assisted laser desorption ionisation (MALDI).

Kaufmann, R.; Spengler, B. and Lützenkirchen, F. (1993) *Rapid Commun. Mass Spec.* **7**, 902–910. Mass spectrometric sequencing of linear peptides by product-ion analysis in a reflectron time-of-flight mass spectrometer using matrix-assisted laser desorption ionization.

Kebarle, P. and Tang, L. (1993) *Anal. Chem.* **65**, 972A–986A. From ions in solution to ions in the gas phase.

Langen, H.; Sander, B.; Vilbois, F. and Lahm, H. W. (1993) *Techniques in Protein Chemistry IV*, ed. by Angeletti, R.H. Academic Press, New York, 47–54. Characterization of the proteins c-kit ligand and DHFR by electrospray mass spectrometry.

Linscheid, M. and Westmoreland, D. G. (1994) *Pure & Appl. Chem.* **66**, 1913–1930. Application of liquid chromatography-mass spectrometry.

Loo, J. A.; Udseth, H. R. and Smith, R. D. (1989) *Anal. Biochem.* **179**, 404–412. Peptide and protein analysis by electrospray ionization-mass spectrometry and capillary electrophoresis-mass spectrometry.

Mann, M.; Meng, C. K. and Fenn, J. B. (1989) *Anal. Chem.* **61**, 1702–1708. Interpreting mass spectra of multiply charged ions.

March, R. E. (1997) *J. Mass Spectrom.* **32**, 351–369. An introduction to quadrupole ion trap mass spectrometry.

Medzihradszky, K. F.; Adams, G. W.; Burlingame, A. L.; Bateman, R. H. and Green, M. R. (1996) *Am. Soc. Mass Spectrom.* **7**, 1–10. Peptide sequence determination by matrix-assisted laser desorption ionization employing a tandem double focusing magnetic-orthogonal acceleration time-of-flight mass spectrometer.

Medzihradszky, K. F. and Burlingame, A. L. (1994) *Methods: A Companion to Methods in Enzymology* **6**, 284–303. The advantages and versatility of a high-energy collision-induced dissociation-based strategy for the sequence and structural determination of proteins.

Meyer, H. E.; Eisermann, B.; Heber, M.; Hoffmann-Posorske, E.; Korte, H.; Weigt, C.; Wegner, A.; Hutton, T.; Donella-Deana, A. and Perich, J. W. (1993) *FASEB J.* **7**, 776–782. Strategies for non-radioactive methods in the localization of phosphorylated amino acids in proteins.

Moritz, B. S. (1996) Diploma thesis, Ruhr-Universität Bochum. Automatisierte HPLC-Tandem MS/MS Sequenzanalyze von Oligopeptiden.

Mortz, E.; O'Conner, P. B.; Roepstorff, P.; Kelleher, N. L.; Wood, T. D.; McLafferty. F. W. and Mann, M. (1996) *Proc. Natl. Acad. Sci USA* **93**, 8264–8267. Sequence tag identification of intact proteins by matching tandem mass spectral data against sequence data bases.

271

Mylchreest, I.; Wheeler, K. and Campbell, C. (1995) *Finnigan MAT Application Report* **240**. Automated scan-mode switching in LC/MS/MS.

Niessen, W. M. A. and Tinke, A. P. (1995) *J. Chrom. A* **703**, 37–57. Liquid chromatography-mass spectrometry: general principles and instrumentation.

Paulus, A. (1997) Kapillarelektrophorese – Chromatographie Würzburger Kolloquium 1997, Bertsch Verlag, Straubing, 155–163. Intergrierte Kapillar-Elektrophorese auf Chips.

Qin, J. and Chait, B. T. (1997) *Anal. Chem.* **69**, 4002–4009. Identification and characterization of post-translational modifications of proteins by MALDI ion trap mass spectrometry.

Ramsey, R. S. and Ramsey, J. M. (1997) *Anal. Chem.* **69**, 1174–1178. Generating electrospray from microchip devices using electroosmotic pumping.

Sander, I.; Raulf-Heimsoth, M.; Siethoff, C.; Lohaus, C.; Meyer, H. E. and Baur, X. (1998) *J. Allergy Clin. Immunol*, in press. Allergy to *Aspergillus*-derived enzymes in baking industry: identification of β-xylosidase from *Aspergillus niger* as new allergen (Asp n 3).

Schindler, P. A.; Van Dorsselaer, A. and Falick, A. M. (1993) *Anal. Biochem.* **213**, 256–263. Analysis of hydrophobic proteins and peptides by electrospray ionization mass spectrometry.

Schnölzer, M.; Jedrzejewski, P. and Lehmann, W. D. (1996) *Electrophoresis* **17**, 945–953. Protease-catalyzed incorporation of ^{18}O into peptide fragments and its application for protein sequencing by electrospray and matrix-assisted laser desorption/ionization mass spectrometry.

Schwartz, J. C. and Jardine I. (1996) *Methods Enzymol.* **270**, Chap. 23, 552–586. Quadrupole ion trap mass spectrometry.

Shevchenko, A.; Chernushevich, I.; Ens, W.; Standing, K.G.; Thomson, B.; Wilm, M. and Mann, M. (1997) *Rapid Commun. Mass Spec.* **11**, 1015–1024. Rapid 'de novo' peptide sequencing by a combination of nano-electrospray, isotopic labeling and a quadrupole/time-of-flight mass spectrometer.

Smith, R. D.; Loo, J. A.; Edmonds, C. G.; Barinaga, C. J. and Udseth, H. R. (1990) *Anal. Chem.* **62**, 882–899. New developments in biochemical mass spectrometry: electrospray ionization.

Smith, R. D.; Wahl, J. H.; Goodlett, D. R. and Hofstadler, S. A. (1993) *Anal. Chem.* **65**, 574A–584A. Capillary electrophoresis/mass spectrometry.

Spengler, B.; Kirsch, D.; Kaufmann, R. and Jaeger, E. (1992) *Rapid Commun. Mass Spec.* **6**, 105–108. Peptide sequencing by matrix-assisted laser-desorption mass spectrometry.

Stahl, D. C.; Martino, P. A.; Swiderek, K. M.; Davis, M. T. and Lee, T.D. (1992) Proceedings of the 40th ASMS Conference on Mass Spectrometry and Allied Topics, Washington, DC, May 31–June 5 1992. Automated LC/MS/MS analysis of peptide mixtures using capillary HPLC and electrospray ionization on a triple sector quadrupole mass spectrometer.

Stimson, E.; Truong, O.; Richter, W. J.; Waterfield, M. D. and Burlingame A. L. (1997) *Int. J. Mass Spectrom. Ion Proc.*, in press. Enhancement of charge remote fragmentation in protonated peptides by high-energy CID MALDI-TOF-MS using 'cold' matrices.

Taylor, J. A. and Johnson, R. S. (1997) *Rapid Commun. Mass Spec.* **11**, 1067–1075. Sequence database searches via de novo peptide sequencing by tandem mass spectrometry.

Valaskovic, G. A.; Kelleher, N. L.; Little, D. P.; Aaserud, D. J. and McLafferty, F. W. (1995*) Anal. Chem.* **67**, 3802–3805. Attomole-sensitivity electrospray source for large-molecule mass spectrometry.

Valaskovic, G. A.; Kelleher, N. L. and McLafferty, F. W. (1996) *Science* **273**, 1199–1202. Attomole protein characterization by capillary electrophoresis-mass spectrometry.

Wachs, T.; Conboy, J. C.; Garcia, F. and Henion, J. D. (1991) *J. Chrom. Science* **29**, 357–366. Liquid chromatography-mass spectrometry and related techniques via atmospheric pressure ionization.

Wilm, M.; Houthaeve, T.; Talbo, G.; Kellner, R.; Mortensen, P. and Mann, M. (1996a) *Mass Spectrometry in the Biological Sciences,* ed. by Burlingame, A.L.; Carr, S.A. Humana Press, Clifton, NJ, 245–265. Approaches to the practical use of MS/MS in a protein sequencing facility.

Wilm, M. and Mann, M. (1996b) *Anal. Chem.* **68**, 1–8. Analytical properties of the nano-electrospray ion source.

Wilm, M.; Shevchenko, A.; Houthaeve, T.; Breit, S.; Schweigerer, L.; Fotsis, T. and Mann, M. (1996c) *Nature* **379**, 466–469. Femtomole sequencing of proteins from polyacrylamide gels by nano-electrospray mass spectrometry.

Wu, J. and Watson, J. T. (1997) *Protein science* **6**, 391–398. A novel methodology for assignment of disulfide bond pairings in proteins.

Yamashita, M. and Fenn, J. B. (1994a) *J. Phys. Chem.* **88**, 4451–4459. Electrospray ion source. Another variation on the free-jet theme.

Yamashita, M. and Fenn, J. B. (1994a) *J. Phys. Chem.* **88**, 4671–4675. Negative ion production with the electrospray ion source.

Yates III, J. R.; Eng, J. K.; McCormack, A. L. and Schieltz, D. (1995) *Anal. Chem.* **67**, 1426–1436. Method to correlate tandem mass spectra of modified peptides to amino acid sequences in the protein database.

Section IV:
Computer Sequence Analysis

Internet Resources for Protein Identification and Characterization

Pierre-Alain Binz, Marc R. Wilkins, Elisabeth Gasteiger, Amos Bairoch, Ron D. Appel and Denis F. Hochstrasser

1 Introduction

The large-scale and rapid identification and characterization of proteins, for example those separated on 2-D gels, represents a central step in the understanding of a proteome. Identification involves the matching of analytical data against protein databases in order to assign a protein as already known or to describe it as novel. Researchers can take advantage of the information contained in databases to predict a number of properties inherent to proteins of interest. The aspects of identification and of structural prediction are complementary and both required in the characterization process of a protein and, more generally, of a proteome. In that respect the development of the World-Wide Web (WWW) is changing the research of all life scientists and in particular of protein chemists and biomedical scientists. The WWW environment allows to search for information in continuously updated databases that are physically located anywhere in the world, to match experimental data with those stored in different and interconnecting databases, to interactively predict protein attributes using intelligent tools, to discuss results and problems with colleagues through newsgroups and other online forums, to share results and suggestions with people in charge of the databases and bioinformatics tools in order to improve the quality of the databases and the power of the tools.

While numerous programs are currently available on the internet for discrete applications, we present here a set of bioinformatics tools particularly adapted to the identification and characterization of proteins from various types of experimentally obtained data. We will first describe a suite of tools provided at the ExPASy WWW server (URL: http://www.expasy.ch) [Appel 1994] developed at the Geneva University Hospital and the University of Geneva. Then we will present sets of similar protein identification and characterization programs, available elsewhere on the WWW. For lack of space, we will not cover all tools available on the Internet. Instead, we will focus on comparable approaches to mass spectrometry issues provided by:

- The ProteinProspector server of the University of California San Francisco Mass Spectrometry Facility
- The PROWL site from the MS groups at Rockefeller and NY Universities

- The PeptideSearch package from the EMBL Protein & Peptide Group
- The MOWSE search program from the Daresbury Laboratory
- The MassSearch tool of the Computational Biochemistry Research Group Server at the ETH Zürich.

2 General Approach

2.1 Experimental Attributes and Choice of Programs

In order to identify and characterize a protein or a complete proteome with the help of bioinformatics tools, the wet-lab scientist first needs to collect experimental data and attributes (Figure 1). This type of information is often deduced from a combination of two-dimensional electrophoresis separation, of proteolytic procedures and/or chromatographic and mass spectrometry analyses. The set of obtained data and known attributes include information about the source of the protein itself (i.e. species, tissue, keyword) and physico-chemical properties of the protein, its peptides or its amino acids. In the second step (i.e. the identification step), the user chooses a program to match his data with those stored in one or more target databases. The outputs of identification programs generally yield lists of proteins ranked by particular scoring methods, with the best matching protein appearing at the top of the list. In the characterization step users complete the description of the protein attributes by the interpretation of data unmatched after the identification. This is done with the help of specific prediction tools.

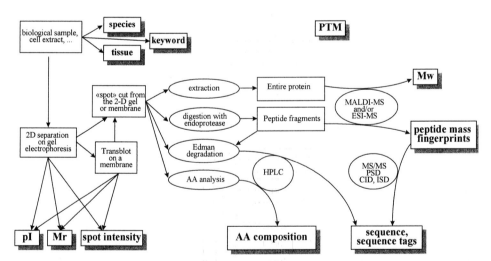

Figure 1. Wet-lab protein attributes for protein identification and characterization. The flow-diagram represents the different protein attributes (shadowed areas) as results of experiments (oval-shaped areas) obtained from various sample preparations (squared areas). These attributes are used by identification and characterization tools as input parameters. Mr: apparent molecular weight on a gel; Mw: MS-measured molecular weight; PTMs: post translational modifications. Since they can be derived from different experimental procedures, the PTMs have not been linked to a single experiment.

All results have to be interpreted carefully to allow a complete characterization of the protein. It is important to note that all identification and characterization programs will always produce a result for any user-specified input data. However, this does not mean that the prediction is always correct. The user must know what the program is able to do. He must be able to appreciate the quality and the effect of the parameters submitted for calculation, and he must be aware of the restrictions specific to the tool. These are indispensable conditions for a correct interpretation of the results.

2.2 Information in Databases

Most of the characterization tools and the identification programs compute and compare user-submitted data with information archived in one or more databases. The ExPASy identification and characterization tools make extensive use of the annotations contained in the SWISS-PROT protein sequence database and its supplement TrEMBL.

2.2.1 The Annotations in SWISS-PROT

SWISS-PROT [Bairoch 1998] (URL: http://www.expasy.ch/sprot/sprot-top.html) is a curated protein sequence database which is particularly well suited to use for the identification and the characterization of proteins. It is completely non-redundant and provides a high level of information (annotations) complementing the primary structure. Discrete protein attributes are located in separate fields of each SWISS-PROT entry and can be retained by searching tools. These are protein name, species, organism classification, organelle, biological source (tissue), keywords, functionality, post-translational modifications (PTMs), known variants, amino acid (amino acid) sequence (see Figure 2). Other information, like pI, molecular weight (MW), amino acid composition, location on a 2-D gel, peptide mass fingerprints and a set of similar sequences can be indirectly deduced using specific tools implemented on ExPASy or on internet sites directly linked to ExPASy.

2.2.2 SWISS-PROT Supplement TrEMBL

TrEMBL **(note 1)** [Bairoch 1998] is a protein sequence database supplementing the SWISS-PROT database. TrEMBL contains the translations of all coding sequences (CDS) present in the EMBL Nucleotide Sequence Database not yet integrated in SWISS-PROT. TrEMBL can be considered as a preliminary section of SWISS-PROT. It includes fewer annotations than SWISS-PROT. SWISS-PROT and TrEMBL can be combined to form a complete and low redundancy protein sequence database updated weekly. Together they contain more than 210 000 entries (about 70 000 in SWISS-PROT release 35 and 140 000 in TrEMBL release 5).

```
ID   TRY1_HUMAN      STANDARD;      PRT;     247 AA.
AC   P07477;
DT   01-APR-1988 (REL. 07, CREATED)
DT   01-APR-1988 (REL. 07, LAST SEQUENCE UPDATE)
DT   01-FEB-1998 (REL. 36, LAST ANNOTATION UPDATE)
DE   TRYPSINOGEN I PRECURSOR (EC 3.4.21.4).
GN   PRSS1 OR TRY1 OR TRP1.
OS   HOMO SAPIENS (HUMAN).
OC   EUKARYOTA; METAZOA; CHORDATA; VERTEBRATA; TETRAPODA; MAMMALIA;
OC   EUTHERIA; PRIMATES.
RN   [1]
RP   SEQUENCE FROM N.A.
RX   MEDLINE; 86221712. [NCBI, Geneva, Japan]
RA   EMI M., NAKAMURA Y., OGAWA M., YAMAMOTO T., NISHIDE T., MORI T.,
RA   MATSUBARA K.;
RL   GENE 41:305-310(1986).
RN   [2]
RP   SEQUENCE OF 16-43.
RX   MEDLINE; 90091010. [NCBI, Geneva, Japan]
RA   KIMLAND M., RUSSICK C., MARKS W.H., BORGSTROEM A.;
RL   CLIN. CHIM. ACTA 184:31-46(1989).
RN   [3]
RP   X-RAY CRYSTALLOGRAPHY (2.2 ANGSTROMS), AND PHOSPHORYLATION.
RX   MEDLINE; 96266496. [NCBI, Geneva, Japan]
RA   GABORIAUD C., SERRE L., GUY-CROTTE O., FOREST E.,
RA   FONTECILLA-CAMPS J.-C.;
RL   J. MOL. BIOL. 259:995-1010(1996).
CC   -!- CATALYTIC ACTIVITY: PREFERENTIAL CLEAVAGE: ARG-, LYS-.
CC   -!- SUBCELLULAR LOCATION: EXTRACELLULAR.
CC   -!- MASS SPECTROMETRY: MW=24348; MW_ERR=2; METHOD=ELECTROSPRAY;
CC       RANGE=24-247.
CC   -!- SIMILARITY: BELONGS TO PEPTIDASE FAMILY S1; ALSO KNOWN AS THE
CC       TRYPSIN FAMILY.
DR   EMBL; M22612; G521216; -. [EMBL / GenBank / DDBJ] [CoDingSequence]
DR   PIR; A25852; A25852.
DR   PDB; 1TRN; 03-JUN-95. [Geneva / Brookhaven]
DR   SWISS-3DIMAGE; TRY1_HUMAN.
DR   GeneCard; PRSS1.
DR   MIM; 276000; -.
DR   PROSITE; PS00134; TRYPSIN_HIS; 1.
DR   PROSITE; PS00135; TRYPSIN_SER; 1.
DR   PRODOM [Domain structure 7 List of seq. sharing at least 1 domain]
DR   PROTOMAP; P07477.
DR   SWISS-2DPAGE; GET REGION ON 2D PAGE.
KW   HYDROLASE; SERINE PROTEASE; DIGESTION; PANCREAS; ZYMOGEN; SIGNAL;
KW   MULTIGENE FAMILY; PHOSPHORYLATION; 3D-STRUCTURE.
FT   SIGNAL        1     15
FT   PROPEP       16     23       ACTIVATION PEPTIDE.
FT   CHAIN        24    247       TRYPSIN I.
FT   ACT_SITE     63     63       CHARGE RELAY SYSTEM.
FT   ACT_SITE    107    107       CHARGE RELAY SYSTEM.
FT   ACT_SITE    200    200       CHARGE RELAY SYSTEM.
FT   DISULFID     30    160
FT   DISULFID     48     64
FT   DISULFID    139    206
FT   DISULFID    171    185
FT   DISULFID    196    220
FT   MOD_RES     154    154       PHOSPHORYLATION.
FT   SITE        194    194       REQUIRED FOR SPECIFICITY (BY SIMILARITY).
SQ   SEQUENCE   247 AA;  26558 MW;  FE7C63A9 CRC32;
     MNPLLILTFV AAALAAPFDD DDKIVGGYNC EENSVPYQVS LNSGYHFCGG SLINEQWVVS
     AGHCYKSRIQ VRLGEHNIEV LEGNEQFINA AKIIRHPQYD RKTLNNDIML IKLSSRAVIN
     ARVSTISLPT APPATGTKCL ISGWGNTASS GADYPDELQC LDAPVLSQAK CEASYPGKIT
     SNMFCVGFLE GGKDSCQGDS GGPVVCNGQL QGVVSWGDGC AQKNKPGVYT KVYNYVKWIK
     NTIAANS
//
```

Figure 2. The attributes in SWISS-PROT used in the identification and characterization tools. This SWISS-PROT entry illustrates the high level of annotation of the proteins included in the database. The lines in a SWISS-PROT entry which are used as arguments for database searches and matching operations in the ExPASy identification and characterization tools are highlighted in bold characters. Additional attributes useful for protein identification but not immediately used by the tools are shown in italic characters. For example, the known mass spectrometric data are annotated in the CC lines, and the pI and Mr coordinates can be deduced from SWISS-2DPAGE which can be accessed through a direct link in the DR line characters) For a detailed description of all codes and abbreviations in SWISS-PROT see [Bairoch 1998] or in the SWISS-PROT user manual (URL http://www.expasy.ch/txt/userman.txt).

3 Identification and Characterization Tools at ExPASy

The ExPASy tools are designed to share a number of goals, i.e.

1) They use the extensive annotation provided in the SWISS-PROT database [Bairoch 1998] whenever possible.
2) They are designed to be cross-linked and transparent to the user.
3) They are all available on the World-Wide Web and are free of use to the academic scientific community.
4) They are "living" programs, and can therefore be adapted and optimized to best serve the users.

The tools are available through the ExPASy World-Wide Web server from the following URL: http://www.expasy.ch/www/tools.html. A mirror of ExPASy tools is located at the URL: http://expasy.bio.mq.edu.au/www/tools.html. Since they are under continuous development these tools are subject to modifications. The "What's New on ExPASy" WWW page located at the URL http://www.expasy.ch/www/history.html provides information about new and updated features. Users can also subscribe to an electronic bulletin called SWISS-Flash, to regularly get by email the news concerning the databases, software and service developments from the ExPASy biocomputing group. and the WWW ExPASy server.\ For details see the URL http://www.expasy.ch/www/swiss-flash.html. SWISS-Flash bulletins by electronic mail. \Feedback is welcome and will help the "Geneva Proteome Project Team" (URL http://www.expasy.ch/www/people.html) to best match the tools with users' requirements.

The tools presented in this section and in section 4 are grouped according to the physico-chemical protein attributes required by their respective algorithms. The number and the variety of the presented identification and characterization programs reflect the wide range of computable protein attributes and experimental procedures that may be undertaken on protein samples. Online documentation is available for each of the ExPASy tools. The exact URLs of all tools can be found in the Appendix.

3.1 Identification Tools

The identification tools in ExPASy are designed to make maximum use of all annotation in the SWISS-PROT database. For example, all tools can process precursor sequences into their mature forms. They can remove signal, transit and propeptides from the database entry. Other annotations from the feature tables such as sequence conflicts, variants, alternative splicing sites and post-translational modifications are also taken into account in a number of these tools. Information from the keyword (KW) and species (OS, OC) lines can be used to restrict the scope of a search.

3.1.1 Identification with Sequence Tags: TagIdent

Description:

TagIdent [Wilkins 1996, 1998a,b] identifies a protein using pI, molecular weight and short sequence tags as input parameters. It searches SWISS-PROT and/or TrEMBL and can:

1) generate of a list of proteins close to a given pI and MW;
2) match a short sequence tag of up to 6 amino acids against proteins in SWISS-PROT/TrEMBL close to a given pI and MW. Results can be shown for *N*-terminal, *C*-terminal and internal tags (see Figure 3);
3) identify proteins by their mass, if this mass has been determined by mass spectrometric techniques.

The intervals of searched pI and MW can be freely chosen. The search can be reduced to a defined range of species.

Comments:

Theoretical isoelectric points and molecular weights are calculated as described for Compute pI/MW.

TagIdent allows a number of proteins to be listed from one or more organisms that may focus within a particular range of pI and MW on a 2-D gel. If available, the additional specification of a short sequence tag up to 6 amino acids drastically narrows the list of matching sequences (see Figure 3). Sequence tags are highly specific and par-

```
Scan in SWISS-PROT database (70787 entries)

496 proteins found in the specified pI/Mw ranges


---
Results with tagging: 1 found
The number before the sequence indicates the position in the
mature protein where your tag SKIV has been found (first occurrence).
If the protein displayed results from the processing of a
precursor, the position of the tag in the precursor polypeptide
will be given in brackets.
The sequence tag itself is printed in lowercase.
---
ENO_ECOLI (P08324)
     ENOLASE (EC 4.2.1.11) (2-PHOSPHOGLYCERATE DEHYDRATASE) (2-PHOSPHO-D-
     GLYCERATE HYDRO-LYASE).pI: 5.32, MW: 45523.75
1    skivKIIGREIIDSRGNPTVEAEVHLEGGFVGMAAAPSGA...
```

Figure 3. Output of TagIdent: an example. Input parameters submitted to TagIdent were: pI = 5.3±1; Mw = 46000±20%; the species chosen as described in OS and OC lines was: ÉCOLÍ; no keywords were specified to restrict the search (KW=ALL); the search is performed using the sequence tag: SKIV; the output has to display the N-terminal sequence for sequences matching the tag only sequence.; Tag = SKIV; Print only the sequences matching your tag 'SKIV' TagIdent reported one single match, i.e. enolase in *E. coli*. The AC and ID (in brackets) codes in SWISS-PROT are displayed together with the description line (DE), and the first 40 AAs of the sequence are shown. The searched tag is highlighted in lower case, and its position in the sequence is given.

ticularly suited to search in organisms with small genomes that are well described at the molecular level. As an illustration, 93 % of the *E. coli* proteins have unique 5 amino acid *N*-terminal sequence tags [Wilkins 1998a] and can thus be unambiguously identified with TagIdent.

3.1.2 Identification with Amino Acid Composition: AACompIdent

Description:

AACompIdent [Wilkins 1996b,c, 1998b] identifies proteins by their amino acid composition. It matches the empirically determined percent amino acid composition of the query protein against the theoretical percent amino acid compositions of proteins in SWISS-PROT and/or TrEMBL. A score is calculated for each protein in the database as the sum of the squared differences between the percent amino acid composition of this protein and that of the query protein. A ranked list is then created where the best matching protein is the one with the lowest score. The output of amino acidCompIdent provides three lists of ranked proteins. The first list yields the result from matching the query amino acid composition against all proteins from a user-specified species or range of species (e.g. "BOVIN" or "MAMMA-LIA"). In the second list the set of target proteins is extended to all species. In the third list, the set of target proteins is restricted to those of the specified species that lie within a specified pI and MW range.

A constellation of empirically quantified amino acids has to be chosen (e.g. Constellation 2 considers all amino acids except Cys and Trp; Asp and Asn are considered together as Asx; Gln and Glu are considered together as Glx).

If the user has calibrated the amino acid analysis with a calibration protein, the experimental amino acid composition of that protein can be entered together with its SWISS-PROT or TrEMBL identification code (e.g. ALBU_BOVIN). AACompIdent will use these values to correct for experimental errors in the determination of the amino acid composition of the unknown protein before matching the composition of that unknown protein against the database.

A *N*- or *C*-terminal sequence tag of maximum 6 amino acids can be used as an additional search parameter, which helps to increase confidence in identification results. In this case the output will contain 40 amino acids of the *N*- or *C*-terminal sequence of each ranked proteins, and the matched tag will be highlighted.

Comments:

A correct identification is assessed to a protein which meets three conditions, as defined by Wilkins et al. [1998b]. "Firstly, the same protein, or type of protein, should appear at the top of the three lists. Secondly, the top-ranked protein in the third list should have a score less than 30 (indicating a "good fit" of the query protein with that database entry). Finally, the third list should show a large score difference between the top-ranked protein and the second ranked protein (indicating a unique matching of the query protein with the top-ranked database entry)."

If the amino acid analysis yields unreliable values for particular amino acids, the "free constellation" can be used to perform matching with another set of amino acids [Golaz 1996].

Cross-species identification can be performed with AACompIdent by restricting the search to species closely related to the species under study or by performing the search against all proteins in the database. Protein MW and amino acid composition are usually well conserved for phylogenetically closely related proteins, which is not the case for pI [Wilkins 1997a]. It has to be noted, however, that higher confidence can be obtained with extensive sequence information and, for very closely related proteins (i.e. for proteins showing high sequence identity), with peptide mass data [Wilkins 1997a].

3.1.3 Identification with Combined Attributes: MultiIdent

Description:

MultiIdent [Wilkins 1998b] uses combinations of parameters like estimated pI and MW, species, keyword, amino acid composition, sequence tag and peptide mass fingerprints to identify proteins. The program can be considered as an extension of AACompIdent. Like AACompIdent, it first generates a list of up to 500 proteins whose amino acid compositions are the closest to that of the query protein. Then the user's experimental masses of peptide fragments are matched against the theoretical fragments calculated from all proteins of this first list. As in AACompIdent the user can focus the search by adding a sequence tag as an additional search parameter. Other parameters that can be used in the identification procedure include the species, the pI and MW ranges, the ion mode ([M]or [M+H]$^+$) and isotopic resolution (mono-

→

Figure 4. Output of MultiIdent: an illustrative example. The query protein (identified as bovine ATPase beta chain) was extracted from a 2-D PAGE, its amino acid composition was quantified using bovine albumine as calibration standard. A sample was submitted to trypsin digestion and the masses of the obtained peptides were measured with mass spectrometry. In this example the amino acid composition search alone with pI and MW did not find ATPB_BOVIN (in italic characters) as the best-matching protein. The peptide mass data provided more specific data for the search and allowed clearly to classify the ATPase beta chains as best-matching among the 200 candidate proteins found by the amino acid analysis. Input parameters used for the calculation, as appearing in the header of the output of MultiIdent: Constellation 2 with peptide mass fingerprinting. Name given to unknown protein: UNKNOWN. Species searched: MAMMALIA. Keyword: ALL. pI: 5.1. Mw: 51000. pI range: 1.00. Mw range: 20 %. Calibration protein: ALBU_BOVIN (P02769). AMINO ACID COMPOSITION FOR CALIBRATION PROTEIN: Asx: 10.40, Glx: 13.90, Ser: 5.40, His: 1.80, Gly: 4.70, Thr: 5.90, Ala: 9.20, Pro: 5.10, Tyr: 3.10, Arg: 4.50, Val: 6.30, Met: 0.40, Ile: 3.50, Leu: 10.70, Phe: 5.50, Lys: 9.40. AMINO ACID COMPOSITION FOR UNKNOWN PROTEIN (in percent): Asx: 8.60, Glx: 13.20, Ser: 5.70, His: 0.70, Gly: 14.70, Thr: 5.40, Ala: 10.60, Pro: 5.40, Tyr: 2.10, Arg: 4.50, Val: 7.30, Met: 0.00, Ile: 5.40, Leu: 9.00, Phe: 2.80, Lys: 4.60. 5.40, His: 1.80, Gly: 4.70, Thr: 5.90, Ala: 9.20, Pro: 5.10, Tyr: 3.10, Arg: 4.50, Val: 6.30, Met: 0.40, Ile: 3.50, Leu: 10.70, Phe: 5.50, Lys: 9.40. \Max. number of proteins generated in list: 200. Max. number of proteins to print: 12. Tag: No_Tagging. Peptide masses for unknown protein: 719.90, 976.20, 1034.00, 1300.10, 1386.40, 1406.60, 1440.20, 1458.60, 1503.10, 1603.10, 1619.20, 1642.40, 1652.20, 1753.30, 1797.80, 1844.20, 1860.50, 1920.40, 1990.00, 2061.90, 2078.90, 2281.90, 2297.90. Using monoisotopic masses of the occurring amino acid residues and interpreting your peptide masses as [M+H]$^+$. The enzyme is Trypsin. Cysteine in reduced form. Peptide mass range: 500 to 3000 Dalton. Tolerance: +/- 2 Dalton. Print the results from: *AAcomposition *Peptide mass fingerprinting *Integrated search for * the closest SWISS-PROT entries for species MAMMALIA and keyword ALL, having pI and MW values in the specified range.

isotopic, average), the enzyme used for the digestion, the type of cysteine modification if any (i.e. treatment with iodoacetic acid, iodoacetamide, 4-vinylpyridine and acrylamide), the presence of oxidized methionine, the maximum number of missed cleavages to be considered and the mass tolerance.

MultiIdent can generate a number of outputs as specified by the user (Figure 4). The first output is a list ranking the proteins according to their amino acid composition

```
SpotNb UNKNOWN
===============

pI:   5.10    Range: (  4.10,    6.10)
Mw:  51000    Range: ( 40800,   61200)
```

The **SWISS-PROT** entries having pI and Mw values in the specified range
for the species **MAMMALIA** and the specified keyword:

```
Rank Score   Protein    (pI      Mw)  Description
============================================================
  1    24   DHA5_HUMAN   6.01   55292  ALDEHYDE DEHYDROGENASE X.
  2    24   NPX2_HUMAN   5.45   45448  NEURONAL PENTRAXIN II.
  3    26   ATPB_BOVIN   5.00   51563  ATPASE BETA CHAIN.
  4    27   ATPB_RAT     4.95   51710  ATPASE BETA CHAIN.
  5    27   UCR1_HUMAN   5.43   49102  UBIQUINOL-CYTOCHROME C REDUCTASE COMPLEX
  6    28   HSR1_HUMAN   4.89   47465  POSSIBLE GTP-BINDING PROTEIN HSR1.
  7    28   ATPB_HUMAN   5.04   53119  ATPASE BETA CHAIN.
  8    28   NPX1_HUMAN   5.84   45393  NEURONAL PENTRAXIN I.
  9    28   MMR1_MOUSE   4.80   47540  POSSIBLE GTP-BINDING PROTEIN MMR1.
 10    29   GLPK_MOUSE   5.47   57458  GLYCEROL KINASE (EC 2.7.1.30) (ATP
 11    29   GSHB_MOUSE   5.56   52247  GLUTATHIONE SYNTHETASE (EC 6.3.2.3)
 12    29   ATPB_MOUSE   4.95   51800  ATPASE BETA CHAIN.
```

PEPTIDE-MASS FINGERPRINTING !

The closest **SWISS-PROT** entries for the species **MAMMALIA** and
keyword **ALL**, having pI and Mw values in the specified range:
The best-matching set of peptide-masses

```
Rank Hits  Protein    Modification         Matched Peptide-masses
============================================================================
  1   10   ATPB_BOVIN  ATPASE BETA CHAIN.  2061.00 / 1988.03 / 1921.97 / 1650.92 / 1601.81 /
                                           1457.84 / 1439.79 / 1406.68 / 1385.71 /  975.56
  2   10   ATPB_HUMAN  ATPASE BETA CHAIN.  2061.00 / 1988.03 / 1921.97 / 1650.92 / 1601.81 /
                                           1457.84 / 1439.79 / 1406.68 / 1385.71 /  975.56
  3   10   ATPB_RAT    ATPASE BETA CHAIN.  2061.00 / 1988.03 / 1921.97 / 1650.92 / 1601.81 /
                                           1457.84 / 1439.79 / 1406.68 / 1385.71 /  975.56
  4    9   ATPB_MOUSE  ATPASE BETA CHAIN.  2061.00 / 1988.03 / 1921.97 / 1842.88 / 1650.92 /
                                           1457.84 / 1439.79 / 1385.71 /  975.56
  5    5   CG2A_MESAU                      2299.11 / 1798.95 / 1754.83 / 1652.86 /  720.40
  6    5   DHAE_ELEED                      1989.94 / 1751.84 / 1617.83 / 1035.47 /  721.42
  7    4   ENOG_HUMAN                      2296.14 / 1858.93 / 1604.78 / 1406.72
  8    4   ENOG_MOUSE                      2296.14 / 1858.93 / 1604.78 / 1406.72
  9    4   GLPK_HUMAN                      1859.01 / 1653.83 / 1504.81 /  977.49
 10    4   K2M2_SHEEP                      2063.03 / 1844.94 / 1407.66 /  974.52
 11    4   PDP_BOVIN  [PYRUVATE DEHYDROGENASE (LIPOAMIDE)]
                                           2298.17 / 1753.86 / 1651.76 /  721.33
 12    4   STCH_HUMAN  MICROSOMAL STRESS 70 PROTEIN ATPASE CORE
                                           1458.74 / 1301.71 / 1034.50 /  721.33
```

The best integrated scores

```
Rank  Int_score  Protein    Modification
==============================================================
  1      2.6     ATPB_BOVIN  ATPASE BETA CHAIN.
  2      2.7     ATPB_RAT    ATPASE BETA CHAIN.
  3      2.8     ATPB_HUMAN  ATPASE BETA CHAIN.
  4      3.2     ATPB_MOUSE  ATPASE BETA CHAIN.
  5      8.5     GLPK_HUMAN
  6      9.0     DHAE_ELEED
  7      9.0     ENOG_HUMAN
  8      9.7     GLPK_MOUSE
  9     10.5     ENOG_MOUSE
 10     10.5     K2M2_SHEEP
 11     11.5     STCH_HUMAN  MICROSOMAL STRESS 70 PROTEIN ATPASE CORE
 12     11.6     CG2A_MESAU
```

scores additional search parameter. The second output ranks the proteins of the first list by the number of peptide hits from the peptide mass fingerprinting analysis. The third output ranks the proteins of the list according to an integrated score derived from both the amino acid composition score and the number of matching peptides in peptide mass fingerprinting.

Comments:

MultiIdent is designed to provide a means of combining many types of analytical information, in order to achieve confident protein identification. It is therefore particularly useful for cross-species protein identification and identification in the case of *N*-terminally blocked proteins.

The list of parameters that can be used for identification encompass those used in TagIdent, amino acidCompIdent and PeptIdent (section 3.1.4). The user can select the degree of detail to be displayed in the output. The result is sent back by email.

3.1.4 Identification with Peptide Mass Fingerprints: PeptIdent

Description:

PeptIdent [Wilkins et al., unpublished data] identifies proteins by peptide mass fingerprinting data. It matches a set of user-specified experimentally measured peptide masses against a database of theoretical peptides. Unlike other peptide mass fingerprinting identification programs, PeptIdent considers some 75 000 known or likely protein processing events and some 5000 known or likely protein post-translational modification events that are documented in SWISS-PROT. Therefore, all proteins in SWISS-PROT are cleaved to their mature forms before peptide masses are calculated, and the program output returns not only a list of likely protein identifications but also any hits with peptides that are known to carry any of more than 20 different types of discrete post-translational modifications and chemical modifications. The program thus offers a degree of protein characterization as part of the identification procedure. The modifications that PeptIdent considers include acetylation, amidation, deamidation, formylation, farnesylation, gamma-carboxy glutamic acid, geranyl-geranylation, hydroxylation, alpha-manno-pyranose, methylation, myristoylation, O-glcNAc, palmitoylation, pyrrolidone carboxylic acid, phosphorylation, and sulfation. Like in MultiIdent other parameters can be used in the identification procedure, i.e. the pI and MW ranges, the ion mode [M+H]$^+$) ænd isotopic resolution, the enzyme used for the digestion, the type of cysteine modification, the presence of oxidized methionine, the maximum number of missed cleavages to be considered and the mass tolerance.

Comments:

The PeptIdent tool does not consider any protein glycosylation apart from O-GlcNAc and C-mannosylation on tryptophan [Doucey 1998]. N-linked and larger O-linked sugar structures are generally of unpredictable mass. Other tools are, however, being developed to address the complexity of glycosylated structures.

PeptIdent does not do the *de novo* prediction of post-translational modifications on proteins, as any modified peptides shown in the results will be the verification of an event documented in SWISS-PROT. However, PeptIdent can match peptides whose modifications are documented in SWISS-PROT as "potential" or "by similarity". PeptIdent thus allows predicted post-translational modifications to be validated.

The FindMod tool (see 3.2.5) can be used subsequent to protein identification with PeptIdent. It allows the *de novo* prediction and discovery of protein post-translational modifications as well as the prediction of potentially mutated amino acid sequences.

It is to be noted that introducing peptide masses known to be autodigestion products of an enzyme (e.g. trypsin) or other artifactual peaks (e.g. matrix peaks or those derived from human keratin) should be avoided as they may give spurious results.

3.2 Characterization Tools

These tools are designed either to predict experimental data from protein attributes (see Figure 1) and/or to help with the interpretation of experimental results. This involves the prediction, the identification and the characterization of PTMs, amino acid substitutions and other physico-chemical parameters of the proteins of interest. They help to complete the characterization of a protein.

3.2.1 Compute pI/MW

Description:
Compute pI/MW [Wilkins 1998b] generates an estimated pI and mass for a given SWISS-PROT or TrEMBL entry or for a user-defined amino acid sequence. If a SWISS-PROT or a TrEMBL entry is chosen, Compute pI/MW reads the feature (FT) lines of the corresponding database entry and then gives the user a choice for which chain or domain of a protein the pI and mass should be calculated. Alternatively, the user can numerically enter the region of a protein for which the parameters should be computed. The pI is calculated from amino acid pK values described in Bjellqvist [1993, 1994]. The mass is calculated from the sum of the average masses for all amino acid fragments in the protein and that of one water molecule.

Comments:
The pI calculation is based on the migration of polypeptides in an immobilized pH gradient (pH 4.5 to 7.3) gel [Bjellqvist 1993, 1994]. This algorithm may therefore be inadequate for pI estimates of highly basic proteins. New experiments are on the way to refine the pI prediction of very basic proteins (D.F. Hochstrasser, personal communication). It should also be noted that since the buffer capacity of a protein can affect the accuracy of its estimated pI, the prediction of pI for small proteins is less accurate. The variation between observed and predicted pI is typically lower than 0.01 for unmodified proteins over 20 kDa, and about 0.04 for smaller proteins [Bjellqvist 1993, 1994].

Post-translational modifications are currently not considered in the mass and in the pI calculations. Therefore the migration of highly modified proteins (particularly glycosylated proteins) can considerably deviate from the pI and mass coordinate computed for the unmodified proteins. Calcium-binding proteins in particular display variable SDS PAGE migration behavior [Sanchez 1997; Dianoux 1992; Bandyo-padhyay 1990]. At the bottom of all SWISS-PROT and TrEMBL entries, the ExPASy server provides a direct link to Compute pI/MW. The entry also contains a link to the corresponding entry in SWISS-2DPAGE [Hoogland 1998], a two-dimensional polyacrylamide gel electrophoresis database accessible on ExPASy. If the protein cannot be found in SWISS-2DPAGE, the region of a 2-D gel where the selected unmodified protein should migrate can be computed and displayed. If phosphorylation and glycosylation modifications are annotated in SWISS-PROT, a region is suggested where the modified protein might be observed (i.e. this region extends towards more acidic pI and higher MW if sialic acids are covalently bound to the polypeptide chain. As an example see the leucine-rich α-2 glycoprotein in the plasma; URL http://www.expasy.ch/cgi-bin/ch2d-compute-map?PLASMA_P02750).

3.2.2 PeptideMass

Description:

PeptideMass [Wilkins 1997b, 1998b] calculates the theoretical masses of peptides generated by the chemical or enzymatic cleavage of a given protein. The protein is submitted either by its SWISS-PROT or TrEMBL code (an identification name ID or an accession number AC) or as a user-specified sequence. The program allows the user-specified sequence to include post-translational modifications, e.g. ILNS-(PHOS)PEKAC is a peptide in which the serine residue in position 4 is phosphorylated. A list of accepted PTMs and their abbreviations is provided. The user then chooses the cleavage agent, the ion mode ([M] or [M+H]$^+$), the isotopic resolution (monoisotopic or average), the maximum number of missed cleavages, any chemical modifications to cysteine and the potential oxidation state of methionine. PeptideMass uses all annotations in SWISS-PROT concerning protein processing and modifications. It thus considers known or potential PTMs, database conflicts, variants (protein isoforms) and varsplices (mRNA splicing variants) annotated in the feature tables of entries in SWISS-PROT. The result appears as a table which contains, for each peptide fragment, the peptide mass, the number of missed cleavages, the mass of the post-translationally modified peptide (if any), the position of the peptide in the sequence, the position and the type of conflicts, variants and varsplices as annotated in SWISS-PROT, and the amino acid sequence. The user can choose to list the peptides according to their masses or to their position in the protein.

Comments:

Signal sequences, propeptides and transit sequences are all removed from proteins before cleavage rules are applied. If a protein entry in SWISS-PROT is known to produce multiple chains or peptides, these are considered separately. The PeptideMass output will therefore contain a separate result list for each of them. The peptide masses related to amino acid substitutions (variants, conflicts or varsplices) are not calculated. However, any peptides that can carry these modifications are highlighted, and a 20x20 substitution table containing the pairwise amino acid mass differences is linked to the output.

3.2.3 ProtParam

Description:
   ProtParam computes various physical and chemical parameters for a protein chosen from SWISS-PROT or TrEMBL or for a user-entered sequence. The computed parameters include molecular weight, isoelectric point, amino acid composition, molar extinction coefficient, estimated half-life, instability index, aliphatic index and grand average of hydropathicity (GRAVY). The molecular weight and the pI are computed as in Compute pI/MW (see section 3.2.1); the extinction coefficient is calculated as described by Gill and coworkers [Gill 1989] phosphate buffer solution at pH 6.5, from the molar extinction coefficients of tyrosine, tryptophan and cystine the half-life is estimated by the N-end rule [Bachmair 1986; Gonda 1989; Tobias 1991]; the instability index is based on a statistical analysis [Guruprasad 1990] where the authors revealed that there are certain dipeptides, the occurrence of which is significantly different in unstable proteins compared with those in stable ones. The authors of this method have assigned a weight value of instability to each of the 400 different dipeptides. The aliphatic index (i.e. the relative volume of a protein occupied by the aliphatic side chains alanine, valine, isoleucine, and leucine) is calculated as proposed by Ikai [1980] and may be regarded as a positive factor for the increase of thermostability of globular proteins; the grand average of hydropathicity (GRAVY) value is computed according to Kyte and Doolittle [Kyte 1982].

Comment:
   The above parameters can be computed for any domain, mature chain or peptide or region of interest as documented in SWISS-PROT feature tables, for a numerically specified sub-sequence or for the full sequence.
   Note that the extinction coefficient is estimated for a protein denatured in 6.0 M guanidine hydrochloride, 0.02 M phosphate buffer solution at pH 6.5, from the molar extinction coefficients of tyrosine, tryptophan and cystine.

3.2.4 ProtScale

Description:
   ProtScale computes and represents series of profiles produced by any amino acid scale on a selected protein. An amino acid scale is defined by a numerical value assigned to each type of amino acid. The most frequently used scales are hydrophobicity or hydrophilicity scales and secondary structure conformational parameter scales, but many other scales exist which are based on different chemical and physical properties of amino acids. This program provides 50 predefined scales entered from the literature. The user specifies a database entry code (AC or ID line from SWISS-PROT or TrEMBL) or a manually entered amino acid sequence, an amino acid scale and a number of statistical parameters to be used for the calculation, i.e. the window size, the relative weight of the window edges, the weight variation model and the normalization scale.

Comments:

As described for Compute pI/MW and ProtParam, the user can restrict the calculation to a specific region of the query protein sequence. The output graphically represents the computed factor as a function of the amino acid position in the protein sequence. The result can be obtained as a gif image, a postscript or a coordinate file.

For details about the different amino acid scales used by ProtScale the reader can consult the separate information sources and literature references provided as links from the ProtScale web page (for URL see Appendix)

3.2.5 FindMod

Description:

FindMod is a program for the *de novo* discovery of protein post-translational modifications and amino acid substitutions [Wilkins et al., unpublished data]. It examines peptide mass fingerprinting results of known proteins for the presence of about 20 types of PTMs of discrete mass. This is done by looking at mass differences between experimentally determined peptide masses and theoretical peptide masses calculated from a specified protein sequence. If a mass difference corresponds to a known PTM not already annotated in SWISS-PROT, "intelligent" rules are applied that examine the sequence of the peptide of interest and make predictions as to what amino acid in the peptide is likely to carry the modification. Potential amino acid substitutions are similarly detected by mass differences.

The user submits a SWISS-PROT or a TrEMBL entry name or an amino acid sequence (Figure 5). In the case of a manually entered sequence, the user is required to specify the biological source of the query protein. This information is used to determine whether certain PTMs are likely to occur in the sequence. In addition to the experimental peptide masses the user enters the ion mode ([M] or [M+H]$^+$) and the isotopic resolution (average or monoisotopic), the digestion agent, any chemical treatment of cysteine, the oxidation state of methionine, the maximum number of missed cleavages to consider and the mass tolerance. The user can also use the program to predict potential PTMs and potential single amino acid substitutions on any peptide.

The results from FindMod are divided into a header and up to three tables (Figure 6). The header contains the submitted protein information and an active link to PeptideMass. The tables report the peptides whose experimental masses match unmodified or modified theoretical digest products of the protein of interest. The first table reports matches to theoretical digest products as unmodified, modified with the

---►

Figure 5. Example of an input form at ExPASy: FindMod tool. All ExPASy input forms are divided into three main sections, i.e. a header containing a short description of the tool and links to detailed information and/or literature references, followed by the form itself, and a lower part including the date of the last modifications to the page, a link to contact people and a link to the ExPASy homepage. In the form section the parameters and experimental data can be introduced using text fields, pull-down menus or clickable buttons. The FindMod form presented here has been filled in for characterization of *E.coli* elongation factor tu digested in-gel with trypsin after 2-D PAGE separation. The masses of its tryptic digests have been measured with a mass spectrometric method.

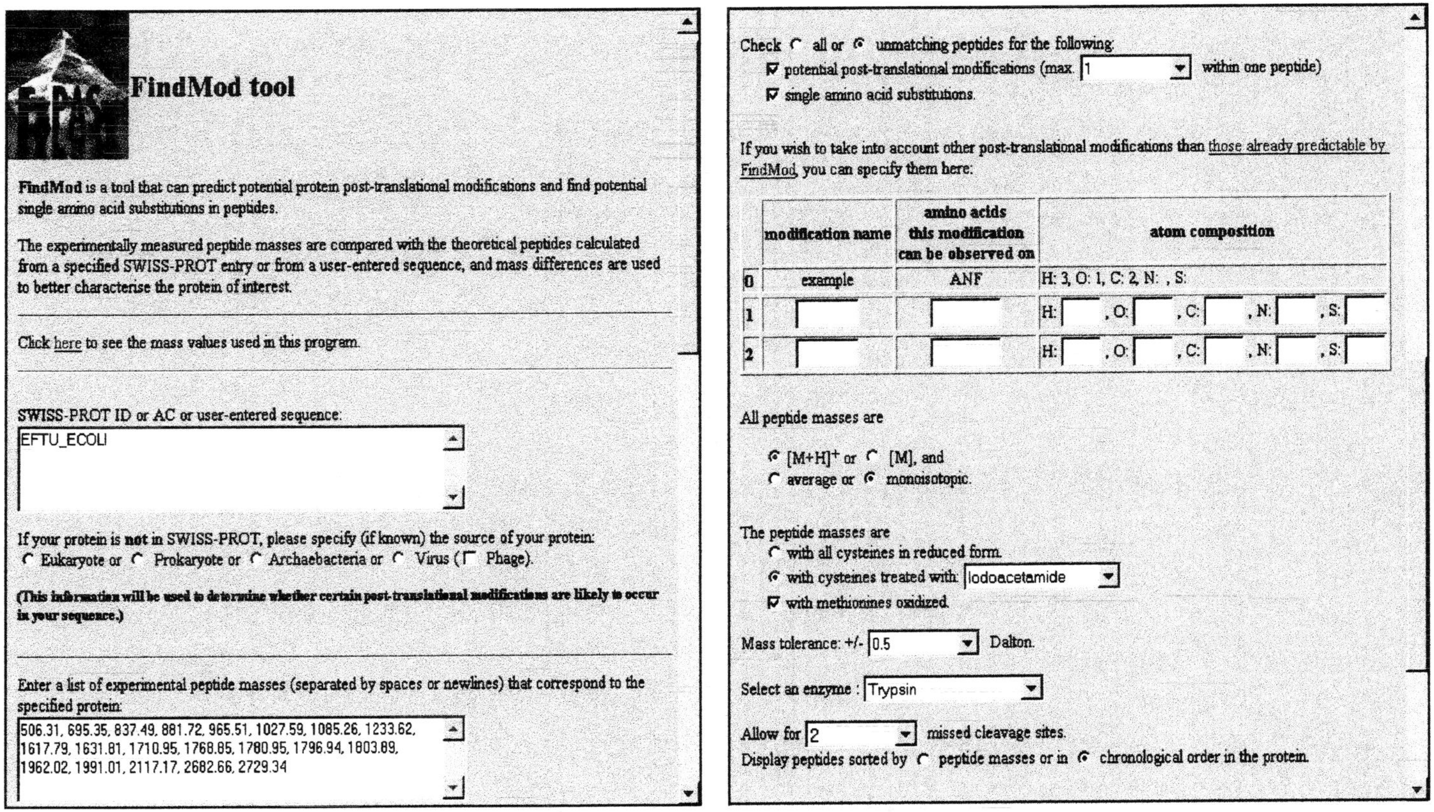

FindMod tool

FindMod is a tool that can predict potential protein post-translational modifications and find potential single amino acid substitutions in peptides.

The experimentally measured peptide masses are compared with the theoretical peptides calculated from a specified SWISS-PROT entry or from a user-entered sequence, and mass differences are used to better characterise the protein of interest.

Click here to see the mass values used in this program.

SWISS-PROT ID or AC or user-entered sequence:
EFTU_ECOLI

If your protein is not in SWISS-PROT, please specify (if known) the source of your protein:
(Eukaryote or (Prokaryote or (Archaebacteria or (Virus (Phage).

(This information will be used to determine whether certain post-translational modifications are likely to occur in your sequence.)

Enter a list of experimental peptide masses (separated by spaces or newlines) that correspond to the specified protein:
506.31, 695.35, 837.49, 881.72, 965.51, 1027.59, 1085.26, 1233.62,
1617.79, 1631.81, 1710.95, 1768.85, 1780.95, 1796.94, 1803.89,
1962.02, 1991.01, 2117.17, 2682.66, 2729.34

Check (all or (unmatching peptides for the following
☑ potential post-translational modifications (max [1] within one peptide)
☑ single amino acid substitutions.

If you wish to take into account other post-translational modifications than those already predictable by FindMod you can specify them here:

modification name	amino acids this modification can be observed on	atom composition
example	ANF	H: 3 O: 1, C: 2, N: , S:
0		
1		H: , O: , C: , N: , S:
2		H: , O: , C: , N: , S:

All peptide masses are
([M+H]$^+$ or ([M], and
(average or (monoisotopic.

The peptide masses are
(with all cysteines in reduced form.
(with cysteines treated with [Iodoacetamide]
☑ with methionines oxidized.

Mass tolerance: +/- [0.5] Dalton.

Select an enzyme: [Trypsin]

Allow for [2] missed cleavage sites.
Display peptides sorted by (peptide masses or in (chronological order in the protein.

FindMod tool

EFTU ECOLI (P02990)
ELONGATION FACTOR TU (EF-TU) (P-43).
ESCHERICHIA COLI.

Theoretical pI/Mw: 5.30 / 43182.39

Calculate   the theoretical masses of peptides generated by the chemical or enzymatic cleavage of this protein using *Petide Mass.*

Click *here* to see the mass values used in this program.

Matching peptides:

User mass	DB mass	Δ mass	#MC	peptide	position	known modifications
965.51	965.506	-0.003	2	SKEKFER	1-7	(ACET: 1)
695.35	695.348	-0.001	0	TYGGAAR	38-44	
1617.79	1617.787	-0.002	1	AFDQIDNAPEEKAR	45-58	(METH: 56)
1803.89	1803.888	-0.001	0	GITINTSHVEYDTPTR	59-74	
1768.85	1768.787	-0.062	0	HYAHVDCPGHADYVK	75-89	(Cys_CAM: 81)
2729.34	2729.342	0.002	0	NMITGAAQMDGAILVVAATDGPMPQTR	90-116	
837.49	837.495	0.005	0	EHILLGR	117-123	
1962.02	1962.022	0.002	0	ILELAGFLDSYIPEPER	188-204	
2117.17	2117.165	-0.004	0	AIDKPFLLPIEDVFSISGR	205-223	
1710.95	1710.95	0	1	LLDEGRAGENVGVLLR	264-279	
1027.59	1027.59	0	0	AGENVGVLLR	270-279	
1991.01	1991.008	-0.001	1	HTPFFKGYRPQFYFR	319-333	
1233.62	1233.617	-0.002	0	GYRPQFYFR	325-333	
1780.95	1780.945	-0.004	0	MVVTLIHPIAMDDGLR	358-373	
1796.94	1796.94	0	0	MVVTLIHPIAMDDGLR	358-373	(1xMSO)
506.31	506.309	-0	0	FAIR	374-377	
881.72	881.45	-0.269	0	TVGAGVVAK	382-390	(PHOS: 382)

Potentially modified peptides, detected by mass difference and conforming to rules (considering only peptide masses that have not matched above):

User mass	DB mass	mass diff.	mod. diff.	Δ mass	potential mod.	#MC	peptide	position	known modifications
1631.81	1603.772	28.038	28.031	-0.006	*DIMETH*	1	AFDQIDNAPEEKAR	45-58	
1631.81	1617.787	14.023	14.016	-0.006	*METH*	1	AFDQIDNAPEEKAR	45-58	(METH: 56)

Potential PTMs detected by mass differences, but not confirmed by rules:

User mass	DB mass	mass diff.	mod. diff.	Δ mass	potential mod.	#MC	peptide	position	known modifications
1085.26	923.495	161.765	162.053	0.288	*CMAN*	2	SKEKFER	1-7	
1631.81	1603.772	28.038	27.995	-0.042	*FORM*	1	AFDQIDNAPEEKAR	45-58	
1631.81	1632.736	-0.925	-0.983	-0.057	*AMID*	1	ETQKSTCTGVEMFR	249-262	(1xMSO)
2682.66	2683.418	-0.757	-0.983	-0.225	*AMID*	2	MVVTLIHPIAMDDGLRFAIREGGR	358-381	(1xMSO)
2682.66	2699.412	-16.751	-17.026	-0.274	*PYRR*	2	MVVTLIHPIAMDDGLRFAIREGGR	358-381	(2xMSO)
1085.26	881.45	203.81	204.188	0.378	*FARN*	0	TVGAGVVAK	382-390	(PHOS: 382)

blue: potential residues carring *modification 1*

Potential single AA substitutions(considering only peptide masses that have not directly matched above):

User mass	DB mass	mass diff.	subst. diff.	Δ mass	potential subst.	*BLOSUM62 score*	#MC	peptide	position	known modifications
1631.81	1603.772	28.038	28.031	-0.006	A->V	0	1	AFDQIDNAPEEKAR*I*	45-58	
1631.81	1603.772	28.038	28.006	-0.031	K->R	2	1	AFDQIDNAPEEKAR	45-58	
1631.81	1603.772	28.038	28.043	0.005	Q->R	1	1	AFDQIDNAPEEKAR	45-58	
1631.81	1617.787	14.023	14.016	-0.006	D->E	2	1	AFDQIDNAPEEKAR	45-58	(METH: 56)
1631.81	1617.787	14.023	14.052	0.029	N->K	0	1	AFDQIDNAPEEKAR	45-58	(METH: 56)
1631.81	1617.787	14.023	14.016	-0.006	N->Q	0	1	AFDQIDNAPEEKAR	45-58	(METH: 56)
1631.81	1711.765	-79.954	-80.036	-0.081	H->G	-2	0	HYAHVDCPGHADYVK	75-89	

annotations in SWISS-PROT and chemically modified as specified in the input form. The second table reports the matches to peptides potentially modified with PTMs. The first part of this table shows matches according to a set of intelligent rules (see comments), whereas the second part lists potential PTMs detected by mass differences, but not confirmed by the rules. The third table shows potential single amino acid substitutions detected by mass difference and a probability score extracted from the BLOSUM62 substitution table [Henikoff 1992].

Comments:

FindMod is a tool which facilitates the characterization of proteins using their peptide mass fingerprint results. It can be best used as a complement of a PeptIdent search to help the interpretation of a peptide mass fingerprinting analysis after protein identification. It allows the screening for unannotated PTMs, for amino acid substitutions and other modifications. In some cases it can also be used in the characterization of a protein identified after cross species identification.

For each PTM a specific prediction rule has been defined by using information in the PROSITE database [Bairoch 1997] examining and by examining all the annotations in SWISS-PROT and relevant information in the literature.

New features of the program are currently under development: for instance Find-Mod will be able to check for peptides that correspond to truncated protein *N*- or *C*-termini and to check for multiply charged peptides in a mass spectrum.

As for PeptIdent, peptide masses known to be autodigestion products of an enzyme (e.g. trypsin) or other artifactual peaks (e.g. matrix peaks) should not be introduced.

Figure 6. Output of a FindMod search: result of the input form of Figure 5 submitted to ExPASy. It first shows a header containing the database identifier of the protein of interest with a link to SWISS-PROT and PeptideMass. Then a number of tables are produced. They report the matching peptides and usually include in separate columns the user specified mass (User mass), the database mass (DB mass) of the matched peptide, the mass difference between query and matched peptide masses (mass diff.), the number of missed cleavages (#MC), the amino acid sequence of the matched peptide (peptide), its start and end positions in the protein sequence (position) and any modifications annotated in the databases for this peptide (known modifications). The first table reports the peptides whose experimental masses match theoretical digest products of the protein of interest. This list includes peptides carrying modifications as annotated in SWISS-PROT as well as peptides containing chemically modified cysteine or methionine residues as specified by the user (i.e. ACET:1 reports an acetylation of the N-terminal serine, METH:56 reports an annotated methylation of the lysine at position 56, Cys_CAM:81 refers to a carboxymethylated cysteine at position 81, 1xMSO suggests that one of the methionines has been oxidized to a sulfoxide, PHOS: 382 reports a phosphorylation of the first threonine of the corresponding peptide). The second table reports potential PTMs for peptides whose masses have not matched above. Potentially modified peptides detected by mass difference with the experimental data and conforming to rules are listed first, followed by peptides with potential PTMs suggested by mass differences, but not confirmed by rules. The amino acid(s) which can carry this modification are highlighted in the peptide sequence. The third table shows potential single amino acid substitutions. For each potential substitution the probabilistic value assessed to this substitution by the BLOSUM62 matrix [Henikoff 1992] (BLOSUM62 score) is shown, and the amino acid(s) concerned by this substitution are highlighted in the peptide sequence. This table has been truncated for easier reading.

IV.1 *P.-A. Binz, M. R. Wilkins, E. Gasteiger, A. Bairoch, R. D. Appel and D. F. Hochstrasser*

4 Other Identification and Characterization Tools on the Internet

This section introduces further identification and characterization tools available on the internet for the study of proteins. The applications of the programs are described in a short form. Since most of them are under continuous development and optimization, only particular features are highlighted. For more detailed information about the input formats, the searchable parameters, the algorithms used, the included choice of peptidases, the program capacities and limitations and other particular questions, the reader should visit the corresponding web sites. Note that unless otherwise mentioned, the programs analyzing molecular masses can compute both monoisotopic and average mass values.

4.1 The ProteinProspector Server (University of California San Francisco Mass Spectrometry Facility)

The facility provides a set of identification and analysis tools. The identification tools are MS-Fit (peptide-mass fingerprinting), MS-Tag (fragment-ion tag search) and MS-Edman (Edman/Peptide Mass search). The analysis tools are MS-Digest (theoretical peptide masses computed from enzymatic digestion of protein sequences), MS-Product (calculation of possible peptide fragment ions) and MS-Comp (theoretical amino acid compositions from peptide parent mass and immonium ions). A detailed description of the programs can be obtained from the URL: http://prospector.ucsf.edu/ or its mirror site at http://falcon.ludwig.ucl.ac.uk/.

- MS-Fit matches peptide mass fingerprints against protein sequence databases. It can optionally search for a limited number of amino acid modifications, i.e. currently protein *N*-terminus modification of Gln to pyroGlu, oxidation of Met, protein N-terminus acetylation and phosphorylation of Ser, Thr or Tyr. It can also search for single amino acid substitutions. In this "homology" mode the amino acid modifications are not considered. Instead, the program uses a mutation mass shift matrix to search for a single amino acid substitution which would transform the mass calculated from a database peptide product to match the experimental mass data.
- MS-Tag fits fragment-ion tag data obtained from a tandem mass spectrometry experiment to a peptide sequence in a sequence database. All fragment-ions in the tag need not be of the same ion series, and amino-acid composition from immonium ions can be added. With the homology mode MS-Tag enables matching to database sequences which contain a single mutation, a cross-species substitution, or a modified amino acid. The Ludwig web site allows users to restrict the searches with a number of parameters, i.e. amino acids known to be absent in the sequence, a number of possibly present amino acid modifications and the addition of a user-specified amino acid defined by its elemental composition.
- MS-Edman performs text-based searches (with or without peptide mass filtering) on different fields in protein sequence databases. The implementation is designed to consider the ambiguity often present in data obtained from an Edman degrada-

tion protein sequencing experiment. MS-Edman can search a sequence database with a sequence including ambiguities ("regular expression") or with a list of similar sequences. If the search has been performed using a regular expression, it can be filtered with peptide mass data.

- MS-Digest performs an in-silico enzymatic digestion of a protein sequence and calculates the mass of each peptides. The user can alternatively specify the accession number from the database, a user-supplied sequence or the MS-Digest index number produced by MS-Fit and MS-Tag (see instruction in the MS-Digest web page for a detailed description of this index number). The program optionally takes into account a number of missed cleavages and the same set of modifications as MS-Fit.
- MS-Product calculates the possible fragment ions resulting from fragmentation of a peptide in a mass spectrometer. Fragmentation possibilities for post-source decay (PSD), high-energy collision-induced dissociation (CID), and low-energy CID processes may be calculated. The syntax used for the query peptide sequence allows to include a number of modified amino acids with a user-specified elemental composition. The user chooses the types of fragments to be computed.
- MS-Comp can be used to complete the possible amino acid composition of a peptide given a parent mass and partial composition from the immonium ion data contained in a user's tandem mass spectrum.

4.2 The PROWL Site from the MS Groups at Rockefeller and NY Universities

PROWL is a project being developed in collaboration between ProteoMetrics and Rockefeller University. Its purpose is to develop WWW accessible resources for protein mass spectrometry [Fenyo 1996]. Two protein identification programs are available at their server (URL: http://prowl.rockefeller.edu/) namely ProFound and PepFrag.

ProFound performs fast database searches combined with Bayesian statistics for protein identification. Masses of peptides from a proteolytic digestion of a protein are compared to the masses of a peptide mass database calculated from the OWL protein database. ProFound can search for a pure sample or a binary mixture. The ranking of the candidate proteins is based on their calculated "posterior probability" [Zhang 1995].

PepFrag attempts protein identification by comparison of the fragmentation pattern of proteolytic peptides to a sequence database. Cleaving enzymes can be endo- and exopeptidases. The user can include a number of phosphorylated Ser, Thr and Tyr in the input parameters. Cysteine modifications are also considered. The search can be further focused by specifying a number of amino acids known to occur in the considered peptide.

4.3 PeptideSearch at the EMBL Protein & Peptide Group

PeptideSearch is an advanced tool for protein database searching by mass spectrometric data, such as peptide mass maps or (partial) amino acid sequences. It has been developed at the EMBL Protein & Peptide Group (URL: http://www.mann.-embl-heidelberg.de/Services/PeptideSearch/PeptideSearchIntro.html).

PeptideSearch uses the non-redundant protein database nrdb updated at EMBL/EBI (European Bioinformatics Institute) as target database. The tool offers three distinct search forms.

The first approach identifies proteins with their peptide mass fingerprints. It can consider oxidation of methionine and a wide range of cysteine modifications. From the output list of matching proteins the user can perform a "second pass search". This second pass search uses the unmatched masses to attempt the detection and possible identification of a second protein in the mixture.

The second variation of PeptideSearch identifies a protein from a peptide sequence tag determined by mass spectrometric measurements. The user submits the mass of a peptide fragment together with a sequence tag obtained from MS fragmentation analysis or Edman degradation. Cleavage specificities and number of allowed errors are included attributes. The search cannot be limited to a taxonomic range.

The third approach identifies a protein by searching an amino acid sequence in the database. The syntax of the query sequence allows ambiguities.

The output of PeptideSearch provides links to the matching protein entries in nrdb and to a virtual digestion program similar to PeptideMass (section 3.2.2) or MS-digest (section 4.1).

4.4 The MOWSE Database Search Tool from the Daresbury Laboratory

The MOWSE search program identifies proteins by their peptide mass fingerprints and optionally sequence information or amino acid composition of protein fragments. The program belongs to the SEQNET project and is located at the Daresbury Laboratory (URL: http://gserv1.dl.ac.uk/SEQNET/mowse.html). Experimental mass values are searched against MOWSE, a molecular weight protein and peptide fragment database derived from the OWL protein database. Only average isotopic masses are considered. The masses in MOWSE are rounded to the nearest integer value.

4.5 The MassSearch Page of the Computational Biochemistry Research Group Server at the ETH Zürich

MassSearch (URL: http://cbrg.inf.ethz.ch/ServerBooklet/MassSearchEx.html) identifies a protein from its peptide mass fingerprints. It was developed at the Computational Biochemistry Research Group Server at the ETH Zürich. The function MassSearch searches the SWISS-PROT database, whereas the function DNAMassSearch locates the best candidates in the EMBL DNA database that would code for a protein fitting the given weights. Results are sent by email.

5 Conclusion

In this chapter we have presented internet tools designed for protein identification and characterization. These programs are all available on the World Wide Web as interactive tools. Scientists now have the possibility of choosing the most appropriate tools or sets of tools for their particular identification and characterization needs. However, rather than using the tools as "black boxes", it is advisable that users spend some time to understand the algorithms, or at least appreciate the possibilities provided by the tools.

The number, the power and the usefulness of these tools will continue to increase in the future. The quality of the results will improve in parallel with the quality and amount of data submitted to the various WWW-accessible databases. Genome sequencing projects will certainly contribute greatly to the rapid growth of sequence information identification tools with increasingly ideal conditions for best matching possibilities. Other information, like PTM, secondary structure, spatial structure and function are more difficult to obtain experimentally and represent therefore a bottleneck in proteome characterization studies. The efficient interpretation of these data in the design of prediction rules will play a central role in the development and the improvement of the characterization tools.

Notes:
Note 1: TrEMBL is split in two main sections, i.e. SP-TREMBL and REM-TREMBL. SP-TREMBL (SWISS-PROT TREMBL) contains the entries which should be eventually incorporated into SWISS-PROT, whereas REM-TREMBL (REMaining TREMBL) contains the entries that will not be included in SWISS-PROT (i.e. Immunoglobulins and T-cell receptors, synthetic sequences, patent application sequences, small fragments shorter than 8 amino acids and CDS not coding for real proteins). In this chapter, TrEMBL will always refer to SP-TREMBL.

6 References

Appel, R. D., Bairoch, A, Hochstrasser, D. F. (1994) *Trends Biochem. Sci. 19*, 258–260. A new generation of information retrieval tools for biologists: the example of the ExPASy WWW server.

Bachmair, A., Finley. D., Varshavsky, A. (1986) *Science 234*, 179–186. In vivo half-life of a protein is a function of its amino-terminal residue.

Bairoch, A., Bucher, P., Hofmann, K. (1997) *Nucleic Acids Res 25 (1)*, 217–221. The PROSITE database, its status in 1997.

Bairoch, A., Apweiler, R. (1998) *Nucleic Acids Res. 26*, 38–42. The SWISS-PROT protein sequence data bank and its supplement TrEMBL in 1998.

Bandyopadhyay, S., Ghosh, S. K. (1990) *J Protein Chem 9 (5)*, 603–611. Goat testis calmodulin: purification and physicochemical characterization.

Bjellqvist, B., Hughes, G. J., Pasquali, Ch., Paquet, N., Ravier, F., Sanchez, J.-Ch., Frutiger, S., Hochstrasser, D. F. (1993) *Electrophoresis 14*, 1023–1031. The focusing positions of polypeptides in immobilized pH gradients can be predicted from their amino acid sequences.

Bjellqvist, B., Basse, B., Olsen, E., Celis, J. E. (1994) *Electrophoresis 15*, 529–539. Reference points for comparisons of two-dimensional maps of proteins from different human cell types defined in a pH scale where isoelectric points correlate with polypeptide compositions.

Dianoux, A. C., Stasia, M. J., Garin, J., Gagnon, J., Vignais, P. V. (1992) *Biochemistry 31 (25)*, 5898–5905. The 23-kilodalton protein, a substrate of protein kinase C, in bovine neutrophil cytosol is a member of the S100 family.

Doucey, M. A., Hess, D., Cacan, R., Hofsteenge, J. (1998) *Mol Biol Cell 9 (2)*, 291–300. Protein C-mannosylation is enzyme-catalysed and uses dolichyl-phosphate-mannose as a precursor.

Fenyo, D., Zhang, W., Chait, B. T., Beavis, R. C. (1996) *Anal Chem 68 (23)*, 721A–726A. Internet-based analytical chemistry ressources: a model project.

Gill, S. C., von Hippel, P. H. (1989) *Anal. Biochem. 182*, 319–326. Calculation of protein extinction coefficients from amino acid sequence data.

Golaz, O., Wilkins, M. R., Sanchez, J.-C., Appel, R. D., Hochstrasser, D. F., and Williams, K. L. (1996) *Electrophoresis 17*, 573–579. Identification of proteins by their amino acid composition: an evaluation of the method.

Gonda, D. K., Bachmair, A., Wunning, I., Tobias, J. W., Lane, W. S., Varshavsky, A. (1989) *J. Biol. Chem. 264*, 16700–16712. Universality and structure of the N-end rule.

Henikoff, S., Henikoff, J. G. (1992) *Proc. Nat. Acad. Sci 89*, 10915–10919. Amino acid substitution matrices from protein blocks.

Hoogland, C., Sanchez, J.-C., Tonella, L., Bairoch, A., Hochstrasser, D. F., Appel, R. D. (1998) *Nucleic Acids Res 26 (1)*: 334–335. Current status of the SWISS-2DPAGE database.

Ikai, A. (1980) *J. Biochem. 88*, 1895–1898. Thermostability and aliphatic index of globular proteins.

Kyte, J., Doolittle, R. F. (1982) *J. Mol. Biol. 157*, 105–132. A simple method for displaying the hydropathic character of a protein.

Sanchez, J.-C., Schaller, D., Ravier, F., Golaz, O., Jaccoud, S., Belet, M., Wilkins, M. R., James, R., Deshusses, J., Hochstrasser, D. H. (1997) *Electrophoresis 18*, 150–155. Translationally controlled tumor protein: a protein identified in several nontumoral cells including erythrocytes.

Tobias, J. W., Shrader, T. E., Rocap, G., Varshavsky, A. (1991) *Science 254*, 1374–1377. The N-end rule in bacteria.

Wilkins, M. R., Gasteiger, E., Appel, R. D., Hochstrasser, D. H. (1996a) *Curr. Biology 6*, 1543–1544. Protein Identification with sequence tags.

Wilkins, M. R., Pasquali, C., Appel, R. D., Ou, K., Golaz, O., Sanchez, J.-C., Yan, J. X., Gooley, A. A., Hughes, G., Humphery-Smith, I., Williams, K. L., Hochstrasser, D. F. (1996b) *Bio/Technology 14*, 61–65. From Proteins to Proteomes: Large Scale Protein Identification by Two-dimensional Electrophoresis and Amino Acid Analysis.

Wilkins, M. R., Ou, K., Appel, R. D., Sanchez, J.-C., Yan, J. X., Golaz, O., Farnsworth, V., Cartier, P., Hochstrasser, D. F., Williams, K. L., Gooley, A. A. (1996c) *Biochem. Biophys. Res. Commun. 221*, 609–613. Rapid protein identification using N-terminal "sequence tag" and amino acid analysis.

Wilkins, M. R.., Williams, K. (1997a) *J Theor Biol. 186 (1)*: 7–15. Cross-species protein identification using amino acid composition, peptide mass fingerprinting, isoelectric point and molecular mass: a theoretical evaluation.

Wilkins, M. R., Lindskog, I., Gasteiger, E., Bairoch, A., Sanchez, J.-C., Hochstrasser, D. F., Appel, R. D. (1997b) *Electrophoresis 18 (3–4)*, 403–408. Detailed peptide characterisation using PEPTI-DEMASS – a World-Wide Web accessible tool.

Wilkins, MR., Gasteiger, E., Tonella, L., Ou, K., Tyler, M., Sanchez, J.-C., Gooley, A. A., Walsh, B. J., Bairoch, A., Appel, R. D., Williams, K. L., Hochstrasser, D. H. (1998a) *J. Mol. Biol.* (in press). Protein Identification with *N*- and *C*-terminal Sequence Tags in Proteome Projects.

Wilkins, M. R., Gasteiger, E., Bairoch, A., Sanchez, J.-C., Williams, K. L., Appel, R. D., Hochstrasser, D. H. Protein Identification and Analysis Tools in the ExPASy Server in: *2-D Protein Gel Electrophoresis Protocols.* (Link AJ ed.) Humana Press, New Jersey. (1998b, in preparation).

Zhang, W., Chait, B. T. (1995) *Proceedings of the 43rd ASMS Conference on Mass Spectrometry and Allied Topics*, Atlanta, Georgia. Protein Identification by Database Searching: A Bayesian Algorithm.

Appendix. Protein identification and characterization programs available on the Internet, and their URL addresses. Programs are grouped according to the protein attributes they use to query databases (modified from [Wilkins 1998a])

Type of Program / Name	Internet URL address
Protein Sequence Tags:	
TagIdent	http://www.expasy.ch/www/guess-prot.html
PeptideSearch	http://www.mann.embl-heidelberg.de/Services/ PeptideSearch/FR_SequenceOnlyForm.html
Peptide Mass Fingerprinting:	
PeptIdent	http://www.expasy.ch/sprot/peptident.html
MassSearch	http://cbrg.inf.ethz.ch/ServerBooklet/MassSearchEx.html
MS-Fit	http://falcon.ludwig.ucl.ac.uk/msfit.html or http://prospector.ucsf.edu/htmlucsf/msfit.htm
PeptideSearch	http://www.mann.embl-heidelberg.de/Services/ PeptideSearch/FR_PeptideSearchForm.html
Peptide Mass Search	http://www.mdc-berlin.de/~emu/peptide_mass.html
ProFound	http://chait-sgi.rockefeller.edu/cgi-bin/prot-id
Peptide Mass Fingerprinting and Sequence Tag Data:	
MS-Edman	http://falcon.ludwig.ucl.ac.uk/msedman.htm or http://prospector.ucsf.edu/htmlucsf/msedman.htm
Mowse	http://gserv1.dl.ac.uk/SEQNET/mowse.htm
Peptide Sequence Tags (from MS/MS or MALDI-PSD):	
MS-Tag	http://falcon.ludwig.ucl.ac.uk/mstag.htm http://prospector.ucsf.edu/htmlucsf/mstag.htm
PepFrag	http://chait-sgi.rockefeller.edu/cgi-bin/prot-id-frag
PeptideSearch	http://www.mann.embl-heidelberg.de/Services/ PeptideSearch/FR_PeptidePatternForm.html
Amino Acid Composition (with or without sequence tag):	
AACompIdent	http://www.expasy.ch/ch2d/aacompi.html
PropSearch	http://www.embl-heidelberg.de/aaa.html
Protein amino acid composition, sequence tag, peptide mass fingerprinting:	
MultiIdent	http://www.expasy.ch/sprot/multiident.html
Programs that Assist Interpretation of Analytical Data:	
Compute pI/MW	http://www.expasy.ch/ch2d/pi_tool.html
FindMod	http://www.expasy.ch/sprot/findmod.html
ProtParam	http://www.expasy.ch/sprot/protparam.html
ProtScale	http://www.expasy.ch/cgi-bin/protscale.pl
PeptideMass	http://www.expasy.ch/sprot/peptide-mass.html
Amino Acid Sequence	http://chait-sgi.rockefeller.edu/cgi-bin/sequence
MS-Comp	http://falcon.ludwig.ucl.ac.uk/mscomp.htm or http://prospector.ucsf.edu/htmlucsf/mscomp.htm
MS-Digest	http://falcon.ludwig.ucl.ac.uk/msdigest.htm or http://prospector.ucsf.edu/htmlucsf/msdigest.htm
MS-Prod	http://falcon.ludwig.ucl.ac.uk/msprod.htm or http://prospector.ucsf.edu/htmlucsf/msprod.htm

Protein Sequences and Genome Databases

Hans-Werner Mewes, Andreas Maierl and Dimitrij Frishman

1 Introduction

Modern research in biology, biotechnology, and medicine draws heavily from the analysis of living material at the molecular level. Although basic principles of energy conversion, information storage, biosynthesis, and metabolism are relatively well known, the details of the highly complex interactions of small, medium, large, and very large biomolecules in time and space are only beginning to be understood. Data concerning the molecules involved in these processes and their properties are a precious commodity, critical to the emergence of these new understandings.

Databases for biological macromolecules are basic resources for research in modern life sciences. Genome analysis, medical diagnostics, environmental research, molecular biology, gene therapy, and biotechnology not only depend on raw data but also on highly structured and interpreted databases for nucleic acid and protein sequences. The use of these databases serves two functions, both essential for the detailed insight into molecular functions and cellular roles of proteins: in practice, the biologist needs access to structured information for the interpretation of experimental data. Knowledge is gained in addition by the analytical methods in bioinformatics that allow for the systematic *in silico* analysis of genomic information. In particular, the interpretation of sequence and sequence-related data became essential for the analysis of complete genomes.

The availability of complete genomes (Table 1) has dramatically changed the view and understanding of living matter. Not only became the complete set of genetic elements from several domains of life available, but also has the systematic comparison of complete genomes in terms of orthologs and paralogs been performed [Tatusov 1997] and opened new insights into the evolution of small organisms [Wolfe 1997; Mewes 1997]. New perspectives and challenges for the systematic functional analysis have been opened by the application of new, sequence based techniques.

Table 1.

Species	Size (kBases)	Date of Release	Reference	Number of ORFs	www-Address
Archeoglobus fulgidus	21800	1997	[Klenk 1997]	2408	www.tigr.org
Bacillus subtilis	4200	1997	[Kunst 1997]	4099	www.pasteur.fr
Borrelia burgdorferi	1440	1997	[Fraser 1997]	850	www.tigr.org
Escherichia coli	4600	1997	[Blattner 1997]	4277	www.wisc.edu/ genetics/
Haemophilus influenzae	1830	1995	[Fleischmann 1995]	1680	www.tigr.org
Helicobacter pylorii	1660	1997	[Tomb 1997]	1590	www.tigr.org
M. thermoautotrophicum	1750	1997	[Smith 1997]	1871	www.genomecorp. com
Methanococcus janaschii	1660	1997	[Clayton 1997]	1735	www.tigr.org
Mycoplasma genitalium	580	1995	[Fraser 1995]	468	www.tigr.org
Mycoplasma pneumoniae	810	1996	[Himmelreich 1996]	677	www.zmbh.uni-heidelberg.de
Pyrococcus horikoshii OT3	1800			1187	www.bio.nite.go.jp
Synechocystis sp.	3570	1996	[Kaneko 1996]	3168	www.kazusa.or.jp
Treponema pallidum	1140	1998		1041	Utmmg.med.uth.tmc. edu/trepone ma
Saccharomyces cerevisiae	13000	1996	[Goffeau 1996]	5880	www.mips.biochem. mpg.de genome-www.stanford.edu

The advent of the Internet as a communication standard was a breakthrough for the dissemination of information and made it easy to create user-friendly interfaces to sequence related data resources. The internet has dramatically changed the availability and the use of biomolecular resources. In particular, access to small, specialized data collections became only practicable through the internet since their cumbersome local implementation is no longer required.

2 Sequence and Sequence-Related Databases

Sequence determination became technically feasible by chemical methods in the early 1950s. Because protein sequencing required the isolation and purification of significant quantities in a difficult, expensive, and time consuming process, the number of protein sequences determined by sequential chemical degradation remained low until the late 1970s. The first systematic analysis of the information available was undertaken in the pioneering work of M. Dayhoff and her colleagues at the National Biomedical Research Foundation in the USA (NBRF). The first editions of the database appeared in printed form as the "ATLAS of Protein Sequence and Structure" [e.g. Dayhoff 1965].

The number of currently available sequence and sequence-related databases is impressive. Table 2 displays a representative, but incomplete selection of the most important data resources.

Historically, the macromolecular database centers have emerged independently; each adopting their own formats for data distribution. The issues of compatibility between data formats has been first discussed by CODATA group resulting in a proposal for a sequence data exchange format [George 1987]. This work was further developed, specifically addressing database semantics, and a more generalized Sequence Database Definition Language SDDL has been developed [George 1993]. As an alternative, the National Center for Biotechnology Information (NCBI) has fostered the use of ASN.1 as a syntax for sequence data representation [Benson 1990]. To overcome the problems in implementing unifying common format specifications, approaches to introduce data warehouse technologies have been proposed [Markowitz 1995; Ritter 1994] but implementations could not keep pace with rapid growth of databases in size and complexity. The idea of federated databases following standards of database definition and interfaces has been discussed [Fasman 1994], but the evident need for a common layer of data exchange has been overruled by the overwhelming success of the Internet technology providing global access to all public database resources.

Table 2.

Database	Subject	Number of entries (Feb. 98)	www-Address[1]	Reference
Complete Genomes/NCBI	Complete Genomes	N/A	www.ncbi.nl.m.nih.gov	[Benson 1998]
MIPS Genomes	Complete Genomes	30931	www.mips.biochem.mpg.de	[Frishman 1997]
TDB	Complete Genomes	N/A	www.tigr.org	[White 1996]
EMBL/EBI	Nucleic Acids	1917868	www.ebi.ac.uk	[Stoesser 1998]
GenBank	Nucleic Acids	2042325	www.ncbi.nlm.nih.gov	[Benson 1998]
PFAM	Protein Domain Families	13816	www.sanger.ac.uk/software/pfam	[Sonnhammer 1998]
BLOCKS	Protein Domains	932	www.blocks.fhcrc.org	[Henikoff 1998]
PRINTS	Protein Fingerprints	750	www.biochem.ucl.ac.uk	[Attwood 1998]
Prosite	Protein Motifs	1335	expasy.hcuge.ch	[Bairoch 1996]
PIR-International	Proteins	105998	www.mips.biochem.mpg.de	[Barker 1998]
SwissProt	Proteins	69113	Expasy.hcuge.ch	[Bairoch 1998]
TREMBL	Protein Translations from EMBL	140555	www.ebi.ac.uk	[Bairoch 1998]
MIPS Yeast	Yeast Database	8448	www.mips.biochem.mpg.de	[Mewes 1997]
YPD	Yeast Database	6027	Quest7.proteome.com/YPDhome.html	[Hodges 1998]

[1] All given addresses must be preceeded by "http://"

An alternative to the interactive, hypertext based use of the Internet for database communication is the employment of the industrial standard as defined in the Common Object Request Broker Architecture (CORBA). CORBA is a middleware system that defines a set of rules for client/server applications and enables transparent access to services provided by independent databases. Part of CORBA is a standardized interface definition language (IDL) allowing for the specification of interfaces and services. New services can be made publicly available by describing the interface using CORBA. Integrated biological applications can be implemented using CORBA connecting the broad variety of data collections existing, while clients are always accessing the most up-to-date data provided by the database centers [Achard 1998].

3 Sequence Data Collection, Distribution and Query

When the technology of DNA sequencing emerged, biological research changed its direction, scope and goals in a revolutionary way. Within a short time, the data flow realized an exponential growth phase that continues today, not only for the DNA sequence data but also for protein sequences that are inferred from them. In the 1980s major centers for data collection and annotation emerged. Nucleic acid sequence data are directly submitted to these centers electronically (EMBL/EBI [Stoesser 1998], GenBank/NCBI [Benson 1998]); the reliance on published literature as a source for this information was entirely replaced by the active submission from the authors.

The macromolecular sequence databases double in size every 2.5 years. As of late 1997 approximately 200 000 protein sequences and more than 500 000 verified nucleic acid sequences have been recorded in these databases (EMBL, Release 53). In addition more than 1.3 million unverified sequences of expressed sequence tags from a variety of organisms are included. The database centers publish updates of the data repositories on a regular basis, daily updates are available through the WWW and are disseminated through mirror sites worldwide (e.g. the nodes of the EMBnet [Harper 1996]).

The early sequence search programs evolved into extensive sequence information retrieval systems which consist of two parts: a data repository module stores the sequence data information and a search engine allows the user to query the information in the data repository. The representation of data in the repository is optimized for fast access through the search engine. Efficient data structures (indexes or trees) are used to achieve fast lookup of data items. As the creation of index structures is computationally expensive, users have no possibility to modify data in the repository. Therefore, the distributed installation of sequence databases has lost impact *versus* the maintenance of database servers that are accessed by several thousand sites daily. The NCBI (National Center for Biotechnology Information, Bethesda, Maryland 20894, USA) offers the ENTREZ software to explore sequence data as well as the literature database retrieval system PubMed [Schuler 1996].

The development of sequence and genome databases must take into account aspects of database technology. Biological data are complex and notoriously inconsistent involving unmerciful problems with respect to syntax and semantics. The latest release of the annual Nucleic Acids Research issue on databases in molecular biology (Volume 26(1), 1998) included a total of 94 contributions. Data collection, administration and manipulation is performed by a variety of methods, depending on the skills and resources of the individual groups maintaining a collection. These methods range from PC based spreadsheets to sophisticated object-oriented databases. Relational database systems power the large centers efficiently but suffer from complex schema development and semantic shortcomings. Relational databases require a linear representation of the hierarchical sequence data. This representation leads to large sets of relational tables. To restore the hierarchical information from the tuples in the tables requires computationally expensive join operations.

For database maintenance, the issues of complex queries and access speed are not of primary importance. However, database access is dominated largely by queries for sequence similarity (see below) or directed to the contents of the sequence associated annotation. The current size of the databases does not allow for linear searches as a result of slow access and data transfer rates from magnetic storage devices, therefore indexing sequence and textual information becomes a necessity.

The limitations of relational database management systems can be overcome by the use of object–oriented ones. However, there are several problems to be solved until object–oriented databases are suitable for a production sequence database environment. First, there are no standards for object–oriented query languages yet. Second, object–oriented databases show poor performance, when used to manage large amounts of sequence data. Currently there exist only prototype databases. The group at MIPS is developing a DBMS using ObjectStore (\cite{object-design-user-guide} and is confident that all crucial issues of sequence database processing can be addressed.

4 Data Processing and Principles of Data Organization

The technology of sequence data processing emerged from the administration of file cabinets and cards. Flat file handling with the help of text editors and deposition of data in a linear order for distribution is still one of the major operating principles for many collections. However, there are basic differences between the operating principles of the nucleic and protein sequence databases. In both cases, the starting point of data processing is a scientific report (either extracted from the literature or directly submitted to the database centers) based on experimental results. The sequence is associated with additional information in the form of text (e.g. the start of the protein coding regions in case of DNA or functional sites in proteins). The sequence database then produces a database entry according to its processing standards. In the case of the nucleic acid sequence databases, the scientific report will persist isolated from any other information related to the particular sequence (e.g. other reports on the same sequence including those generated from the same laboratory). The goal of the protein sequence databases is to merge information from different source reports that can be identified as related to the same sequence (which is defined as an unique gene product from one particular location in the genome). During this process, discrepancies among the reports are detected and recorded (and resolved if possible) resulting in a much more reliable, comprehensive, and complete representation of the information.

In addition a significant level of redundancy in the data is eliminated [Stoesser 1998; Benson 1998; Barker 1998; Bairoch 1998].

5 Sequence Data and Related Information

Data in the biological sequence database are generally characterized as: (1) facts and (2) associated, interpretive information. The so–called facts (the sequence data themselves) are inferences on experimental results and are subject to experimental errors and errors in interpretation; they are not immutable and may be modified at any later time as errors are detected and resolved. The second class of information is derived from a very broad range of experimental techniques and inference chains; these data are generally denoted as annotation; they describe the biological properties of the sequences. The annotation is a difficult task and can only be properly performed by technically–trained biological research scientists familiar with the data and the experimental techniques employed in their determination. Adequate representation of the information requires an understanding of the current, rapidly changing state of biological knowledge as well as an understanding of computer science.

For example, 20 years ago it was only beginning to be recognized that protein coding regions in higher organisms are often interrupted by untranslated segments. Now a large fraction of the protein sequence data are inferred based on knowledge of known biological translation mechanisms. This new understanding fundamentally changed the data in the protein sequence database. Any model for representing biological data must be highly extensible and sufficiently robust to be adapted rapidly to fundamental changes in biological understanding.

Despite the fundamental progress in genome and proteome analysis, the current biological knowledge is still incomplete and many important properties of the molecules are yet to be discovered. Others are predicted based on empirical observations and comparative methods. The data inferred by the interpretation of sequence relations are subject to the limitations of these methodologies [Galperin 1998]. Upcoming techniques such as quantitative expression analysis and the detection of protein/protein interactions will contribute large amounts of data in relation to the sequence and genome databases scaffold. As more is learned from information based on experimental evidence, former annotation must be continually reassessed for compliance with new facts. Any scheme developed for processing these data must be highly dynamic. In particular, the rapid progress in the systematic sequencing of complete genomes makes the data analysis the critical, time-limiting step that is often only insufficiently represented in updates of the databases.

6 Sequence Information Retrieval

Molecular sequence databases provide an extraordinary resource to gain insight into biological function. They serve as data repositories for experimental results as well as reference compendia, summarizing the current state of biological knowledge. In principle, they are structured in the form of separate entries, representing unbranched chains of nucleic acid or amino acid residues.

A powerful tool to access the sequence related information has been developed by Etzold et al. [Etzold 1996]. The SRS system links heterogeneous data collections by common indices and cross-references. The heterogeneous data formats are digested by a sophisticated parsing engine (ICARUS) providing flexible access to information scattered over several databases. SRS is installed on a number of public sites world-wide serving a large user community.

A primary use of the sequence database is sequence data analysis by homology searches. A large part of the information available has been deduced from sequence data analysis. The quality of sequence annotation is a critical factor for the interpretation of database information. Unfortunately misassignments tend to propagate into sequence related database entries. Critical inspection of sequence analysis results is obligatory [Galperin 1998]. An experimentally determined sequence can be scrutinized against the most up-to-date information of the sequence databases to answer the following questions:

- Is the sequence already known in the database?
- Do significantly similar sequences exist in the database?
- Is any additional information related to the similar sequences found in the database available (e.g., functional assignment, 3D structure)?

These questions are typically addressed first by searching against the sequence portion of a database. Most database searching programs do not employ the text portion of the database directly in resolving such queries, rather the text is made available after the database is searched to aid in the interpretation of the results. Because of the size of the current databases (20–200 million residues), even simple searches require several minutes for read and compare operations. Hence the emphasis is on pre-structuring the data and/or limiting the operations necessary during the initial examination [Gonnet 1992; Mewes 1995].

Homology searches are conducted as complex string comparisons. For proteins, the alphabet of matching characters consists of the set of 20 naturally occurring amino acids. Based on similar physico–chemical properties and the redundancy of the genetic code, similarities between pairs of amino acids are quantified by the use of comparison matrices [Henikoff 1993; Taylor 1991]. The optimal alignment of a sequence pair is the arrangement of two sequences (including gaps) that exhibits the maximal sum of the scores of matching amino acid pairs. Although rigorously proven algorithms have been developed to compute such optimal alignments, the methodology is fundamentally limited by the nature of the problem. The essential question is whether two (or more) sequences specify proteins that fold into similar structures and/or that exhibit similar biological functions. Because the structure/function problem remains unsolved, it is not possible to construct an effective measure for this propensity. Hence, sequence comparison methods can only suggest (rather than prove) that the observed sequence similarity is indicative of similarity of structure or function.

Early analytical methods focused on the global comparison of two complete sequences along their entire length [Needleman 1970; Waterman 1983]. Given a comparison matrix, the algorithms always find an optimal alignment. The algorithms have been tuned for the number of steps that are necessary to compute an alignment: The number of residue comparisons is proportional to the product of the length of the compared sequences. Internally the algorithm allocates an amount of memory that is proportional to the length of the shorter sequence.

Sequence relations as reflected in sequence homology are based on evolutionary inheritance. The classical concept of a protein superfamily was built on the assumption that point–mutational events related members of a superfamily to their common ancestors. The PIR–International Protein Sequence Database was originally designed based on this concept. The term homology is used to indicate that pairs or sets of sequences are of common evolutionary origin. Provided that the sequences are closely related, the identification of a query sequence as matching a well described family is straightforward [Barker 1996].

7 Common Algorithms for Homology Search

Sequence similarity searches are used as versatile tools to identify homology related entries in the database and to provide clues for protein function prediction. Sensitivity and selectivity, in particular the choice of the right scoring matrices are important issues to be considered in the analysis of newly discovered sequences against the sequences stored in the databases [Henikoff 1993].

Exhaustive comparisons are computationally demanding. To obtain reasonable response times, either special parallel computer architectures [Brutlag 1993] or heuristic methods are applied. The most advanced algorithms to scan the databases, such as FASTA and BLAST, use algorithms to explore only a selected subset of candidates. The heuristic strategies work in three steps. First they identify all substrings of sequences in the database that directly match a substring of the query sequence without insertions or deletions. This step uses specific data structures and fast algorithms. In a second step the significance of each database hit is evaluated. Significant subsequences are extended in both directions using the computationally expensive exact matching algorithm. The extended subsequences are returned as the result of the database query in the third step. The optimization results from the reduction of the number of sequences that must be compared using the expensive optimal alignment algorithm.

The FASTA program [Lipman 1985] focuses on regions of the database sequence that show a high density of direct matches with the query sequence. FASTA identifies regions of identity using a lookup table. A parameter allows to set a threshold for the number of subsequent expansions.

The BLAST (Basic Local Alignment Search Tool) [Altschul 1990] algorithm constructs for a given query sequence a set of similar sequences. The similar sequences are retrieved using an algorithm that does not recognize mismatches and gaps. The BLAST algorithm cuts the query sequence from left to right into overlapping substrings of a fixed length (k–tuples). For each substring all similar substrings up to a certain threshold are generated. For the total set of substrings all matching subsequences in the database are retrieved. This is done using a look up table or a finite state machine.

The FLASH algorithm (Fast Look–up Algorithm for String Homology [Califano 1993] is based on a probabilistic indexing framework. It creates an index that already contains deletions by splitting the database sequence into k–tuples. FLASH adds for each k–tuple a set of n–tuples (n <= k) to the index. The n–tuples are formed by deleting residues from the k–tuple. The FLASH algorithm regards point mutations in amino acid sequences by forming classes of amino acids. Position tree variants have also been successfully applied to protein and genome sequences [Gonnet 1992; Mewes 1997].

The evaluation of the sequence comparison results is often time–consuming and statistical evaluations are subject to misinterpretation [for a review see Altschul 1994]. There is an evident tradeoff between sensitivity and selectivity. The sensitivity of a sequence comparison algorithm is measured by the ratio of the number of occurrences of a substring in the database to the number of the reported correct hits. The selectivity is measured as the ratio of the number of correct reported hits to the number of totally reported hits (including false positives). The following observations must be taken into account whenever sequences of weak similarity are compared:

- Scores associated with query/randomized sequences are not normally distributed but more closely follow the extreme value distribution. Probabilities calculated in the form of standard deviations from the average matching score against all sequences in a database (which assume a normal distribution) are underestimates.
- Many naturally occurring sequences show strong deviations from random distribution in their amino acid compositions. Standard sequence comparison methodology is not applicable in these cases. Compositionnaly biased regions must be masked.
- It has been shown that the length of an aligned domain influences the significance of the comparison, i.e., longer stretches of the same statistical score are more reliable [Doolittle 1990].
- Evolutionary modification of sequences occurs continuously in time. At a given moment in the evolutionary process, one may observe examples of any evolutionary distance from near identity to statistically insignificant similarity. Thus, no mathematical algorithm *per se* is able to distinguish between related and unrelated sequences as long as the three dimensional structure is unknown.

8 Exhaustive Methods for Genome Analysis

Interest in large-scale sequence analysis was triggered by the advent of high-throughput DNA and cDNA sequencing and practical implementation of the first-ever whole genome sequencing projects in the beginning of this decade. Although approximately 150 000 protein sequences were known by that time, each of them was typically analyzed in a distributed fashion by individual researchers and subsequently submitted to public protein sequence databanks. The 14 genome sequencing projects completed so far generated over 30 000 protein sequences. The necessity to cope with large batches of sequences simultaneously made manual application of common sequence analysis packages impractical. However, the new challenges go far beyond merely increased data volumes. The availability of representative genomic fragments or even complete genomes substantially changed the very nature of sequence analysis work. The new tasks specific for computational genomics include:

- Creating possibly complete functional catalogues of gene products based on experimentally determined facts as well as computer predictions, [e.g. Mewes 1997].

- Examining the general organization of the complement of genes, such as gene order [Mushegian 1996] and operon architecture [Koonin 1997].
- Deriving paralogous relationships for completely sequenced organisms through exhaustive all-against-all comparisons of gene products; assessing the redundancy of genetic information [Brenner 1995; Labedan 1995; Mewes 1997].
- Making definitive conclusions about the presence of absence of certain proteins (and hence functions) in a cell; documenting non-orthologous gene displacement, i.e. the cases when non-homologous proteins in two different organisms fulfil the same function [Koonin 1996].
- Studying the complete set of relationships between the gene products from genomes of pathogenic organisms to the entire spectrum of known antigenic, membrane, drug resistance and virulence proteins from other species and thus obtaining valuable leads for drug discovery and vaccine development [see, e.g., Huynen 1997].
- Conducting cross-genome comparisons to obtain valuable insights into features characteristic for particular organisms [Smith 1997; Tatusov 1997]. For closely related species (e.g., pathogenic vs. non-pathogenic strains) subtle variations of gene content can shed light on the origins of their physiological dissimilarity. Complicated genomic events, such as gene shuffling, can be studied in detail. A minimal set of physiological functions necessary for a free living cell can be deliniated based on similarity relationships between multiple genomes [Mushegian 1996] and made more precise as new genomic data become available.

The effect of these emerging tasks has been twofold. First, they stimulated the development of many novel computationally efficient algorithmic approaches designed to address the specific problems of sequence analysis on genomic scale. For example, programs for DNA-protein comparisons [Gish 1993; Huang 1996] allow to obtain a sneak preview of the genome functional repertoire before the complete sequence has been assembled and verified. They combine conceptual translation of the query sequence in all six reading frames and database search into one step and are able to merge marginally significant database hits from the same strand into longer and more reliable similarity regions.

Secondly, genome analysis efforts underscored the need for integrated software tools capable of applying a whole spectrum of computational methods to genomic sequences, storing, generalizing, and visualizing the results in a systematic way. The first substantial piece of genomic DNA subjected to massive sequence analysis was the yeast chromosome III [Bork 1992]. Based on this early experience, GeneQuiz (www.sander.ebi.ac.uk/genequiz), the first system for automatic genome analysis was developed [Scharf 1994; Ouzounis 1995]. The main focus of this set of programs is reliable assignment of biochemical functions to the protein sequences encoded in each particular genome. This system performs sensitive similarity searches followed by automatic evaluation of results by a rule-based expert system. MAGPIE (multipurpose automated genome investigation environment; [Gaasterland 1996] http://www-c.mcs.anl.gov/home/gaasterl/magpie.html) is a genome annotation system that attempts to incorporate human logic into the process of decision on gene functional assignment. Different types of data obtained from local and remote sequence analysis services are represented as facts and analyzed through a system of logical rules imple-

Protein Extraction, Description, and ANalysis Tool

Computational analysis of genomic sequences

- *Saccharomyces cerevisiae*
- *Mycoplasma pneumoniae*
- *Synechocystis sp.*
- *Escherichia coli*
- *M. thermoautotrophicum*
- *Bacillus subtilis*
- *Archaeoglobus fulgidus*
- *Pyrococcus horikoshii OT3*

- *Mycoplasma genitalium*
- *Methanococcus jannaschii*
- *Haemophilus influenzae*
- *Helicobacter pylori*
- *Rhizobium sp. (plasmid)*
- *S. solfataricus (fragment)*
- *Borrelia burgdorferi*
- *Treponema pallidum*

Figure 1. PEDANT home page (http://www.mips.biochem.mpg.de).

mented in Prolog programming language. The system is designed to be adaptive and can change its behavior and analysis parameters dependent on the particular stage of a sequencing project. MAGPIE assigns confidence levels to multiple features established for each ORF and provides links to associated information, such as MedLine bibliographic references and relevant metabolic pathway database entries. A recently developed program and Web resource, PEDANT (Protein Extraction, Description, and ANalysis Tool; http://pedant.mips.biochem.mpg.de/ frishman/pedant.html) by Frishman and Mewes [Frishman 1997] makes available exhaustive functional and structural characterization of proteins coded in all publicly released genomes using a variety of bioinformatics methods. Figure 1 shows the current lists of genomes available. PEDANT makes it possible to create a list of gene products from a given organism belonging to a particular category, e.g., membrane proteins or proteins involved in amino acid metabolism, and then obtain detailed reports on each sequence summarizing all known and predicted features.

9 Future Development of Sequence Databases

Large sequencing projects have major impact on the future development of the macro-molecular sequence databases. The number of complete genome sequences will increase rapidly. Small model genomes from distant positions in the taxonomic kingdom are selected as prototypes for the systematic evaluation of proteins and their functional roles in the cell.

These projects, together with the ongoing sequencing of *A. thaliana* [Bevan 1998], *O. sativa* and human genome project will fundamentally change the quality of sequence data available for processing by the sequence databases. There are two directions to support scientists that process the data: The integration of databases

311

into a workgroup computing environment. The environment contains a set of tools that allow semi–automatic processing of sequence data. It guides the scientist through the annotation process and is able to execute simple mechanical tasks like the classification of sequences that are very similar to sequences already in the database. To provide the scientist control over the annotation process, tools for the visualization of the data are essential (e.g. the display of chromosomes, Figures 2 and 3).

Current public sequence databases display a high level of redundancy. The nucleic acid sequence databases contain a large number of independent sequencing reports describing the same sequence; discrepancies are often observed among these various overlapping reports. The emphasis has been on assembling large collections of data at relatively low cost with limited concern over data reliability. Current estimates indicate that as many as 10 in every 10 000 bases deposited in the nucleic acid sequence databases may be erroneous, suggesting that more than 50 % of the translations into open reading frames are subject to error. Proper assessment of the problem is complicated by the widely varying degrees of reliability found among the currently employed sequencing techniques. A crucial reinvestigation will be required as soon as appropriate and inexpensive technologies are available for the determination of molecular sequences within well–defined error limits. However, the issue to update databases with respect to the current research-driven biological knowledge is most relevant for the database user.

Simple tasks like the detection of new database hits could be taken over by software agents that are substitutes for human database annotators. Agents are autonomous

Graphical Display of Chromosome X

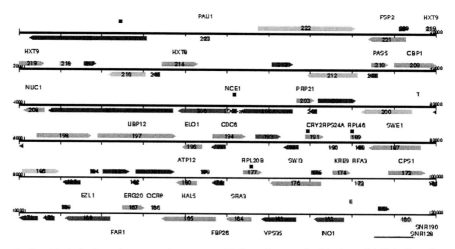

Figure 2. Graphical display of y yeast chromosome X (http://www.mips.biochem.de) linked to the yeast sequence database.

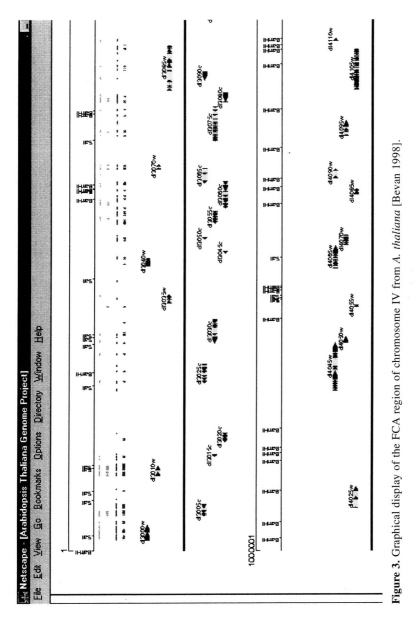

Figure 3. Graphical display of the FCA region of chromosome IV from *A. thaliana* [Bevan 1998].

programs that are able to generate and execute plans in order to achieve a goal. They communicate with each other and with human database users.

Another strategy is to use active database systems. Active databases are designed to support the processing of data items able to recognize predefined patterns in their database content and to react by modifying the data. Active databases could help to reduce redundancy and to enforce correctness of the data. Advanced data structures should be suitable for knowledge extraction and data mining, i.e. the nontrivial extraction of implicit, previously unknown, and potentially useful information from data [Frawley 1992].

A uniform access scheme could enhance the interoperability between the databases. Efforts in software design could be improved, because tools can be optimized to be used for multiple databases. The integration of information available from different biological databases presents still a major challenge.

The development of industrial standards in computing like programming languages (e.g. Java), database management systems, and wide area network communication software will have its impact on the future of the genome related databases. Techniques building on genome information such as expression analysis [DeRisi 1997], proteome analysis [Fey 1997] and systematic efforts in functional analysis will generate massive amounts of data. For yeast, these efforts are in progress (Eurofan), other model organisms will soon follow. The post genome era will not only augment the volume but also the complexity of the sequence related databases.

Modern software for navigation among data and informatics tools for their analysis and manipulation must be developed. However, such tools alone cannot compensate for the redundancy, incompleteness, inconsistency, and outright errors found within today's sequence databases. Effectively organizing the sheer volume of data elucidated will require the underlying concepts employed in the databases to be more precisely defined and eventually will point out the shortsightedness of accumulating all data without regard to quality, correctness, accuracy, or semantic consistency.

10 References

Achard, F., Barillot, E. (1998) *Nucleic Acids Res 26*, 100–101. Virgil: a database of rich links between GDB and GenBank.

Altschul, S. F., Boguski, M. S., Gish, W., Wootton, J.C. (1994) *Nature Genetics 6*, 119–129. Issues in searching molecular sequence databases.

Altschul, S. F., Gish, W., Miller, W., Myers, E. W., Lipman, D. J. (1990) *J Mol Biol 215*, 403–410. Basic local alignment search tool.

Attwood, T. K., Beck, M. E., Flower, D. R., Scordis, P., Selley, J. N. (1998) *Nucleic Acids Res 26*, 304–308. The PRINTS protein fingerprint database in its fifth year.

Bairoch, A., Apweiler, R. (1998) *Nucleic Acids Res 26*, 38–42. The SWISS-PROT protein sequence data bank and its supplement TrEMBL in 1998.

Bairoch, A., Bucher, P., Hofmann, K. (1996) *Nucleic Acids Res 24*, 189–196. The prosite database its status in 1995.

Barker, W. C., et al. (1998) *Nucleic Acids Research 26*, 27–32. The PIR-International Protein Sequence Database.

Barker, W. C., Pfeiffer, F., George, D. G. (1996) *Methods Enzymol 266*, 59–71, Superfamily classification in PIR-International Protein Sequence Database.

Benson, D. A., Boguski, M. S., Lipman, D. J., Ostell, J. (1990) *Genomics 6*, 389–391. The national center for biotechnology information.

Benson, D. A., Boguski, M. S., Lipman, D. J., Ostell, J., Ouellette, B. F. (1998) *Nucleic Acids Res 26*, 1–7. GenBank.

Bevan, M., et al. (1998) *Nature 391*, 485–488. Analysis of 1.9 Mb of contiguous sequence from chromosome 4 of Arabidopsis thaliana.

Blattner, F. R., et al. (1997) *Science 277*, 1453–1462. The complete genome sequence of Escherichia coli K-12.

Bork, P., Ouzounis, C., Sander, C., Scharf, M., Schneider, R., Sonnhammer, E. L. (1992) *Nature 358*, 287. What's in a genome?

Brenner, S., Hubbard, T., Murzin, A., Cothia, C. (1995) *Nature 378*, 140. Gene duplications in *H. influenzae*.

Brutlag, D. L., Dautricourt, J.-P., Diaz, R., Fier, J., Moxon, B., Stamm, R. (1993) *Comput Chem 17*, 203–207. Blaze (tm): an implementation of the smith-waterman sequence comparison algorithm on a massively parallel computer.

Califano, A., Rigoutsos, I. (1993) *Ismb*-9256–64. FLASH: a fast look-up algorithm for string homology.

Clayton, R. A., White, O., Ketchum, K. A., Venter, J. C. (1997) *Nature 387*, 459–462. The first genome from the third domain of life.

Dayhoff, M. O. (1965) National Biomedical Research Foundation, Maryland. Atlas of protein sequence and structure.

DeRisi, J. L., Iyer, V. R., Brown, P. O. (1997) *Science 278*, 680–686. Exploring the metabolic and genetic control of gene expression on a genomic scale.

Doolittle, R. F. (1990) *Meth Enzymol 183*, 99–110. Searching through sequence databases.

Etzold, T., Ulyanov, A., Argos, P. (1996) *Methods Enzymol 266*, 114–128, SRS: information retrieval system for molecular biology data banks.

Fasman, K. H. (1994) *J Comput Biol 1*, 165–172. Restructuring the genome data base: a model for a federation of biological databases.

Fey, S. J., et al. (1997) *Electrophoresis 18*, 1361–1372. Proteome analysis of Saccharomyces cerevisiae: a methodological outline.

Fleischmann, R. D., et al. (1995) *Science 269*, 496–512. Whole-genome random sequencing and assembly of *Haemophilus influenzae* Rd.

Fraser, C. M., et al. (1997) *Nature 390*, 580–586. Genomic sequence of a Lyme disease spirochaete, *Borrelia burgdorferi*.

Fraser, C. M., et al. (1995) *Science 270*, 397–403. The minimal gene complement of Mycoplasma genitalium.

Frawley, W. J., Piatetsky-Shapiro, G., Matheus, C. J. (1992) *Aaai* 57–70. Knowledge Discovery in Databases: An Overview.

Frishman, D., Mewes, H. W. (1997) *Trends Genet. 13*, 415–416. PEDANTic genome analysis.

Gaasterland, T., Sensen, C. W. (1996) *Trends Genet 12*, 76–78. MAGPIE: automated genome interpretation.

Galperin, M. Y., Koonin, E. V. (1998) *In Silico Biology 1*, 0007. Sources of systematic error in functional annotation of genomes: domain rearrangement, non-orthologous gene displacement, and operon disruption.

George, D. G., Mewes, H. W., Kihara, H. (1987) *Protein Seq. Data Anal. 1*, 27–39. A Standardized Format for Sequence Data Exchange.

George, D. G., Orcutt, B. C., Mewes, H.-W, Tsugita, A. (1993) *Protein Seq Data Anal 5*, 357–399. An object-oriented sequence database definition language (sddl).

Gish, W., States, D. J. (1993) *Nature Genetics 3*, 266–272. Identification of protein coding regions by database similarity search.

Goffeau, A., et al. (1996) *Science 274*, 546–567. Life with 6000 genes.

Gonnet, G. H., Cohen, M. A., Benner, S. A. (1992) *Science 256*, 1433–1445. Exhaustive matching of the entire protein sequence database.

Harper, R. A. (1996) *Trends Biochem Sci 21*, 150–2. EMBnet: an institute without walls.

Henikoff, S., Henikoff, J. G. (1993) *Proteins 17*, 49–61. Performance evaluation of amino acid substitution matrices.

Henikoff, S., Pietrokovski, S., Henikoff, J. G. (1998) *Nucleic Acids Res 26*, 309–12. Superior performance in protein homology detection with the Blocks Database servers.

Himmelreich, R., Hilbert, H., Plagens, H., Pirki, E., Li, B. C., Herrmann, R. (1996) *Nucleic Acids Research 24*, 4420–4449. Complete sequence analysis of the genome of the baterium *Mycoplasma pneumoniae*.

Hodges, P. E., Payne, W. E., Garrels, J. I. (1998) *Nucleic Acids Res 26*, 68–72. The Yeast Protein Database (YPD): a curated proteome database for Saccharomyces cerevisiae.

Huang, X. (1996) *Microbial and Comparative Genomics 1*, 281–291. Fast comparison of a DNA sequence with a protein sequence database.

Huynen, M. A., Diaz-Lazcoz Y., Bork P. (1997) *Tends Genet. 13*, 389–390. Differential genome display.

Kaneko, T., et al. (1996) *DNA Res 3*, 109–136. Sequence analysis of the genome of the unicellular Cyanobacterium Synechocystis sp. Strain PCC6803. II. Sequence determination of the entire genome and assignment of potential protein-coding regions.

Klenk, H. P., et al. (1997) *Nature 390*, 364–370. The complete genome sequence of the hyperthermophilic, sulphate- reducing archaeon Archaeoglobus fulgidus.

Koonin, E. V., Galperin, M. Y. (1997) *Curr Opin Genet Dev 7*, 757–763. Prokaryotic genomes: the emerging paradigm of genome-based microbiology.

Koonin, E. V., Mushegian, A. R., Bork, P. (1996) *Trends in Genetics 12*, 334–336. Non-orthologous gene displacement.

Kunst, F., et al. (1997) *Nature 390*, 249–256. The complete genome sequence of the Gram-positive bacterium *Bacillus subtilis.*

Labedan, B., Riley, M. (1995) *Mol. Biol. Evol. 12*, 980–987. Gene products of Escherichia coil: sequence comparisons and common ancestries.

Lipman, D. J., Pearson, W. R. (1985) *Science 227*, 1435-1441. Rapid and sensitive protein similarity searches.

Markowitz, V. M., Ritter, O. (95) *J Comput Biol 2*, 547-556. Characterizing heterogeneous molecular biology database systems.

Mewes, H. W., et al. (1997) *Nature 387*, 7–65 The yeast genome directory.

Mewes, H. W., Heumann, K. (1995) Combinatorial Pattern Matching (Galil Z., U.E., Eds.), Vol. 937, pp. 261–285, Springer Verlag, Berlin-Heidelberg. Genome Analysis: Pattern Search in Biological Macromolecules.

Mewes, H., Alberman, K., Heumann, K., Liebl, S., Pfeiffer, F. (1997) *Nucl. Acids Res. 25*, 28–30. MIPS: a database for protein sequences, homology data and yeast genome information.

Mushegian, A. R., Koonin, E. V. (1996) *Trends Genet 12*, 289–290. Gene order is not conserved in bacterial evolution [letter].

Mushegian, A. R., Koonin, E. V. (1996) *Proc.Nat.Acad.Sci. 93*, 10268–10273. A minimal gene set for cellular life derived by comparison of complete bacterial genomes.

Needleman, S. B., Wunsch, C. D. (1970) *J Mol Biol 48*, 443–453. A general method applicable to the search for similarities in the amino acid sequence of two proteins.

Ouzounis, C., Bork, P., Casari, G., Sander, C. (1995) *Protein Science 4*, 2424–2428. New protein functions in yeast chromosome VIII.

Ritter, O., Kocab, P., Senger, M., Wolf, D., Suhai, S. (1994) *Comput Biomed Res 27*, 97–115. Prototype implementation of the integrated genomic database.

Scharf, M., Schneider, R., Casari, G., Bork, P., Valencia, A., Ouzounis, C. (1994) ISMB, pp. 348–353, AAAI Press, Menlo Park, CA.

Schuler, G. D., Epstein, J. A., Ohkawa, H., Kans, J. A. (1996) *Meth. Enzymol. 266*, 141–162. Entrez: molecular biology database and retrieval system.

Smith, D. R., et al. (1997) *Genome Research 7*, 802–819. Multiplex sequencing of 1.5 Mb of the *Mycobacterium leprae* genome.

Smith, D. S., et al. (1997) *J. Bacter. 179*, 7135–7155. Complete genome sequence of Methanobacterium thermoautotrophicum DH: Functional analysis and comparative genomics.

Sonnhammer, E. L., Eddy, S. R., Birney, E., Bateman, A., Durbin, R. (1998) *Nucleic Acids Res 26*, 320–322. Pfam: multiple sequence alignments and HMM-profiles of protein domains.

Stoesser, G., Moseley, M. A., Sleep, J., McGowran, M., Garcia-Pastor, M., Sterk, P. (1998) *Nucleic Acids Res 26*, 8–15. The EMBL nucleotide sequence database.

Tatusov, R. L., Koonin, E. V., Lipman, D. J. (1997) *Science 278*, 631–637. A genomic perspective on protein families.

Taylor, W. R., Jones, D. T. (1991) Curr Opinion Struct Biol 1, 327–333. Templates, consensus patterns and motifs.

Tomb, J. F., et al. (1997) *Nature 388*, 539–547. The complete genome sequence of the gastric pathogen *Helicobacter pylori.*

Waterman, M. S. (1983) *Proc Natl Acad Sci USA 80*, 3123–3124. Sequence alignments in the neighborhood of the optimum with general application to dynamic programming.

White, O., Kerlavage, A. R. (1996) *Meth Enzymol 266, 27–40*, TDB: new databases for biological discovery.

Wolfe, K. H., Shields, D. C. (1997) *Nature 387*, 708–713. Molecular evidence for an ancient duplication of the entire yeast genome.

Index